Knowledge Computing and Its Applications

S. Margret Anouncia · Uffe Kock Wiil
Editors

Knowledge Computing and Its Applications

Knowledge Manipulation and Processing Techniques: Volume 1

 Springer

Editors
S. Margret Anouncia
Computer Science and Engineering
VIT University
Vellore, Tamil Nadu
India

Uffe Kock Wiil
The Maersk Mc-Kinney Moller Institute
University of Southern Denmark
Odense
Denmark

ISBN 978-981-10-6679-5   ISBN 978-981-10-6680-1   (eBook)
https://doi.org/10.1007/978-981-10-6680-1

Library of Congress Control Number: 2017962106

Printed on acid-free paper

This Springer imprint is published by Springer Nature
The registered company is Springer Nature Singapore Pte Ltd.
The registered company address is: 152 Beach Road, #21-01/04 Gateway East, Singapore 189721, Singapore

Dedicated to
The
Almighty God

Foreword

Knowledge Computing plays a key role in the current information age. Processing of huge amounts of data and extracting relevant knowledge is a central part of data science today.

A glance into this book and its organization shows that it provides a comprehensive collection of contributions covering current directions, challenges, and applications in the field of knowledge computing. The contributions deal with different dimensions of knowledge processing exemplified by a broad spectrum of applications. The tools and methodologies underlying the contributions provide a state-of-the-art glimpse into modern knowledge computing.

The book consists of five parts divided into two volumes. Volume I: (1) knowledge manipulation techniques for Internet technologies, (2) knowledge computing in large organizations, (3) knowledge computing and competency development. Volume II: (4) knowledge processing tools and techniques in specific domains, and (5) application of knowledge engineering process in information management. Taken together these five parts (including a total of 29 chapters) provide an excellent overview of current tools, methodologies, and practice in the field of knowledge computing.

The book can serve as important inspiration for academics, industry experts, business professionals as well as students in their future research endeavours.

Enjoy!

Vijayawada, India

Dr. Suresh Chandra Satapathy
LMCSI, Senior Member IEEE, Professor & Head
Department of CSE
PVP Siddhartha Institute of Technology

Preface

Information society is rapidly migrating to the knowledge economy where knowledge management is renovating to the knowledge ecosystem. Several entities may not provide convenient capabilities directly but can enable capabilities of other products, services or users. Knowledge economy is concerned with substituting knowledge-based products and services in knowledge markets which are built on supportive mechanisms to help mobilization,allotment or exchange of knowledge among providers and users.

This Handbook of Knowledge Computing and its applications is a edited book contributed by Seventy number of researchers' who work in the field of Knowledge Engineering. The scope of the book includes knowledge manipulation techniques for Internet Technologies, Knowledge Processing tools and techniques in Specific Domains, knowledge computing in Large and international Organizations, application of knowledge engineering process in information Management, Skill and Competencies Development.

The book is edited as two volumes comprising fourteen and fifteen chapters respectively.

The first volume focuses on *Knowledge Manipulation and Processing Techniques* comprising three parts such as: *Knowledge Manipulation Techniques for Internet Technologies* that includes articles dealing with knowledge manipulation, representation techniques that are useful for recent Internet technologies such as Big Data and IoT. The second part on *Knowledge Computing in Large Organizations* focusing the various organizational issues sorted out through knowledge computing methods. The third part deals with the *Knowledge Computing and Competency Development*.

The second Volume titled *Knowledge Computing in Specific Domains* has two parts. The first part of volume two delineates on *Knowledge Processing Tools and Techniques in Specific Domains* with articles associating knowledge processing in domains such as health care, manufacturing, multimedia and remote sensing.

The second part deals with the articles concentrating on the *Application of knowledge engineering process in Information Management*. With the drastic growth of knowledge collections in varieties of forms and in diversified fields, the demand for the knowledge processing and representation increases. Consequently, the claim for the usage of knowledge tools and techniques arises. With this motivation, the handbook is planned and is intended for a spectrum of people spanning from academicians, researchers, industry experts and students. The book also can be a beneficial to business managers and entrepreneurs.

The salient feature of this hand book includes:

- Covers knowledge processing in the developing domains such as Big Data, IoT, health care and multimedia.
- It also presents methods associated with the knowledge computing and its applications.
- Contributors of the book are researchers and practitioners from various reputed academia and industry within and outside country.

We thank all the contributors for their relentless effort. Without their support, this book would never come to realization. We express our gratitude to Springer Team for their support throughout the project.

Vellore, India   S. Margret Anouncia
Odense, Denmark   Uffe Kock Wiil

Acknowledgement

First and foremost, the editors would like to wholeheartedly thank the Lord, from the bottom of their hearts, who has been the enduring source of strength and comfort for the completion of this book project. You have given us the power to believe in our passion and pursue our dreams.

The editors would like to acknowledge the help of all the people involved in this project and, more specifically, the authors and reviewers who took part in the review process. Without their support, this book would not have become a reality.

The editors would like to thank each one of the authors for their contributions. Our sincere gratitude goes to the chapter's authors who contributed their time and expertise to this book.

The editors wish to acknowledge the valuable contributions of the reviewers regarding the improvement of quality, coherence and content presentation of chapters. They also thank the reviewers for their thoughtful, detailed and constructive comments.

The editors' heartfelt thanks go to all the friends and colleagues who assisted them in all endeavours of assembling this book and their outstanding and generous help in providing recommendations and observations which has been an inspiration for this book.

Finally, the editors would like to acknowledge with gratitude the support and love of their family members for compiling this book. This book would not have been possible without their support.

Dr. S. Margret Anouncia
Dr. Uffe Kock Wiil

Contents

About the Editors

S. Margret Anouncia is a Professor at Vellore Institute of Technology (VIT University) in India. She received her bachelor's degree in Computer Science and Engineering from Bharathidasan University (1993), Tiruchirappalli, India, and a master's degree in Software Engineering from Anna University (2000), Chennai, India. She was awarded a doctorate in Computer Science and Engineering at VIT University (2008). Her areas of interest include digital image processing, software engineering and knowledge engineering.

Uffe Kock Wiil received his M.Sc. (1990) and Ph.D. (1993) degrees from Aalborg University, Aalborg, Denmark. He has been a Professor of Software Engineering at the University of Southern Denmark, Denmark, since 2004. His primary research interests are health informatics (clinical decision support systems; patient empowerment) and security informatics (analysis and visualization of crime-related data).

Abbreviations

ABE	Attribute-Based Encryption
ABPL	Average Backbone Path Length
ADHD	Attention Deficit/Hyperactivity Disorder
AES	Advanced Encryption Standard
AF	Age Factor
AIC	Akaike Information Criterion
AIG	Adjusted Information Gain
ANFIS	Adaptive Neuro-Fuzzy Inference System
ANN	Artificial Neural Network
APA-DSM	American Psychiatric Associations Diagnostic and Statistical Manual of Mental
API	Application Programming Interface
APL	Average Path Length
AR	Adjusted Rand Index
AR	Association Rule
ARI	Adjusted Rand Index
ASCII	American Standard Code for Information Interchange
ASE	Average Squared Error
AUC	Area Under Curve
AUE	Accuracy Updated Ensemble
AWE	Accuracy Weighed Ensemble
BM	Benchmark
BMI	Body Mass Index
BNC	British National Corpus
BPNN	Back Propagation Neural Network
CAD	Computer-Aided Diagnosis
CART	Classification and Regression Technique
CBR	Case-Based Reasoning
CDS	Connected Dominating Set
CDSS	Clinical Decision Support System

CE	Cross Entropy
CFG	Context Free Grammar
CIA	Central Intelligence Agency
CKS	Contextual Knowledge Structure
CLI	Command Line Interface
COPD	Chronic Obstructive Pulmonary Diseases
CQL	Cassandra Query Language
CRCDD	Cross Recurrence Concept Drift Detection
CSP	Context Search Phrase
CT	Clustering Trees
DBN	Deep Belief Network
DCT	Discrete Cosine Transform
DDoS	Distributed Denial of Service
DGB	Disk Graph with Bidirectional links
DLP	Discrete Logarithm Problem
DM	Decision-Making
DNA	Deoxyribonucleic Acid
DNN	Deep Neural Network
DOS	Denial-of-Service Attacks
DRM	Digital Rights Management
DST	Dempster–Shafer Theory
DT	Distance Transform
DWT	Discrete Wavelet Transform
ECAES	Elliptic Curve Authentication Encryption Scheme
ECDLP	Elliptic Curve Discrete Log Problem
FD	Fractional Differential
FEAC	Fast Evolutionary Algorithm for Clustering
FFT	Fast Fourier Transform
FLC	Fuzzy Logic Controller
FPR	False-Positive Rate
GA	Genetic Algorithm
GCS	Geographic Coordinate System
GIS	Geographic Information System
GL	Grunwald–Letnikov
GLCM	Gray Level Co-occurrence Matrix
GRNN	General Regression Neural Network
GRR	Google Rapid Response
GUI	Graphical User Interface
GWS	GateWay-based Strategic
GWS-CDS	GateWay-based Strategic Connected Dominating Set
HAS	Harmony Search Algorithm
HCI	Human–Computer Interaction
HDFS	Hadoop Distributed File System
HI	Huberts Index
HOG	Histogram of Oriented Gradients

HTTP	Hypertext Transfer Protocol
IBE	Identity-Based Encryption
ICA	Independent Component Analysis
ID	Intrusion Detection
IDS	Intrusion Detection System
II	Intensity of Interaction
IMDB	Internet Movie Database
IoT	Internet of Things
IPS	Internet Protocol Security
ISODATA	Interactive Self-Organizing Data Analysis Technique
JT/RM	Job Trackers/Resource Manager
K-ANMI	K-means and Mutual Information
KB	Knowledge Base
KNN	K-Nearest Neighbor
LDA	Latent Dirichlet Allocation
LDA	Linear Discriminant Analysis
LET	Link Expiration Time
LHS	Left-Hand Side
MAC	Mandatory Access Control
MAE	Mean Absolute Error
MANET	Mobile Ad hoc Network
MaxD	Maximum Density
MB	Measure of Belief
MCDA	Multicriteria Decision Analysis
MCDS	Minimum Connected Dominating Set
MD	Measure of Disbelief
MDD	Major Depressive Disorder
MDL	Minimum Description Length
MDM	Mobile Device Management
MDP	Missed Detection Rate
MGR	Mean Gain Ratio
MLP	Multilayer Perceptron
MLR	Multinomial Logistic Regression
MM	Mammographic Mass
MMHCI	Multimodal Human–Computer Interaction
MPANN	Memetic Pareto Artificial Neural Network
MRI	Magnetic Resonance Imaging
MTFA	Mean Time between False Alarms
MTP	Mean Time Detection
MWMCDS	Maximal Weight Minimum Connected Dominating Set
NB	Naive Bayes
NBP	Neighbor-based Binary Pattern
NECAS	National Epidemiological Catchment Area Survey
NIDS	Network Intrusion Detection System
NIMH	National Institute of Mental Health

NLP	Natural Language Processing
NMI	Normalized Mutual Information Matrix
NN	Neural Network
NOS	Not Otherwise Specified
NSA	National Security Agency
OCR	Optical Character Recognition
OE-FV	Online Ensemble based on Feature Vector
PAM	Physical Activity Monitoring
PAM	Pluggable Authentication Module
PCA	Principal Component Analysis
PFD	Partial Fractional Derivative
PM	Proposed Methodology
PNN	Probabilistic Neural Network
PP	Prepositional Phrase
QIS	Query Indexed Structure
RBF	Radial Basis Function
RBFNN	Radial Basis Function Neural Network
RDBMS	Relational Database Management System
RFE	Recursive Feature Elimination
RFID	Radio Frequency Identification
RHS	Right-Hand Side
RMSE	Root-Mean-Square Error
ROC	Receiver Operating Characteristic
RT	Radon Transform
SAN	Storage Area Network
SCCDD	Stable Clustering Concept Drift Detection
SFPM	Soft Frequent Pattern Mining
SLPP	Simple Spatial Preposition and Prepositional Phrase
SNA	Social Network Analysis
SOM	Self-Organizing Map
SSE	Searchable Symmetric Encryption
SSL	Secure Sockets Layer
SSO	Single Sign On
ST-MaxD	Strategic Maximum Density
STS	Sentiment Training Set
ST-SN	Strategic Strong Neighborhood
SUNDR	Secure Untrusted Data Repository
SVM	Support Vector Machine
TDM	Term-Document Matrix
TLS	Transport Layer Security
TM	Twitter Monitor
TNDR	Threshold Neighbor Distance Ratio
TPR	True Positive Rate
TT/NM	Task Tracker/Node Manager
UE	User Experience

UI	User Interface
UIML	User Interface Markup Language
UIMS	User Interface Management System
URL	Uniform Resource Locator
USIXML	USer Interface eXtensible Markup Language
VP	Verb Phrase
VPN	Virtual Private Network
WBCD	Wisconsin Breast Cancer Dataset
WCFPOF	Weighted Closed Frequent Pattern Outlier Factor
WD	Weight Decay
WSD	Word Sense Disambiguation
WSN	Wireless Sensor Network
XIML	eXtensible Interface Markup Language

Part I
Knowledge Manipulation Techniques for Internet Technologies

Implementation of Connected Dominating Set in Fog Computing Using Knowledge-Upgraded IoT Devices

V. Ceronmani Sharmila and A. George

Abstract Wireless sensor networks (Wsns) have a worldwide attraction because of its increasing popularity. The key enablers for the Internet of Things (IoT) are WSN, which plays an important role in future by collecting information through the cloud. Fog Computing, the latest innovations, connects sensor-based IoT devices to the cloud. Fog Computing is a decentralized computing infrastructure in which the data, compute, storage, and applications are distributed efficiently between the data source and the cloud. The main aim of Fog Computing is to reduce the amount of data transported to the cloud and hence increase the efficiency. The knowledge-upgraded IoT devices will be embedded with a piece of software into it, which can able to understand the Distributed Denial of Service (DDoS). Such attacks are not forwarded to the cloud and thus the cloud server down problem is avoided. The IoT devices enabled with such knowledge is connected together to form a Connected Dominating Set (CDS). The data routed through only such IoT devices will be directly connected to the cloud. The CDS-based approach reduces the search for a minimum group of IoT devices called nodes, thus forming the backbone network. Various CDS algorithms have been developed for constructing CDSs with minimum number of nodes. However, most of the research work does not focus on developing a CDS based on application and requirement. In this chapter, a Gateway-based Strategic CDS (GWS-CDS) is constructed based on strategy and communication range. Here, any node in the network assigned a critical communication range, which is in a strong neighbourhood and which is within the communication range of more than one network, will be selected as the starting node, instead of the node with maximum connectivity. If a node is not within a critical communication range, then the following factors will be increased:

V. Ceronmani Sharmila (✉)
Department of Information Technology, Hindustan Institute
of Technology and Science, Chennai, Tamilnadu, India
e-mail: ceronvlsi@gmail.com

A. George
Department of Mathematics, Periyar Maniammai University,
Thanjavur, Tamilnadu, India
e-mail: amalanathangeorge@gmail.com

© Springer Nature Singapore Pte Ltd. 2018
S. Margret Anouncia and U. K. Wiil (eds.), *Knowledge Computing and Its Applications*, https://doi.org/10.1007/978-981-10-6680-1_1

the number of nodes that locally compete over a shared channel, access delay, network throughput and network partitioning. The other nodes for CDS construction are selected based on density and velocity. The focus of this research work was to construct a CDS in heterogeneous networks. The algorithm was tested with respect to three metrics—average CDS node size, average CDS Edge Size and average hop count per path. Simulation results showed that the proposed algorithm was better when compared to the existing algorithms.

Keywords Wireless sensor networks · IoT · Fog Computing · Communication range · CDS · Strategy · Density · Velocity and gateway nodes

1 Introduction

Fog Computing and Internet of Things (IoT) are the two evolving highest technological prototypes that to date have a larger influence in the recent world [1]. However, because of their equivalent features, their combination can raise a number of computing and network-demanding universal applications under the incoming dominion of the upcoming Internet. Fog Computing has developed as an upcoming technology that can bring cloud applications closer to the physical IoT devices at the network edge. One of the first attempts to define a fog node was made by Cisco, as a 'mini-cloud' situated at the edge of the network and implemented through a variety of edge devices, interconnected by wireless communication technologies [2]. Internet of Things (IoT), one of the main research topics in current years, composed with notions from Fog Computing, conveys speedy advancements in Smart City, industrial control, monitoring systems, transportation and other fields [3]. The background of these applications meeting big data needs has to be processed on clouds instead of specially relying on computational resources and limited storage of small devices. But due to the growth in the number of Internet-connected devices, the improved request of low-latency, real-time services is proving to be a challenging task for the outdated cloud computing context [4]. These technologies can be incorporated to the wireless sensor networks (WNSs) to support cloud applications.

An ad hoc wireless network can be symbolized by a unit disk graph [5], where each node is connected with a disk focused at this node with the same communication range (also called radius). Any two nodes are neighbours if and only if they are within the communication range of each other. For example, both the nodes q and r are neighbours of node p because they are within the communication range of p as shown in Fig. 1a. But the nodes q and s are not neighbours since they do not share their communication range as shown in Fig. 1b. Gateway node is a network node that serves as an entrance to another network. Doubly cycled nodes such as p and r are called gateway nodes.

Wireless sensor networks (WSNs) are made up of small devices deployed in large scale that run autonomously. These small devices are embedded with sensors

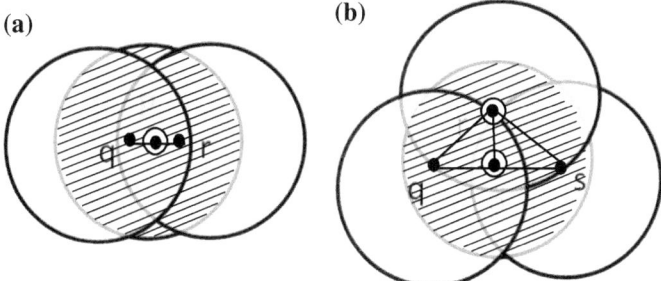

Fig. 1 Example for ad hoc networks

termed as 'nodes' with capability to feel the real world. The main task of WSN is to sense and monitor the physical phenomenon in the environment, reporting back the same to base station for further processing. The base station (BS) is a node that controls and coordinates the activities of the networks and takes decisions, assigns tasks and also can query the network for data or any information. The size and design of the node pose some potential energy issues with communication capability, computation capacity, storage capacity and battery operated. The significant characteristics of WSNs are fast deployment, self-configuration, distributed operation, multi-hop routing, autonomous terminal, lightweight terminals and shared physical medium. The challenges faced by wireless sensor networks are Denial of Service attacks (DOS), data centric, low data rate and application specific. With the growth of mobile technology and wireless communication, WSN has been utilized in most civilian applications such as tactical operations, rescue missions, battlefield.

Multi-hop routing mostly achieves a high degree of network connectivity, as the broadcast range of the nodes in WSN is usually lower than the span of such networks. However, owing to the changing network topology and mobility of nodes, the wireless connections are often broken and new connections are established. Fixing the communication range for nodes plays a major role in improving the network connectivity. If the communication range of the node is more, the conditions are as follows: decreased number of hops, decreased network capacity, increased number of nodes that are nearby contend over the shared channel and increased access delay. If the communication range of the node is less, the conditions are as follows: improved frequency reuse, extend battery lifetime, increased throughput and increased network partitioning. Different authors [6–8] have presented solutions for calculating the minimum value of nodes' communication range called as the critical communication range, which avoids frequent network partitioning.

The Connected Dominating Set (CDS) approach can be used to develop a critical communication range-based algorithm for finalizing the nodes responsible for transmitting the data in heterogeneous networks. The traditional protocols developed for constructing the CDS are all parameter-based. The parameters can be degree, density, routing cost, strong neighbour, identity, location, interference, energy and reliability. In [9], four CDS algorithms were built based on strategy that the starting node can be selected within the network provided there is a need and application for that particular node. The proposed communication range-based algorithm constructs a CDS with the strategy that any node in the network assigned to the critical communication range in a strong neighbourhood and in the communication range of more than one network (gateway node) will be nominated as the starting node rather than the node with maximum connectivity. The other nodes of the CDS are selected based on density and velocity. The algorithm was tested with respect to three metrics—average CDS node size, average CDS Edge Size and average hop count per path. Simulation results showed that the proposed algorithm was better in terms of the size of the CDS.

The section distribution of the paper is as follows: Sect. 2 deals with the review of related works. Section 3 describes the construction of GWS-CDS algorithm with examples. Section 4 discusses the simulations and results. The conclusion is outlined in Sect. 5.

2 Related Work

Multi-hop routing related tasks are very difficult in WSN because of the absence of predefined infrastructure. Normally, broadcasting is the routing methodology used for data communication. The simple broadcasting leads to flood where each node recommunicates the broadcasting message that it receives. Such recommunications lead to broadcast storm problems [10, 11]. Extensive research has been carried out to construct a virtual backbone that will remove unnecessary recommunication links by selecting the required nodes for communication. A backbone is a subset of nodes which will be able to complete special tasks to assist the other nodes in the network. If the backbone of the network is connected, it enables the backbone nodes to communicate with other nodes easily. The special tasks that can be performed by the backbone nodes are broadcasting, routing, etc. Connected Dominating Set (CDS) has been suggested to serve as a virtual backbone to secure good connectivity and to decrease routing overhead in a WSN. Having such a CDS simplifies routing by restricting the main routing tasks to the dominators only. Given an undirected graph $G = (V, E)$, a CDS is a set P of vertices with two properties:

- Any node in P can reach any other node in P by a path that stays entirely within P.
- Every vertex in G either belongs to P or is adjacent to a vertex in P.

The dominating-set-based routing is a promising approach among the existing routing protocols. The major task of this approach is to construct and maintain the CDS. It is advantageous to construct a small CDS (least number of nodes in the CDS) without compromising the efficiency and functionality of the WSN. The two important necessities for a CDS-based routing protocol are: the CDS formation should be localized for less overhead and fast convergence. Inappropriately, finding a minimum CDS is NP-complete for most of the graphs.

The CDS construction algorithms can be divided into two groups: centralized and decentralized algorithms. The centralized algorithms depend on the entire network information, whereas the decentralized algorithms depend on the local information only. Centralized algorithms construct a CDS with a high maintenance cost and smaller CDS than decentralized algorithms. Das et al. [12] introduced a centralized algorithm to find a small CDS. This centralized algorithm utilized the Guha and Khuller's first approximation algorithm [13] for CDS formation. In [13], a spanning tree T is grown in numerous consecutive iterations. In the first iteration, a node with maximum degree is selected as the root of the tree T. In the succeeding iterations, a node n_1 in T which has the maximum number of neighbours not in T is selected. Finally, a spanning tree was built with non-leaf vertices forming a CDS. Decentralized algorithms are further divided into localized and distributed algorithms. The localized algorithms are distributed, and the assessment process requires a constant number of iteration rounds. The assessment process is serialized and decentralized in case of distributed algorithms. These algorithms can be utilized in *homogeneous* networks, *heterogeneous* networks and *gateway*-based networks, based on the network models.

2.1 Homogeneous Networks

The Connected Dominating Sets (CDS) are widely used to model a homogeneous network, in which all the nodes have same communication range. The CDS problem has been studied in-depth in unit disk graph [14], where the communication range of all the nodes is one unit. The minimum CDS (MCDS) issue in UDG has been examined and shown to be NP-hard [15]. To build a CDS in homogeneous networks, most of the present algorithms first select a parameter based on which the CDS nodes are identified. In [16], two centralized and one distributed algorithm were developed based on the parameters diameter and identity. The first algorithm builds a Breadth First Search (BFS) Tree initially, then a maximal independent set (MIS) was constructed based on the BFS tree and MIS nodes were connected to form a CDS. In the second algorithm, an MIS was computed initially and MIS nodes were connected by constructing a spanning tree. The hop distance from the root node to each of the nodes in the network is manipulated in the algorithm. As a result, the second algorithm has a lesser performance ratio for size and a bigger one for diameter than the first algorithm. The third algorithm is the distributed version of the second algorithm. After knowing the hop distance from every node to the root node, each node decides whether it can be an MIS node based on the information about its neighbours. Theoretical analysis of the algorithms was carried out to prove that the performance ratio of the above algorithms was better when compared to other related algorithms. The parameters used for simulation are CDS size, diameter, Average Backbone Path Length (ABPL) and network lifetime. The results demonstrated that the above algorithms outperformed the others.

Meghanathan and Terrell [17] proposed a stable Connected Dominating Set for mobile ad hoc networks using strong neighbourhood. The strong neighbourhood of a node is defined using the parameter Threshold Neighbour Distance Ratio (TNDR) that can be at most 1. The TNDR was defined as the ratio of physical Euclidean Distance and the communication range. The construction of CDS starts with the inclusion of the node with maximum number of stronger neighbours into the CDS. The consecutive nodes are selected based on strong neighbour parameter and if there is, a tie, one among the contended nodes is selected arbitrarily. The CDS constructed was a long-living stable CDS without acquiring a significant increase in the size of CDS. The parameters used for simulation are CDS node size, CDS Edge Size, hop count per path and CDS lifetime. The results demonstrated that the CDS has better balance in load handling, thus improvising the fairness of node usage.

Misra and Mandal [18] proposed an algorithm for identifying a smallest Connected Dominating Set using the collaborative cover heuristic for ad hoc networks. The collaborative cover heuristic was based on two principles: (1) domatic number of a connected graph is at least two, which is defined as the maximum number of disjoint dominating sets. (2) optimal substructure with a common connector, which is defined as the highest weight independent set. Initially a partial Steiner tree was constructed during the formation of independent set (dominators). Finally, a post-processing step identified the Steiner nodes, when the Steiner tree for the independent set was formed. The results demonstrated that collaborative heuristic was able to give a marginally better bound when the distribution of nodes was uniform.

Meghanathan [19] proposed a Maximum Density-based CDS (MaxD-CDS) algorithm for mobile ad hoc networks (MANET). This algorithm generated a Minimum Connected Dominating Set (MCDS) in a static graph. The MaxD-CDS algorithm selected the nodes with larger number of uncovered neighbours for constructing the CDS. The construction of the MaxD-CDS started with the inclusion of the node having the highest density (uncovered nodes) into the CDS. Once a node was added to the CDS, all its neighbours were said to be covered. The node with the highest density in the coveted list was selected as the next node for inclusion into the CDS. This process was repeated until all the nodes in the network were covered. The parameters used for evaluation were CDS node size and CDS Edge Size. Simulation results showed that the proposed algorithm achieved an average number of nodes and edges in the CDS.

Ceronmani and George [9] proposed four Strategic CDS algorithms for mobile ad hoc networks. In the algorithms proposed by the authors the strategy used to be, to select any node in the network as the starting node for constructing the CDS, provided the selected node have a specific application and need. The four algorithms proposed were Strategic Maximum Density CDS (ST-MaxD CDS), Strategic Strong Neighbourhood CDS (ST-SN CDS), Strategic Minimum Velocity Minimum Density CDS (ST-MV-MD CDS) and Strategic Minimum Density Minimum Velocity CDS (ST-MD-MV-CDS). The construction of the above algorithms started with any node in the network into the CDS. Once a node was added to the CDS, all its neighbours were said to be covered. The node with highest density or strong neighbourhood or minimum velocity (in case of a tie a node with minimum density) or minimum density (in case of a tie a node with minimum velocity) in the coveted list was selected as the next node for inclusion into the CDS. This process was repeated until all the nodes in the network were covered. The parameters used for evaluation were CDS node size, CDS Edge Size and CDS circuit size. On comparing the four algorithms, ST-MD-MV-CDS was found to have a moderate number of CDS nodes, less number of CDS edges and no circuits.

Bannoura et al. [20], proposed a wake-up dominating set problem for unit disk graphs, where the CDS was used for waking up the nodes. In this algorithm, an online variant was considered and starting from an initial node all nodes have to wake-up. But the online variant algorithm knows only the nodes awakened so far and has no information about the count and location of the sleeping nodes. The problem was solved by setting a worst-case competitive ratio by using the number of wake-up signals and the size of minimum CDS. The CDS was developed using dense random uniform placement and restricted adversary placement of CDS nodes. Simulation results deterministic wake-up algorithm and competitive epidemic algorithm produced reasonable results, which allow its application in the real world.

2.2 Heterogeneous Networks

Even though many researchers have developed algorithms for homogeneous networks, in practice the communication range of all nodes is not necessarily equal. So, CDS was constructed for heterogeneous networks, where each node could operate at a different communication range. In [21], the network was modelled as a disk graph and two efficient approximation algorithms are introduced to obtain a minimum CDS. A size relationship between an MIS and a CDS was provided in disk graph with bidirectional links (DGB). The upper bound of the number of independent neighbours of any node in DGB was also presented in the algorithm. It was observed that the performance ratios of the algorithms are constant when the communication range was bounded. Through theoretical analysis and simulations, it was demonstrated that, in order to reduce the size of a CDS, a Steiner tree with minimum number of Steiner nodes could be interconnected with the maximal independent set.

Xiaohong et al. [22] proposed a distributed lifetime-extended CDS algorithm (TCDS) based on node weight function. The imbalance of node energy consumption in heterogeneous networks was measured, and a novel node weight function was constructed to solve the joint optimization problem. The node weight function was developed, considering the network lifetime and node degree. The joint optimization problem was developed to handle the trade-off between minimizing the CDS node size and prolonging network lifetime. TCDS involves three phases. In the first phase, a BFS tree was built and neighbour information was exchanged. In the second phase, a dominating set was constructed. In the third phase, the connector nodes were selected. The results demonstrated that the lifetime-extended CDS algorithm prolonged the network lifetime more efficiently than others did. Deng et al. [23] proposed an analytical model based on the energy utilization model and a two-dimensional Poisson node distribution, to examine the optimal value for communication range. The analysis of the research work showed that the node density, rather than the network coverage area, influenced the optimum communication radius more. Simulation results were cross-checked with analytical results to approve the optimality of the per-hop communication range.

2.3 Gateway-Based Networks

Wu and Li [24] proposed a modest and competent distributed algorithm for calculating the CDS for ad hoc networks. The CDS was constructed based on the geographical distances between the nodes. The various ways to reconstruct the CDS due to the mobility of the nodes were presented in this paper. The proposed algorithm outperformed the classical algorithm such as the Das' algorithm [25] in terms of size of the CDS and complexity. This algorithm calculated the CDS in $O(\Delta^2)$ time with 2-distance neighbourhood information, where Δ was the maximum node degree in the graph. Wu et al. [26] proposed a power-aware Connected Dominating Set to extend the lifespan of each CDS node during bypass traffics. The nodes chosen for CDS were altered for matching the energy utilization in the network. The nodes for CDS construction were chosen based on the node degree and energy level. The results were verified through theoretical and simulation analysis.

Yan et al. [27] proposed a simple, efficient and heuristic algorithm for finding a Maximal Weight Minimum Connected Dominating Set (MWMCDS). Every node in the network n was assigned a weight w_n (a real number ≥ 0), in order to reflect the influence of gateway node's power. The weight-based approach of choosing the gateway nodes ensured that the most appropriate nodes were chosen for the role of gateway nodes so that they can properly communicate with all the other non-gateway nodes. The simulation results demonstrated that the proposed algorithm confirmed the gateway nodes quickly, ensuring the maximality of sum of CDS's weight and minimality of CDS' size. Gandhi and Parthasarathy [28] proposed two randomized distributed algorithms such as CDS-Colour and CDS-Top based on wireless interference and the subsequent loss of messages during the execution of the algorithm. The above two algorithms have produced a CDS of constant size and constant stretch ratio with high probability and converge within the poly-logarithmic running time. These algorithms were identified to be the first interference-aware distributed virtual backbone algorithms which have broken the linear barrier time.

3 Gateway-Based Strategic Connected Dominating Set (GWS-CDS)

The Gateway-based Strategic Connected Dominating Set (GWS-CDS) finds a minimal CDS in a WSN in a distributed manner. The proposed algorithm finds a Strategic CDS in a heterogeneous network based on the communication range of the node. Santi [7] proved the critical communication range (CCR) to be,

$$CCR_{GM} = c\sqrt{\frac{\ln g_n}{\pi g_n}} \tag{1}$$

where c is a constant ≥ 1, CCR_{GM} is the CCR in the presence of M-like hood gateway node mobility and g_n is the number of gateway nodes. M is an arbitrary model such that

(1) M is obstacle free
(2) The gateway node is allowed to move within a certain bounded area.

This bounded area can be linked to the Link Expiration Time (LET) as discussed by the authors Fly and Meghanathan [29]. The neighbourhood of two nodes can be predicted by LET. Let two nodes A and B (as shown in Fig. 2) be within the communication range of each other. Let (x_1, y_1) and (x_2, y_2) be the coordinates of the nodes A and B, respectively. Let v_1 and v_2 be the velocities and Θ_1, Θ_2 where $(0 \leq \Theta_1, \Theta_2 < 2\pi)$ indicate the direction of movement of nodes A and B, respectively.

The quantity of time the two nodes A and B will stay connected, $T_{A\ B}$, can be predicted using the following equation:

$$T_{A-B} = \frac{-(ab + cd) + \sqrt{(a^2 + c^2)r^2 - (ad - bc)^2}}{a^2 + c^2} \tag{2}$$

Fig. 2 Relative angular velocity in a plane

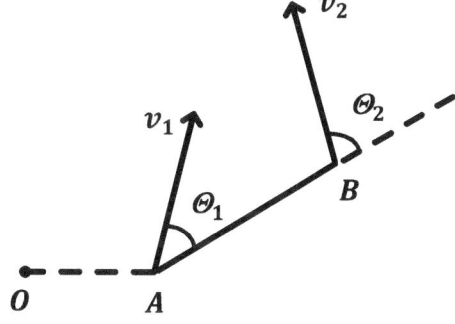

where:
$$a = v_1\cos \Theta_1 - v_2\cos \Theta_2; b = x_1 - x_2; c = v_1\sin \Theta_1 - v_2\sin \Theta_2; d = y_1 - y_2$$

The x and y coordinates of the gateway nodes must lie within the critical communication range of the network, given by the following equation of circle:

$$x^2 + y^2 \leq \text{CCR}_{\text{GM}} \tag{3}$$

Using Eq. (2), the gateway node is allowed to move a distance D_{A-B} within a certain bounded area and is given by the following equation:

$$D_{A-B} = T_{A-B} * \dot{\theta}_{A-B} \tag{4}$$

where $\dot{\theta}_{A-B}$ is the relative angular velocity between the two nodes A and B moving in a circular plane.

Result Relative angular velocity is given by the following equation:

$$\dot{\theta}_{A-B} = \frac{bc - ad}{b^2 + d^2} \tag{5}$$

where
$$a = v_1\cos \Theta_1 - v_2\cos \Theta_2; b = x_1 - x_2; c = v_1\sin \Theta_1 - v_2\sin \Theta_2; d = y_1 - y_2$$

Proof Let A and B be two nodes moving in a plane. If their velocities are v_1 and v_2 making angles Θ_1 and Θ_2 with AB as shown in Fig. 2, then the component, in the direction perpendicular to AB, of the velocity of A relative to B is $v_1\sin \Theta_1 - v_2\sin \Theta_2$ because the velocity components of A and B in this direction are $v_1\sin \Theta_1$ and $v_2\sin \Theta_2$.

The angular velocity of A relative to B is given by the following equation:

$$\frac{v_1\sin \Theta_1 - v_2\sin \Theta_2}{AB} \tag{6}$$

Also $\overrightarrow{AB}$ is the position vector of A with reference to B. So, if v_1 and v_2 are the velocity vectors of A and B, then the velocity of A relative to B is $v_1 - v_2$ and consecutively the angular speed θ of A about B is obtained by the following equation:

$$\dot{\theta} = \frac{\left| \overrightarrow{AB} \times (\overrightarrow{v_1} - \overrightarrow{v_2}) \right|}{(AB)^2} \tag{7}$$

$$\dot{\theta} = \frac{\left| (\overrightarrow{OA} - \overrightarrow{OB}) \times (\overrightarrow{v_1} - \overrightarrow{v_2}) \right|}{(AB)^2} \tag{8}$$

$$\dot{\theta} = \frac{\left| \left[(x_1\hat{i} + y_1\hat{j}) - (x_2\hat{i} + y_2\hat{j}) \right] \times \left[(v_1\cos\Theta_1\hat{i} + v_1\sin\Theta_1\hat{j}) - (v_2\cos\Theta_2\hat{i} + v_2\sin\Theta_2\hat{j}) \right] \right|}{\left[(x_1\hat{i} + y_1\hat{j}) - (x_2\hat{i} + y_2\hat{j}) \right]^2}$$

(9)

$$\dot{\theta} = \frac{\left| \left[(x_1 - x_2)\hat{i} + (y_1 - y_2)\hat{j} \right] \times \left[((v_1\cos\Theta_1 - v_2\cos\Theta_2)\hat{i} + (v_2\sin\Theta_1 - v_2\sin\Theta_2)\hat{j}) \right] \right|}{\left[(x_1 - x_2)\hat{i} + (y_1 - y_2)\hat{j} \right]^2}$$

(10)

$$\dot{\theta} = \frac{\left| \left[(x_1 - x_2)\hat{i} + (y_1 - y_2)\hat{j} \right] \times \left[(v_1\cos\Theta_1 - v_2\cos\Theta_2)\hat{i} + (v_2\sin\Theta_1 - v_2\sin\Theta_2)\hat{j} \right] \right|}{(x_1 - x_2)^2 + (y_1 - y_2)^2}$$

(11)

$$\dot{\theta} = \frac{bc - ad}{b^2 + d^2}$$

(12)

where:

$a = v_1\cos\Theta_1 - v_2\cos\Theta_2; b = x_1 - x_2; c = v_1\sin\Theta_1 - v_2\sin\Theta_2; d = y_1 - y_2$

Consider a WSN network consisting of set V vertices and E edges between the nodes, i.e. (V, E). Assume that the WSN area is a cluster of networks with some minimum number of nodes per network. The node that has at least one or two neighbours (i.e. node degree is 2) in more than one network can be the strategic node. The mobility of the strategic node is restricted using the critical communication range and strong neighbourhood of the nodes. The network cluster structure consists of a strategic node, CDS nodes and non-CDS nodes. The communication range of the strategic node is different from the communication range of all the other nodes. The nodes within the network cluster are selected to form a graph, in which the existence of an edge is ensured if and only if two nodes are connected within one hop (i.e. these two wireless nodes can always receive the signal from each other directly). The main design objective of the GWS-CDS approach is to construct a CDS with strong neighbourhood and maximum lifetime. The minimal lifetime of CDS leads to frequent destruction of CDS. Therefore, in this paper, an algorithm is developed which integrates multiple factors like LET, critical communication range, strategy, density and velocity for deriving the Gateway-based Strategic CDS (GWS-CDS).

3.1 Data Structures

The following data structures are used in the GWS-CDS algorithm.

(i) *Uncovered-Nodes-List*: It is a list which includes the nodes that are so far not covered by the CDS nodes.

(ii) *CDS-Node-List*: It is a list which includes the selected CDS nodes in each round of the GWS-CDS algorithm.
(iii) *Covered-Nodes-List*: It is a list of nodes that are in *CDS-Node-List* and the nodes covered by the CDS nodes.
(iv) *CDS-Edge-List*: It is a list which includes the edges that exists between any two CDS nodes in the *CDS-Node-List*.

3.2   Construction of Gateway-Based Strategic Connected Dominating Set (GWS-CDS)

The WSN network area is divided into various networks based on the homogeneous communication range for normal nodes and critical communication range CCR_M for gateway nodes. The normal nodes are placed randomly within the WSN network area, whereas the gateway nodes are placed using the distance parameter D_{i-j}. A scenario for a sample network is shown in Fig. 3. Here, four networks are created with different communication ranges. The black nodes represent the normal nodes, and the colour nodes represent the gateway nodes. The grey colour nodes are within the communication range of any two networks.

The green-coloured node is within the communication range of three networks, i.e. Network A, Network B and Network D. The yellow-coloured nodes are within the communication range of three networks, i.e. Network B, Network C and Network D. The red-coloured node is within the communication range of four networks—Network A, Network B, Network C and Network D. The best option for choosing the strategy node is the red gateway node because it is a connector to three other networks.

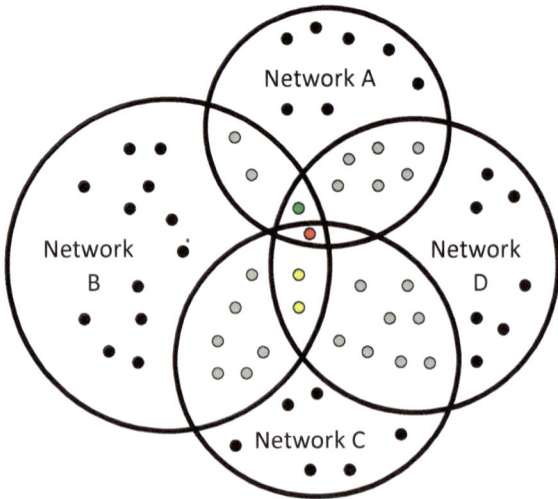

Fig. 3 Sample network with normal and gateway nodes

The assumptions for constructing the GWS-CDS algorithm are listed below:

- The partition of the WSN network area is assumed in such a way that each network will be having sufficient number of nodes and at least one node will be present in the communication range of more than one network.
- The graph is connected and each node knows its adjacent neighbours.
- The movement of the gateway nodes is restricted within the bounded area.
- Normal nodes are assigned with homogeneous communication range, and gateway nodes are assigned with critical communication range CCR_{GM}.

Algorithm for constructing GWS-CDS algorithms is listed below:

1. Input: An undirected graph $G = (V, E)$ //V, E-set of vertices and edges.
2. Output: A pipelined strategic Connected Dominating Set, a sub graph of G.
3. Randomly assign a unique identifier (ID) to every node in the network.
4. Include all the nodes in *CDS-Uncovered-List.*
5. Compute *Edge-List* of all the nodes so that each node knows the neighbour information.
6. Select any node in the network as the start node, which satisfies the basic criterion
   The basic criteria for selecting the strategic node are listed below:

   - The communication range of the node must satisfy the Eq. (1).
   - It should be within the strong neighbourhood network, i.e. it is allowed to move only within a certain bounded area, as per Eq. (3).
   - If there are more than one gateway nodes, a node that is maximally connected to other networks is to be selected. If there is a tie, the selection can be random.

7. Remove the selected node from *CDS-Uncovered-List.*
8. Include the selected strategic node into the *CDS-Node-List* and *Covered-Nodes-List.*
9. Include all the one-hop neighbours of the selected node (which are said to be covered) into the *Covered-Nodes-List.*
10. Set the density of the start node as 0.
11. Compute the density of covered nodes.
12. Choose the next node from the *Covered-Nodes-List* **by checking** the following conditions,

    - A node not included in *CDS-Node-List.*
    - A node included in *Covered-Nodes-List.*
    - A node with at least one uncovered neighbour.
    - A node with density greater than one. If there is a tie or the density of all the nodes included in the Covered-Nodes-List is less than two, choose a node with minimum velocity.
    - If still there is a tie, choose randomly.

13. Add the node chosen (from step 12) to the *CDS-Node-List*.
14. Add the entire one-hop neighbours to the *Covered-Nodes-List*.
15. Remove the chosen node (from step 12) and the entire one-hop neighbours from *CDS-Uncovered-List*.
16. Change the density of the selected node as 0 and recompute the density of all the other nodes (except for the CDS nodes and the nodes whose density is 0).
17. Go to step 12 and continue the above process until the *CDS-Uncovered-List* is empty.
18. Compute the final *CDS-Node-List* and *CDS-Edge-List*.

4   Simulations and Results

A sample network in which the WSN network area is divided into five scattered networks using different communication range is shown in Fig. 4. The communication range of Network A is four, Network B is two, Network C is five, Network D is three and Network E is one. The red-coloured node is in the communication coverage of Network B, Network C and Network E, i.e. it is the maximally network-connected gateway node. Even though the red-coloured node is connected to only three other nodes, it is in the network range of B, C and E. All the gateway nodes (grey-coloured nodes) have different communication range, ranging from 4 to 12, whereas the non-gateway nodes (black-coloured nodes) have the homogeneous communication range of one.

    The input to the algorithm is a snapshot of the network as shown in Fig. 5a. Here the circles represent the nodes, the density of the node represents the number in numerical form inside the circle, and the ID of the node represents the number in

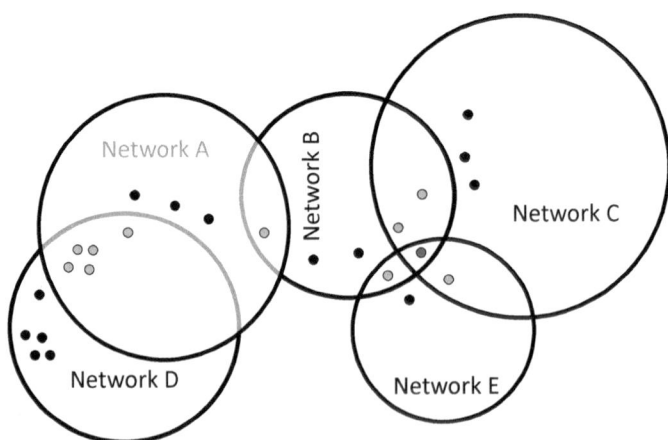

Fig. 4 Example network with normal and gateway nodes

numerical form outside the circle. Table 1 tabulates the values of density, identity and velocity for the initial static graph with 25 nodes and 44 edges. In the following figures, nodes indicated as dotted circles with a plus notation are included into the *CDS-Node-List*, the edges between any two CDS nodes in the *CDS-Node-List* are included into the *CDS-Edge-List* and the nodes indicated as dotted circles are included into the *Covered-Nodes-List*. Initially, all the nodes are included into the *Uncovered-Nodes-List*. The maximally network-connected gateway node (node 19) with density five and velocity 0.9 is selected as the strategic node as shown in Fig. 5b. It is included in the *CDS-Node-List* and *Covered-Nodes-List*. The node 19's covered neighbours (node 16, 17, 18, 21, 22) are included into the *Covered-Nodes-List*. The node 19 and its covered neighbours are excluded from the *Uncovered-Nodes-List*. The density of the strategic node (node 19) is changed to 0 and the density of all other nodes is changed as shown in Fig. 5b. Further the node with lowest density has to be selected for the next iteration. But two nodes (node 16 and 18) have the same density as one. As per the algorithm, if there is a tie, a node with minimum velocity has to be selected. Hence node 18 is included into the *CDS-Node-List* and *Covered-Nodes-List* as shown in Fig. 5c. The process is continued, and covered nodes are included into the *Covered-Nodes-List*. The next node to be selected into the *CDS-Node-List* is node 16 as shown in Fig. 5d. The process is continued until the *Uncovered-Nodes-List* is emptied. The final iteration is shown in Fig. 5e. The number of CDS nodes in *CDS-Node-List* is 12 and number of CDS edges in *CDS-Edge-List* is 12 as shown in Fig. 5f.

The GWS-CDS algorithm is implemented on an ad hoc peer-peer network by placing nodes in an M × M area. Nodes are moving randomly and the proposed algorithm was performed on a network with varying number of nodes and density. The *CDS-Node-List* and *CDS-Edge-List* are examined using the Breadth First Search (BFS) algorithm to find whether the basic CDS is connected or not. The connectors between the CDS nodes are checked using the BFS algorithm. A new CDS is constructed if and only if at least one node is not connected.

The simulations are performed in Visual C++ Express Tool. The dimensions of the network topology are 1000 m × 1000 m. The model works as follows: each node starts moving from a random location to an arbitrary destination with a random speed in the range of $[v_{min} \ldots v_{max}]$. After reaching the destination, the node stops there for a pause time before continuing on to another arbitrarily selected destination with a new randomly selected speed. The minimum velocity per node, v_{min} as well as the pause time of the nodes is set to 0. The values of v_{max} are varied between 5 and 15 m/s to represent levels of low and high mobility, respectively. Under each of the above simulation conditions, mobility trace files are generated for a simulation time of 500 s. The network topology is sampled every 0.2 s. The communication range of normal nodes is fixed as 250 m. The communication range of the gateway nodes is 150 ~ 250 m. The communication range of other normal nodes is fixed at 100 m. If a CDS of a cell breaks at a particular time instant, the appropriate CDS formation is reinitiated.

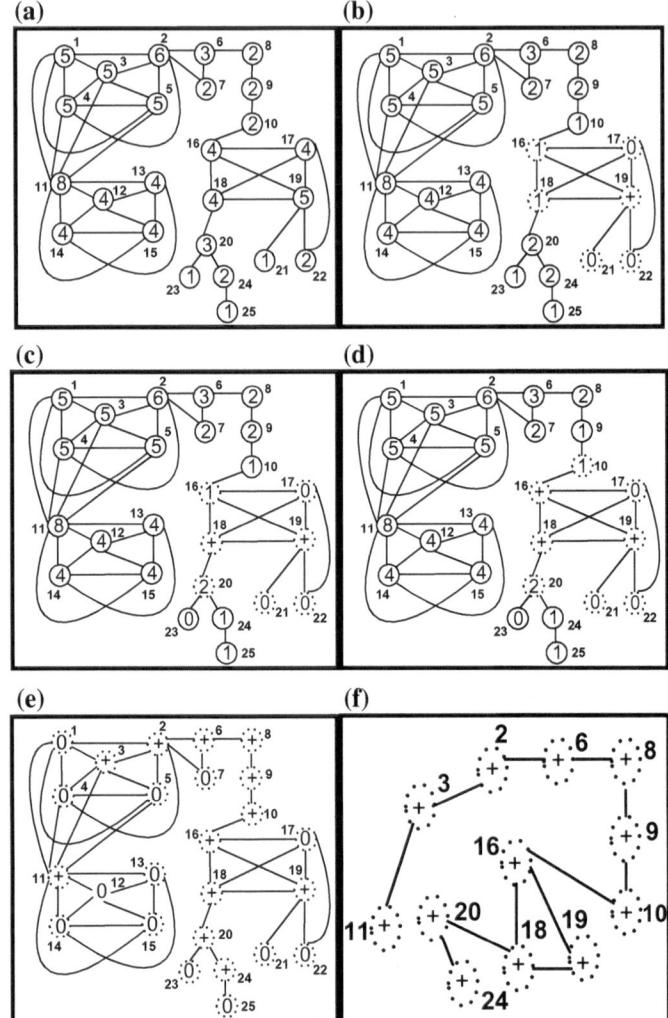

Fig. 5 Example to construct GWS-CDS **a** Initial static network graph, **b** Round # 1, **c** Round # 2, **d** Round # 3, **e** Round # 12 (last iteration), **f** CDS nodes and CDS edges list

4.1 Performance Metrics

The following performance metrics were measured in the simulations. The proposed approach was compared with MaxD-CDS [19], ST-MD-MV-CDS [9] and TCDS [22]. Each data value in Figs. 6, 7, 8, 9, 10 and 11 is an average computed over 10 mobility trace files and 10 randomly chosen source-destination (r-s) pairs from each file.

Table 1 Sample values for parameters

Identity	Density	Velocity
1	5	1.2
2	6	0.5
3	5	0.8
4	5	1.3
5	5	1.6
6	3	1.1
7	2	7.2
8	2	5.6
9	2	2.1
10	2	0.8
11	8	6.4
12	4	2.2
13	4	4.9
14	4	6.8
15	4	1.9
16	4	1.9
17	4	5.6
18	4	0.5
19	5	0.9
20	3	0.9
21	1	8.9
22	2	4.3
23	1	0.8
24	2	0.3
25	1	1.6

Fig. 6 CDS node size during low mobility network

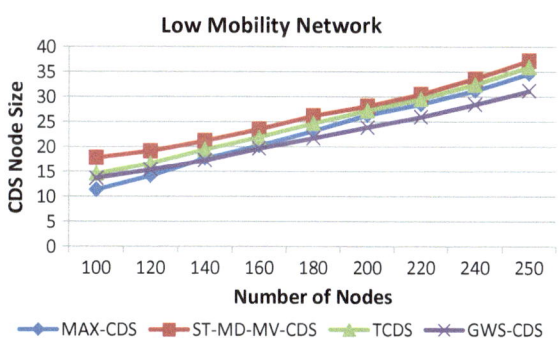

- CDS Node size: It is the number of nodes included in the CDS. A smaller sized CDS suffers less from interference problem, but ends up in the frequent CDS

Fig. 7 CDS node size during
high mobility network

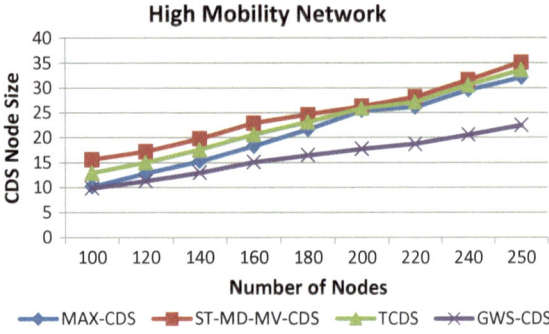

Fig. 8 CDS Edge Size
during low mobility network

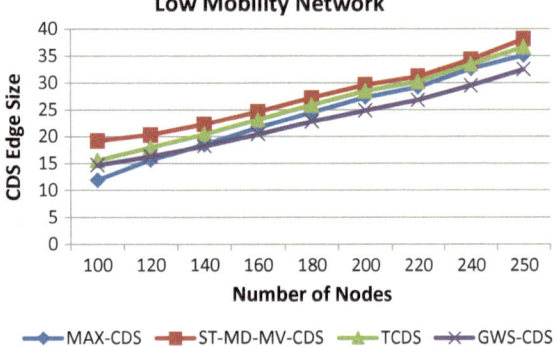

Fig. 9 CDS Edge Size
during high mobility network

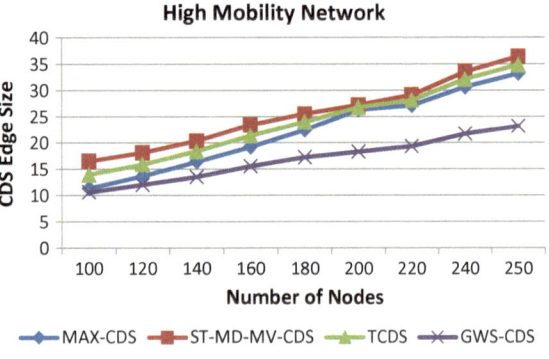

reconstruction. Hence, a CDS with a slight increase in size is constructed, which helps in maintaining the CDS easily.

- CDS Edge Size: It is the number of edges connecting the nodes that are included into the CDS.

Fig. 10 Average path length
during low mobility network

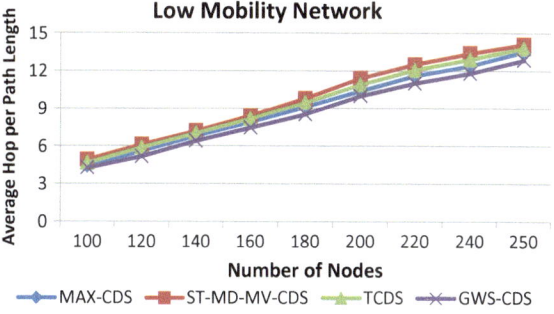

Fig. 11 Average path length
during high mobility network

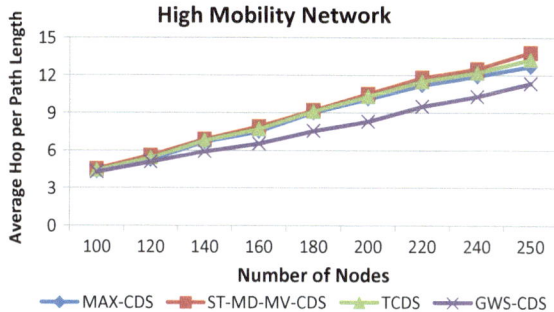

- Average path length (APL): It is the time-averaged value of the distance trav-
  elled by a packet before reaching its destination, over the entire duration of the
  simulation. Diameter is therefore the worst-case of APL.

4.2   CDS Node Size

The GWS-CDS includes less nodes than the MaxD-CDS, ST-MD-MV-CDS and
TCDS algorithms (refer to Figs. 6 and 7). The aim of the GWS-CDS is to
approximate a minimum CDS (MCDS) within the WSN. From the results, it is
observed that the CDS node size of the MaxD-CDS is 1 (at low mobility) to 1.3 (at
high mobility) times larger than the node size of the GWS-CDS. The CDS node size
of the ST-MD-MV-CDS is 1.2 (at low mobility) to 1.5 (at high mobility) times
larger than the node size of the GWS-CDS. The CDS node size of the TCDS is 1.1
(at low mobility) to 1.4 (at high mobility) times larger than the node size of the
GWS-CDS. A CDS with minimum number of nodes is constructed using the
GWS-CDS, which results in maximized energy conservation of the network.
The CDS constructed in this way is a strong neighbourhood CDS so that the
lifetime of the CDS is prolonged.

4.3   CDS Edge Size

The nodes of the MaxD-CDS are more likely to be far apart, hence the number of edges between the MaxD-CDS nodes tend to be very low (an edge can only exist between two nodes if they are within each other's communication range). The GWS-CDS, however, incurs a less CDS node size and subsequently incurs a correspondingly smaller number of edges between CDS nodes (refer to Figs. 8 and 9). From the results, it is observed that the CDS Edge Size of MaxD-CDS is 1.0 (at low mobility) to 1.3 (at high mobility) times larger than the CDS Edge Size of the GWS-CDS. The CDS Edge Size of ST-MD-MV-CDS is 1.2 (at low mobility) to 1.5 (at high mobility) times larger than the CDS Edge Size of the GWS-CDS. The CDS Edge Size of TCDS is 1.1 (at low mobility) to 1.4 (at high mobility) times larger than the CDS Edge Size of the GWS-CDS.

4.4   Average Path Length (APL)

The GWS-CDS contributes to lowest average path length (refer Figs. 10 and 11) between a r-s pair when compared to MaxD-CDS and ST-MD-MV-CDS algorithms. From the results, it is observed that the APL of the MaxD-CDS is 1.0 (at low mobility) to 1.1 (at high mobility) times longer than the APL of the GWS-CDS. On the other hand, the APL of ST-MD-MV-CDS and TCDS is 1.1 (at low network density) to 1.2 (at high network density) times longer than the APL of the GWS-CDS. The lower path length between r-s paths using the GWS-CDS algorithm is due to the selection of CDS based on the critical communication range, strong neighbourhood, density and velocity.

5   Conclusion

Simulation results showed that the GWS-CDS could be more stable than MaxD-CDS, ST-MD-MV-CDS and TCDS. The increased stability is due to the selection of gateway nodes within the strong neighbourhood and critical communication range, which can bring a direct improvement in performance to the CDS as well as the entire WSN. The GWS-CDS algorithm included a relatively lesser number of nodes, edges and has a lesser average path length. The minimized average path length increased the possibility of finding the shortest routes between the nodes in the network. The GWS-CDS algorithm constructs a CDS by considering the homogeneous and heterogeneous communication range of the nodes that rules out the assumption of having same communication range for all the nodes. Practically, the communication range of all the nodes cannot be identical for specific applications. The methodology of choosing a gateway node with critical communication range and the mobility restricted to the strong neighbourhood results in an efficient CDS construction when compared to existing algorithms. The selection and usage of gateway nodes are well analysed in this paper. In the future, as an extension to this work, selection of gateway nodes in a disk graph with bidirectional links (DGB) within a heterogeneous network can be analysed.

The proposed methodology will also be a solution to the following problem happened during 21 October 2016 where an army of million hacked IoT devices broke the Internet through a massive DDoS attack against Dyn, a major Domain Name System (DNS) provider. A significant outage to a ton of websites and services, including Twitter, GitHub, PayPal, Amazon, Netflix and Spotify, was experienced by the users. The IoT manufactures focus on the usability and performance of IoT devices, but neglects the encryption mechanisms and security measures of the devices. Thus, the IoT devices are frequently hijacked and used as weapons in cyber attacks.

A huge army of hacked IoT devices was the reason for the problem. Such issues may also affect the cloud services, thereby the fame of the cloud provider is brought down instantly. A piece of software embedded into the IoT devices connected to the cloud will never allow such attacks to be forwarded to the cloud. This software can able to understand the malware which will be in the form of bots and avoid the unnecessary routing to the cloud.

Acknowledgement The authors gratefully acknowledge the use of services and facilities of the Centre for Networking and Cyber Defense (CNCD) at Hindustan Institute of Technology and Science, Chennai, India.

References

1. Baccarelli, Enzo, et al. (2017). Fog of everything: Energy-efficient networked computing architectures, research challenges and a case study. *IEEE Access, 5,* 9882–9910.
2. Marín-Tordera, E., et al. (2017). Do we all really know what a fog node is? Current trends towards an open definition. *Computer Communications, 109,* 117–130.
3. Verba, N., et al. (2016). Platform as a service gateway for the fog of things. *Advanced Engineering Informatics,* Article in press.
4. Arkian, H. R., Diyanat, A., & Pourkhalili, A. (2017). MIST: Fog-based data analytics scheme with cost-efficient resource provisioning for iot crowdsensing applications. *Journal of Network and Computer Applications, 82,* 152–165.
5. Wu, J., Lou, W., & Dai, F. (2006). Extended multipoint relays to determine connected dominating sets in manets. *IEEE Transactions on Computers, 55*(3), 334–347.
6. Sanchez, M., Manzoni, P., & Haas, Z. J. (1999). Determination of critical transmission range in ad-hoc networks. *Multiaccess, Mobility and Teletraffic in Wireless Communications, 4,* 293–304.
7. Santi, P. (2005). The critical transmitting range for connectivity in mobile ad hoc networks. *IEEE Transactions on Mobile Computing, 4*(3), 310–317.
8. Deng, J., et al. (2007). Optimal transmission range for wireless ad hoc networks based on energy efficiency. *IEEE Transactions on Communications, 55*(9), 1772–1782.

9. Sharmila, C., & George, A. (2014). Construction of strategic connected dominating set for mobile ad hoc networks. *Journal of Computer Science, 10*(2), 285–295.
10. Akbari Torkestani, J., & Meybodi, M. R. (2010). An intelligent backbone formation algorithm for wireless ad hoc networks based on distributed learning automata. *Computer Networks, 54*(5), 826–843.
11. Hussain, S., Shafique .M. I., & Yang. L. T (2010).Constructing a CDS-based network backbone for energy efficiency in industrial wireless sensor network. In *Proceedings of HPCC* (pp. 322–328).
12. Das, B., Sivakumar, R., & Bhargavan, V. (1997). Routing in ad hoc networks using a spine. In *International Conference on Computer Communications and Networks* (pp. 1–20).
13. Guha, S., & Khuller, S. (1998). Approximation algorithms for connected dominating sets. *Algorithmica, 20*(4), 374–387.
14. Clark, Brent N., Colbourn, Charles J., & Johnson, David S. (1990). Unit disk graphs. *Discrete Mathematics, 86*(1-3), 165–177.
15. Garey, M. R., & Johnson, D. S. (1979). *Computers and intractability*. San Francisco: W. H. Freeman, Print. Print.
16. Kim, D., et al. (2009). Constructing minimum connected dominating sets with bounded diameters in wireless networks. *IEEE Transactions on Parallel and Distributed Systems, 20*(2), 147–157.
17. Meghanathan, N., & Terrell, M. (2012). An algorithm to determine stable connected dominating sets for mobile ad hoc networks using strong neighborhoods. *International Journal of Combinatorial Optimization Problems and Informatics, 3*(2), 79–92.
18. Misra, R., & Mandal, C. (2010). Minimum connected dominating set using a collaborative cover heuristic for ad hoc sensor networks. *IEEE Transactions on Parallel and Distributed Systems, 21*(3), 292–302.
19. Meghanathan, N. (2012). Graph theory algorithm for mobile ad hoc networks. *Informatica, 36*, 185–200.
20. Bannoura, A., et al. (2016). The wake up dominating set problem. *Theoretical Computer Science, 608*, 120–134.
21. Thai, M. T., et al. (2007). Connected dominating sets in wireless networks with different transmission ranges. *IEEE Transactions on Mobile Computing, 6*(7), 721–730.
22. Xiaohong, L. I., et al. (2014). A lifetime-extended size-bounded construction algorithm for connected dominating sets in heterogeneous wireless sensor networks. *Journal of Computational Information Systems, 10*(16), 6973–6981.
23. Deng, J., et al. (2007). Optimal transmission range for wireless ad hoc networks based on energy efficiency. *IEEE Transactions on Communications, 55*(9), 1772–1782.
24. Wu J., & Li H. (1999). On calculating connected dominating set for efficient routing in ad hoc wireless networks. In *Proceedings of the 3rd International Workshop on Discrete Algorithms and Methods for Mobile Computing and Communications* (pp. 7–14). August 1, 1999 ACM.
25. Das, B., Sivakumar, E., & Bhargavan, V. (1997) Routing in ad-hoc networks using a virtual backbone. In *Proceedings of the 6th International Conference on Computer Communications and Networks (IC3N'97)* (pp. 1–20). September 1997.
26. Wu, J., et al. (2002). On calculating power-aware connected dominating sets for efficient routing in ad hoc wireless networks. *Journal of Communications and Networks, 4*(1), 59–70.
27. Yan, X., et al. (2004). A heuristic algorithm for minimum connected dominating set with maximal weight in ad hoc networks. In *Grid and Cooperative Computing, Springer Berlin Heidelberg* (pp. 719–722).
28. Gandhi, R., & Parthasarathy, S. (2007). Distributed algorithms for connected domination in wireless networks. *Journal of Parallel and Distributed Computing, 67*(7), 848–862.
29. Fly, P., & Meghanathan, N. (2010). Predicted link expiration time based connected dominating sets for mobile ad hoc networks. *International Journal of Computer Science and Engineering, 2*(6), 2096–2103.

Developing Security Intelligence in Big Data

Hardik A. Gohel and Himanshu Upadhyay

Abstract In today's world, as the volume of digitized data grows exponentially, the need and the ability to store and computationally analyze large datasets are growing along with it. The term "big data" refers to very large or complex datasets, such that classical data processing software applications are insufficient to manage. A great example of a company that symbolizes the modern mass data-driven world is Google. It is possibly the most successful IT company in the world as well as the largest data processing company of modern times. In April 2004, Larry Page and Sergey Brin wrote their first and now famous "Founders Letter" to their employees which stated "Google is not a conventional company. We do not intend to become one." Twelve years down the line, with a change in leadership, incoming CEO Sundar Pichai wrote a letter to employees in 2016 and concluded it with "Google is an information company. It was when it was founded, and it is today. And it's what people do with that information that amazes and inspires me every day." There are many challenges in the analysis of large volumes of data, including data capture and storage, data analysis, curation, searching, sharing and transfer-ring, data visualization, data inquiry and updating, among others. However, the biggest challenge is information security and privacy of big data [29]. A lack of securi-ty around big data can lead to great financial losses and damage to the reputation for the company. Security threats and attacks are becoming more active in violating cyber rules and regulations. These attacks also affect big data and the information contained in it. Attackers target personal and financial data, or a company's confidential intellectual property information, which greatly affects their competitiveness. The biggest threat is when attackers target personal or consumer financial information stored in big data. Although there are rules and regulations in place to protect data, there are still vulnerabilities in big data that are serious enough to warrant substantial concern. In a recent and highly publicized incident, WikiLeaks released a huge trove of alleged internal documents from the US Central Intelligence Agency (CIA). It is by far the

---

H. A. Gohel (✉) · H. Upadhyay
Florida International University, Miami, FL, USA
e-mail: hgohel@fiu.edu

H. Upadhyay
e-mail: upadhyay@fiu.edu

© Springer Nature Singapore Pte Ltd. 2018
S. Margret Anouncia and U. K. Wiil (eds.), *Knowledge Computing and Its Applications*, https://doi.org/10.1007/978-981-10-6680-1_2

25

largest leak of CIA documents in history. There are thousands of pages describing sophisticated software tools and techniques used by the agency to break into smartphones, computers, and even Internet-connected televisions. Both government and corporate leaks have been made possible due to the ease of downloading, storing, and transferring millions of documents in a very short time. With this state of affairs in mind, there needs to be a comprehensive examination of these threats and attacks on big data, and a study of novel approaches to defend it. This chapter presents an in-depth look into the threats and attacks on big data and inspects the methods of defense and protection. We discuss the vulnerabilities of modern big data systems, and the characteristic methods of intrusion, and unauthorized seizure of data. We present a few case studies of big data weaknesses and their exploitation by attackers. The information offered here is very useful in building proper defenses against potential malicious incidents. We also discuss the specific security demands of big data environments in government and medical sectors.

Keywords Big data · Security threats · Cryptography · Digital footprints Machine learning

1 Introduction of Big Data Security

Big data is a term for datasets that are so huge or complex that usual data processing software is insufficient to manage them. The possible challenges of big data are to capture, store, analyze, exchange, curate, share, search, visualize, update, and query data. The expression "big data" normally refers to the utilization of predictive analysis, client behavior analysis, or other data analytics strategies that concentrate on extraction of value from information, and occasionally to a specific size of the dataset [2, 28]. There is little uncertainty that the amount of information now accessible is expansive, yet that is not the most relevant feature for this new information ecosystem [3]. Analysis of datasets can discover new connections to figure out business patterns, avert diseases, fight crime, and so on [27]. Business executives, scientists, medical practitioners, the media, and government agencies routinely face challenges with vast datasets in realms such as Internet searches as well as urban and business informatics. Researchers experience several restrictions in e-Science work, including genomics [4], meteorology, complex simulations, connectomics, and biological and ecological research [21].

Datasets can develop quickly since they can be progressively assembled by numerous cheap data-detecting Internet of things (IoT) gadgets, for example, cell phones, aerial(remote sensing), cameras, software logs, amplifiers, radio frequency identification(RFID) devices, and wireless sensor networks (WSNs), [24]. The world's technical per capita ability to store data has almost doubled every 40 months since 1980 [9] with almost 2.5 exabytes (2.5×10^{18} bytes) each day. One question for major organizations is to determine who ought to possess the big data activities that influence the whole organization [20].

Relational database management systems (RDBMS) and desktop statistics and visualization application packages experience issues with managing big data. This task may need massive parallel programs executing on a great multitude (almost one thousand) of servers (Villanova University). What can be considered as big data differs, relying on the abilities of the users as well as their tools, and extending capacities make defining big data a moving target. For a few organizations, managing gigabytes of information may in fact trigger the need to rethink the options for data administration. For others, however, several terabytes may be gathered before the size of data would need to be considered [8].

Big data is thus a term that depicts the huge volume of data—structured, as well as unstructured—that the business experiences every day. However, it is not the specific measure of this data which is imperative; it is, in fact, what the firms do with it that matters. Big data could be broken down for any intelligence that could prompt better choices and key business decisions.

While the expression "big data" is comparatively new, the collaboration of socially available data and its further analysis has long been practiced. However, the extent and scope were limited. This idea became important in the mid-2000s when industrial analysts contemplated the now-standard meaning of big data through the following three Vs:

- Volume—Organizations gather information from an assortment of sources, including business exchanges, online networking, and the data from sensors or even machine-to-machine information. Storage would have been an early issue although new advancements (for example, Hadoop) have reduced this weight.
- Velocity—Data rushes in at enormous speeds and needs to be managed instantaneously. RFID labels, sensors, and smart meters drive the need to manage torrential data in real time.
- Variety—Data comes in a wide range of structures and formats, including an unstructured document containing text or a set of organized numeric information stored in conventional databases. Unstructured data could even extend to forms like e-mail, audio, video, stock ticker information, and monetary transactions.

Another V, Veracity, has been added by a few firms to depict it (Villanova University), with the frequency of revisions being tested by industrial authorities [8]. Further, the 3Vs have been extended to other corresponding qualities of big data.

Digital footprints: Big data usually is a free-of-cost result arising due to the digital interactions (Source: www.bigdataparis.com).

Machine learning: Big data with respect to machine learning doesn't requires reason to identify patterns. In other words, we don't need to worry about the reason to discover patterns and correlations in the data which offers novel and great insights [15].

Big data is gathered from several sources. Sensors used to assemble atmospheric data, social media posts, digital images and video, transaction records (usually purchases), as well as wireless GPS signals, are some examples. On account of cloud computing and Internet socialization, several *peta*bytes of unformatted information are generated on the Web every day and most of this data would have an inherent business value on the off chance that it could be recorded and analyzed.

Mobile communication firms, for instance, gather information from cellular towers; gas companies gather information from seismic investigations and refinery sensors; electricity utilities collect data from not just power plants, but also circulation grids. Organizations gather a huge amount of user-generated data from clients and prospects including debit/credit card numbers, social security numbers, information on purchasing propensities, and utilization patterns of customers. This ingestion of big data and the requirement to circulate it all through the firm has actually provided a potential focus area for cyber-criminals and hackers. This information, which was earlier unusable by the firm, is now extremely important and, being liable to security laws and compliance conditions, needs to be secured.

All in all, as we can see, big data is a very well-known feature; so, what are we truly examining? Here, we look into the security issues, which include two distinct focal points: securing the firm and its clients' data in a big data setting and utilizing big data methods to examine, and even anticipate, security lapses. Be that as it may, security and protection issues are amplified by the velocity, volume, and veracity of big data, for example, massive cloud reserves, diversity of information sources and data formats, streams of data being acquired, and high-volume cloud-based migration. Conventional legacy systems for security, which are customized to smaller static data, are thus deficient.

2 Big Data Architecture Vulnerabilities (www.cisoplatform.com)

2.1 Existing Big Data Architecture

Big data is basically quite different from conventional relational databases in aspects of requisites and architecture. As we have seen, big data is normally described by 3Vs (volume, velocity, and variety). Some of the basic differences of the big data architecture include:

- **Distributed architecture**: The architecture for big data is largely distributed, scaling up to almost 1000 s of storage and data processing nodes. The data is partitioned horizontally, duplicated, and then shared across the available data nodes. Consequently, the big data architecture is quite resilient as well as fault tolerant.
- **Real-time computations**: The processing of such data needs to be continuous in nature, supporting real-time computations, which is expected to succeed the current batch processing supported by Hadoop.
- **Ad hoc queries**: Big data permits data analysts to extract optimal results by executing appropriate queries to analyze the big data sources.

- **Powerful, parallel programming language**: Extremely complex and largely parallel calculations, which are more computation intensive, need to be performed in big data rather than the usual PL/SQL and SQL queries. Hadoop, for example, uses the MapReduce framework, usually written in Java, in order to perform calculations on data processing nodes.
- **Easier code relocation**: With big data, it is easier to move and relocate the code rather than the data.
- **Non-relational data**: Big data is largely non-relational, as compared to the usual databases that follow the traditional relational approach. The primary benefit of data being stored in a non-relational form is in its ability to accept and hold huge volumes of data that exhibit considerable variety.
- **Automatic tiers**: The most frequently accessed (hottest) big datasets are tiered into high performing media, whereas the coldest ones are accordingly sent to cheaper, high-volume disks. Consequently, it is very difficult to precisely know where the data would be placed among the possible data nodes.
- **Data input from various sources**: Data collected from a variety of sources, for example system logs, social media, end-to-point devices, comprises big data.

Keeping these features of big data in mind, we could outline the following vulnerabilities that could render the existing big data architecture inadequate.

I. **Insecure computation**

An insecure program could pose as a major security challenge for any big data solution. Particularly, these would include any insecure program that could:

- Access sensitive or confidential data including personal particulars, age information, and credit card details.
- Corrupt the data, causing results to be incorrect.
- Present denial-of-service to the solutions proposed to the big data, in turn inflicting financial loss.

II. **End-point validation of input/filtering**

Big data, as specified earlier, gathers data from a huge variety of sources. As a result, two major challenges arise during the process of data collection:

- **Data filtering**: This approach aims to categorically filter the data which is suspected to be malicious or rogue data.
- **Input validation**: Another mechanism includes a clear understanding of what data can be trusted and what cannot. Thus, there would be efforts made to identify whether the data has been received from valid sources or not.

The massive amount of data being collected in big data renders it impossible to filter or validate the input data. An additional feature of the data (i.e., its behavior) creates another challenge for input data filtering and validation. Usual signature-based filtering of data may not provide a complete solution to the data filtering and input validation problem. A malicious or rogue data source could, for instance, insert large volumes of legitimate although incorrect figures into the system to alter the expected results.

III. Granular access control

Current big data solutions have been designed by keeping the performance and scalability aspects in mind, without focusing much on security. This is quite in contrast to existing relational database management systems that have quite considerable security features such as access control at various levels, such as user, table, row, and even up to cell levels. However, several constraints limit the provision of comprehensive-level access controls for a big data store, such as:

- Big data security is still an ongoing research concern.
- The non-relational character of data does not conform to the traditional realms of access control, viz., tables, rows, and cell levels. Present databases such as NoSQL depend on third-party packages or middleware applications to provide suitable access control solutions.
- Unplanned queries create added challenges regarding access control. For instance, the end user may have submitted SQL queries that would be valid for relational databases.
- Further, access control by default is disabled. This gives a suitable explanation for the practical problems encountered when trying to provide access controls at a global level to the big data.

IV. Insecure storage and communication of big data

The storage and communication of big data pose several additional security challenges:

- **Distributed data nodes**: Big data, in order to optimize its data storage, utilizes a large collection of data nodes that may even be distributed. This introduces a great challenge for the authorization, authentication, and encryption of the data at every node.
- **Auto-tiered data**: The tiering of data, as we saw earlier, is a method to optimize the process of data storage and retrieval. Such a procedure, being performed automatically, could incorrectly save very sensitive data on low-cost, less-sensitive media.
- **Real-time analysis and continuous calculations**: The effectiveness of analytical processing would require low latency to be maintained for the execution of queries. Therefore, the steps of encryption, and further decryption, could inflict overheads to system performance.
- **Transactional logs**: The transactions performed on the big data would create huge volumes of transaction logs, which could be considered as another big data that should be protected in the same way as the data.

V. Invasion of privacy by data mining and analytics

Big data is being visualized as a ready source of data for many kinds of studies, surveys, etc. Such monetization of these big data sources would in turn involve data mining as well as analytics on a large scale. However, this raises several concerns regarding the security of this data, which could in fact be drilled into and misused, thus creating a possible invasion of privacy into the data belonging to the user without their knowledge and consent. Similarly, the related issues of invasive marketing and disclosing of sensitive details need to be addressed. For instance, when AOL released unspecified search logs to be used for academic purposes, the users could be easily identified by the people performing searches. Netflix faced similar issues when the users of their unspecified datasets could be identified by their IMDB scores being correlated with respective Netflix movie scores.

3 Big Data Security Techniques

Numerous organizations now utilize big data for promotion and research, yet might not have the basics right—especially from a security point of view. Big data ruptures can be huge as well, with the potential for causing significantly more reputational harm and legitimate repercussions than at present. An increasing number of organizations are utilizing the innovation to store and dissect petabytes of information including Web logs, click stream information, and online networking substance to increase their knowledge about their clients and their businesses. In this section, we look at the security techniques proposed by federal agencies (www.splunk.com) as well as several corporate firms (www.utdallas.edu) for securing big data.

a. How government IT agencies can counter security threats by analyzing big data (www.splunk.com)

Corporations and government agencies are under almost constant attacks as cyber-gangs, as well as nation states, regularly troll to obtain valuable information. Moreover, the inherent complexity of enterprise IT infrastructure, as well as cyber-threat techniques, makes the detection of such attacks extremely overwhelming. It is of course harder for government agencies, where expertise in information security is usually thin while budgets are tight; for instance, the figures obtained from the White House depict a total growth of just about 1.5% per year in federal IT expenditures since 2009. Thus, a survey of around 300 cyber-security professionals, spread across federal as well as state/local agencies, discovered that it usually takes 16 days on average to identify the threats after intrusions into their systems and networks. Although the majority of the threats are caught quite quickly, the more sophisticated attacks could need several weeks to uncover the malicious plans. Moreover, the most fatal cyber-attacks are those undetected threats that might never be known.

Security research has provided warnings over the years regarding the growing complexity and fatality of advanced cyber-threats. The latest generation of malware contains tricky and highly evasive methods devised to exploit the uncertainties in Internet's core technology, defects in network software stacks, and constraints of security devices. Recent infringements have rested any doubts whether these attacks could be sustained for a substantial amount of time, while collecting and distributing sensitive data that often gets used to commit blackmail, if not being resold, as is usual in most cases. Attackers searching internal networks and systems usually function beneath the purview of security devices and usual security event monitoring systems (SIEMs). For instance, the attack which siphoned almost 80 million social security numbers and the associated health records off the Anthem system took around nine months to be discovered. Similarly, an attack made on a key retailer was not noticed for many days, while it sucked up customers' credit card details during the chain's busiest shopping season, mocking the almost $1.5 million spent by the firm on installing malware detection systems. Similarly, the IRS would need several months, as well as many millions of dollars, to try to recover the damages inflicted when taxpayer data were stolen to file fake returns (Fig. 1).

- Survey Results Show Government IT is Not Ready for Security Challenges

MeriTalk [16] recently conducted a survey of government security professionals, according to which more than 75% consider that their data security team is rather reactive, largely due to the sheer volume of security-relevant data that is coming into the systems. Prevention is always better than cure, thus without significant change in their strategies and processes, these public sector agencies will never be well prepared.

Organizations with traditional security systems usually learn the hard way, since the evidence of any threat, as well as the incriminating data, is not directly obtainable from the security devices. The survey found that the government IT agencies are brimming with data. Vulnerability scans from the intrusion detection systems (IDS), logs from different servers like e-mail servers, VPN, DHCP, proxy servers, show that a huge proportion of such big data often goes unexamined.

Fig. 1 Big data analytics—
The missing piece to the
cyber-security puzzle. *Source*
www.splunk.com

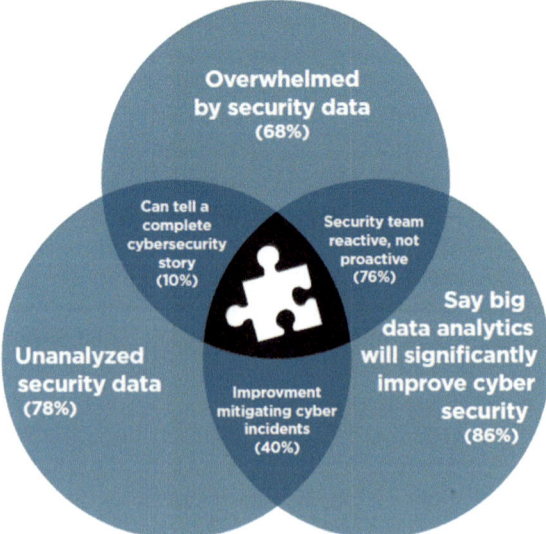

Barely 10% of the respondents claim to be getting a total security profile from the data analysis, while 78% mention that some security data always remains unchecked either owing to limited time or the lack of capable security analysts.

Government security professionals, however, see a possibility to fix these security issues. The survey results suggest that applying big data analytics to cyber-security issues would generate a big effect (as in Fig. 2). More than 60% say that it would detect breaches in real time, while almost half believe that big data analytics would also allow them to determine the probable causes of a breach. Yet, the survey suggests a cognitive discord (Fig. 3). Although most respondents considered the strategy of analyzing the security data comprehensively, only 28% of

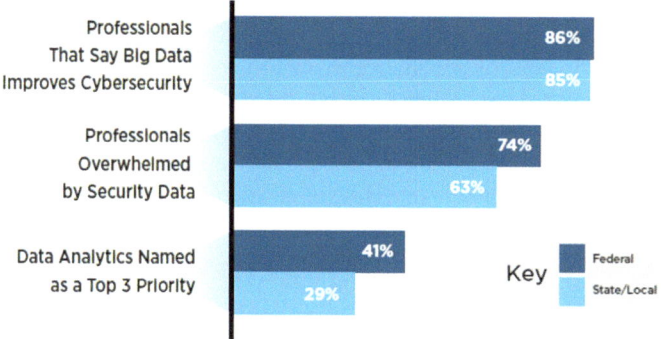

Fig. 2 Big data analytics is not seen as a priority despite the opportunity it provides to improve security. How Government IT Can Counter Security Threats by Analyzing Big Data MeriTalk Survey Analysis

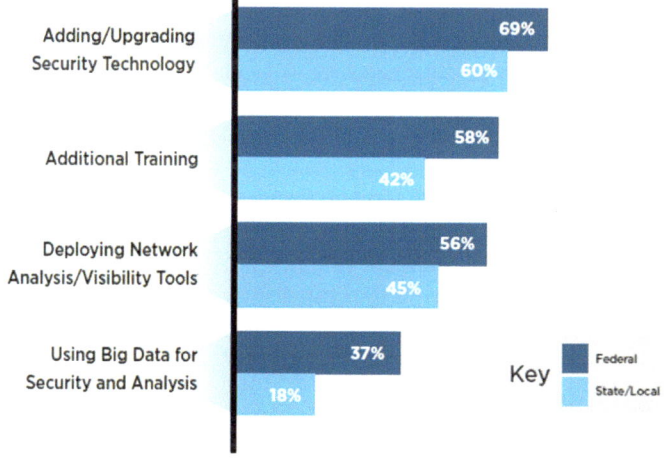

Fig. 3 Breakdown of public sector currently improving security. How Government IT Can Counter Security Threats by Analyzing Big Data MeriTalk Survey Analysis

the organizations want to actually use it to try and connect, as well as correlate, the information. This is evidence of the constricted government IT budgets, making it hard to defend new projects, though it is a myopic view.

The MeriTalk survey discovers that most of the government organizations spend a large portion of their security budget on basics, including upgrading existing systems and hardware deployment for enhanced network visibility, as well as on training the security personnel on the latest threats and associated technologies, as shown in Fig. 3. We see that less than one-third of the respondents hire new security employees. Thus, this lack of proper security measures actually reinforces the need for big data security analysis, as it is a well-proven means to achieve greater efficiency, also allowing the experts, however limited, to be much more effective.

All these issues raise doubts for the confidence people have on public sector data security. In this context, just 40% of government IT agencies can confidently say that they could stop 90% of possible cyber-incidents before the damage occurs. On the inclusion of big data analytics, results of the survey show an increase in effectiveness to almost 60%. The extent of such an improvement in cyber-detection, as well as prevention, would, however, depend on the size of the incident, as well as the information targeted, though recent research approximates the average breach to be costing almost $4 M. Thus, the government agencies should choose if they wish to resolve just one such incident or implement suitable platforms, like Splunk, to provide comprehensive operational intelligence to the organizational big data.

These platforms derive meaning from the machine data that is created by the systems, apps, and security software and applications. Such a comprehensive approach makes this platform well suited to solve problems of cyber-security data analysis for preventive, as well as forensic, uses. Such software tools would ingest, analyze, and review system data, holding terabytes of new data each day, and supporting petabytes of historical searches, like a typical network of cyber-security applications and servers would produce. The key to identifying and revealing cyber-threats is to correlate the events arising from multiple sources. Present attacks are multipronged, as well as distributed, so as to elude traditional security mechanisms and secretly mix with usual network traffic and system activity. They can only be detected after joining snippets of unusual behavior over a period of time across several systems. Typical attacks launch a start through malware which invades an unsuspecting user's PC through Trojan e-mail attachments or through a compromised Web link. After entering the network, this precursor for the multiphase attack begins to download more complex programs which can examine the internal network, launch attacks on systems displaying vulnerabilities, and return the targeted data to their selected destination. Individual events would probably be detected by the existing mechanisms, though they would not be able to trigger an alert.

- Government Action Plan

Government security professionals are now digging into the large stores of their under-utilized data and still continue to be reactive in their approach, rather than proactively trying to resolve problems before they mature into a crisis situation. The need, however, is to aggregate all existing data sources into big data, at a single place, and to further analyze and correlate the data. They could, in fact, filter and examine the data, so that it could search for ad hoc security analysis of all incidents. The MeriTalk survey displays that at a rate of almost 90%, a big data analysis system such as Splunk would significantly enhance cyber-security, thus making it easy for government IT organizations to improve their security position.

b. Strategies by corporate firms to counter threats for big data security (www. utdallas.edu), (in.pcmag.com)

Each business organization is making attempts to gather large amounts of business intelligence (BI), as much as their executives, marketing professionals, and other department personnel can collect. But, once accumulated, the striking problem is not just how to analyze and derive valuable information from such massive data, but also how to secure this big data. So, in addition to executing analytics and data visualization programs on this trove, the firm also needs to ensure that there are no spillages or leaking spots in the data store. As a solution, the Big Data Working Group of the Cloud Security Alliance (CSA) has released a handbook to avoid issues in privacy as well as security. It lists the following 10 best practices in order to secure big data that utilize a cache of storage, data encryption, monitoring, governance, and, of course, security techniques.

1. Safeguarding Distributed Programming Frameworks

The philosophy of having distributed big data stores as well as programming frameworks like Hadoop brings along a strong possibility of data being leaked. There is a probable issue termed untrusted mappers as the data being obtained from various sources might generate erroneous aggregated results.

The CSA suggests that firms should first create trust by employing methods like Kerberos authentication and also ensure conformity to existing security regulations. Then, we must un-normalize the data, by disassociating it from all the possible personal identification information (PII), thus ensuring that the privacy of the persons is not compromised. Further, we implement a predefined security policy in order to authorize access to the files, and in turn ensure that there is no leakage of the information through any of the system resources. This is done by the use of mandatory access control (MAC), for example, the sentry tool for Apache HBase. Then, once the tough part is over, what is left is to just prevent any data leaks through timely maintenance. IT department officials must keep checking the worker nodes as well as mappers into the cloud-based virtual environment, and be vigilant for fake nodes as well as modified duplicate data.

2. Securing Big Non-Relational Data

Non-relational databases, for instance NoSQL, are quite vulnerable to common attacks like NoSQL injection. The CSA provides a range of preventive counter-measures. We could begin with encryption or hashed passwords, as well as ensuring end-to-end encryption using algorithms like advanced encryption standard (AES), Rivest, Shamir, Adleman (RSA), and secure hash algorithm 2 (SHA-256). Other mechanisms, like transport layer security (TLS) and secure sockets layer (SSL), are useful encryption methods as well. Beyond those usual measures, additional layers including data tagging and object level security could be used to secure the non-relational big data, by a scheme called pluggable authentication modules (PAM). This provides a flexible means to authenticate users while ensuring the safety of log transactions by using tools like NIST log. Lastly, there are also *fuzzing methods* that uncover vulnerabilities injected, through cross-site scripting, between NoSQL and HTTP protocols when using automated means for data input at the data node, protocol, and application level of this distribution.

3. Secure Big Data Store and Transaction Logs

Storage management is a vital portion of the big data security concern. The CSA suggests the use of signed message digests (MDs) to assign digital identifiers to every digital document or file, as well as to use a method named secure untrusted data repository (SUNDR) to identify unauthorized modifications to files through malicious server agents. There are also several other techniques, such as key rotation, lazy revocation, broadcast and policy-based data encryption, as well as digital rights management (DRM). Yet, it is of course best to try and securely build the personal cloud storage atop the existing infrastructure.

4. End-Point Filters and Validation

End-point security is of utmost importance to the business. The organization could begin with utilizing trusted certificates, performing resource testing, as well as connecting only trusted devices to the network by using a mobile device management (MDM) solution (in addition to the antivirus and other malware protection software). Further, the firm could use similarity, as well as outlier detection techniques, to filter the inputs into proper and malicious, thus safe-guarding the system against ID-spoofing, or Sybil attacks, wherein one entity masquerades as having multiple identities.

5. Real-Time Compliance and Security Monitoring

Compliance is always an issue for firms, and more so when actually dealing with a continuous overflow of data. This can be best tackled with real-time analytics as well as security at each level of the storage stack. The CSA suggests that firms apply big data analytics through the use of tools like Kerberos, secure shell (SSH), and Internet protocol security (IPS) to be able to control big data in real time. Once that is done, further steps include mining event logs, implementing prime security systems for the routers as well as application-level firewalls, and also starting to implement comprehensive security control, at not just the cloud, but even at cluster and application levels. The CSA also warns enterprises to be aware of evasion attacks that try to bypass the big data security infrastructure, commonly known as "data poisoning" attacks.

6. Preserving Data Privacy

Sustaining data privacy in ever-increasing datasets is truly hard. According to the CSA, the key is to exhibit *scalability* and *composability*, by the implementation of methods like **differential privacy**—to maximize the accuracy of query results with minimal identification of records—and **homomorphic encryption** for storing and processing the encrypted information into the cloud. Further, the CSA recommends the incorporation of employee awareness programs that focus on imminent privacy regulations, and to maintain the software infrastructure by employing proper authorization methods. Finally, best practices encourage the implementation of an approach called "privacy-preserving data composition" that controls leakage of data from multiple databases by monitoring and reviewing that infrastructure which links the databases.

7. Cryptography for Big Data

Cryptography has become more advanced. Using a simple mechanism like the searchable symmetric encryption (SSE) protocol, corporate firms can in fact execute Boolean queries on encrypted big data, by just creating a system to be able to search and then filter the encrypted data. The CSA further recommends several cryptographic techniques. For example, relational encryption allows encrypted data to be compared without sharing encryption keys. This is done by matching the identifiers and their attribute values. Another mechanism called identity-based encryption (IBE) makes it easier for key management in public key systems. This is done by letting plaintext be encrypted for any given identity values. Attribute-based encryption (ABE) could be used to integrate all access controls into a comprehensive encryption scheme. Moreover, there is the converged encryption method where encryption keys could assist cloud providers in identifying duplicate data.

8. Granular Access Control

Access control comprises two core steps: restricting and granting user access. A policy to choose the right level of control in any particular situation needs to be built and implemented. The setup of such granular access controls requires the following:

- Normalize elements that are mutable, and de-normalize others.
- Administer secrecy needs and guarantee proper execution.
- Monitor access labels.
- Observe admin data.
- Use single sign-on (SSO), as well as
- Use a proper labeling scheme to uphold appropriate data federation.

9. Auditing

Auditing, especially at a granular level, is a must for big data security, especially after the system has been attacked. It is recommended that corporate firms create a consistent audit analysis following any attack, as well as to be sure to provide a complete audit trail. However, this should not hamper performance, as it is necessary to guarantee quick access to data to reduce response time. The confidentiality and integrity of audit information are also equally essential. Information related to the audit shall be separately stored, with granular protection in the form of user access control as well as regular monitoring. It is also possible to use open source audit layering or even a query coordinator like ElasticSearch that could make the entire process simpler.

10. Data Provenance

Contextual to the requirements, data provenance could have several meanings. Here provenance refers to the origin of the metadata being generated by usual big data applications. Such data truly needs significant protection. This would need to first develop an infrastructure authentication protocol, which would control the access, and also set up regular status updates as well as data integrity verification by using methods like *checksums*. Additionally, to secure the big data at the source itself, we need to implement scalable, dynamic, granular access control while implementing encryption methods. Just one approach may not be suitable to provide security to the big data across all levels of the organization and so we need to secure every level of the big data infrastructure as well as the application stack.

4 Threats and Attacks Against Big Data

Considering the security and privacy aspects, we need to visualize big data security from several viewpoints. We need to think not just of how to protect the data primarily, but also the processing of big data, and of course the output of this big data processing. Kim et al. [13] provide three major concerns where security is needed in big data: access control, data security, and information security.

Xu et al. [30] have presented a big data security model that considers the user role of security in each of these phases of the big data process. Actually, to secure the big data environment, it is vital to identify the possible threats, and attacks, that big data could experience during its lifecycle. Such an identification of the threats, as well as attacks, would help the security groups to strengthen the defense against such threats. Alshboul et al. [1] present a big data lifecycle model consisting four phases: data collection, data storage, data processing, and knowledge creation. Further, they integrated their model with possible threats and attacks to provide a security threat model, which could further be used to secure the big data infrastructure. They have identified four types of roles the users portray in the big data environment, including data provider, data collector, data miner, and decision maker [30].

Alternatively, various threats as well as attacks threaten the big data technology. Dev et al. [6] have explained a data mining-based threat that utilizes data mining methods to extract important information as well as sensitive data. Big data is also threatened by the aspect of data privacy. This is because the declarations of such sensitive data may tarnish the reputation of individuals or organizations. These include threats of re-identification and wrong results [10]. Wu and Guo [29] have also maintained that information assurance and privacy are major concerns for big data environments where the extraction of personal or sensitive data could harm individuals as well as corporate organizations that could lead to numerous business problems.

Big data technology, as we saw earlier, faces several security threats and attacks that usually are drawn from the usual features of big data technology, in turn relying on data analytics methods and data mining algorithms. Moreover, attackers could also make use of data mining procedures to locate sensitive data and further release it to the public, causing a data breach. Threats as well as attacks to big data can thus be classified in terms of the four phases of the big data lifecycle, which have been depicted in Table 1.

Each of these phases has peculiar characteristics and is allocated different tasks, making each of the phases susceptible to particular threats and attacks. For example, the data collection phase experiences attacks such as phishing and spoofing, which target people engaged in the process of data gathering and distribution. Better awareness and proper compliance to security procedures and policies are one of the ways to enhance security for this phase. Data mining-based attacks, commonly performed by hackers, can be characterized by the use of these extraction techniques to retrieve sensitive data, which is then used illicitly. Such attacks could be restricted by the fragmentation of the datasets either horizontally or vertically, as well as by adopting means for non-centralized storage of data. The risk of physical threats like theft or unauthorized access, however, needs to be dealt with in the usual ways.

The data analytics phase poses a risk to attacks that may result in either the release of sensitive data or in harming the data process. Data mining-based attacks may occur to discover and provide vital information, or associative techniques

Table 1 Threats model (Alshboul, Wang, Nepali)

Phases	Threats and attacks	Description	Suggested defense
Data collection	Phishing Spamming Spoofing	These attacks are hacking data provider and collector to get an access to the data in the collection phase	Security awareness program
Data storage	Data mining-based attacks	Targeted datasets to extract knowledge [6]	Divide datasets (vertically and horizontally) and noncentral data storage framework
	Attacks on data storage devices	Stealing hard disks or make images of them	Physical security measures noncentral data storage framework
	Unauthorized data access	People access data illegally	Access control
Data analytics	Data mining-based attacks	Using data mining methods to extract sensitive knowledge	Divide datasets (vertically and horizontally) and use access control
	Re-identification threat	Identification threats of personal information [10]	Core attribute encryption
	Wrong result threat	Using incorrect analysis process, which leads to incorrect results [10]	Follow correct analysis procedures and document, audit, and review the process
Knowledge creation	Privacy threats	Releasing the resulted knowledge (e.g., rival competitors)	Adopt encrypt the resulted knowledge and adopting access control strategy
	Phishing and spoofing	Decision makers are targeted	Security awareness programs

could help to locate personal attributes which would actually have an impact on data privacy. To protect the big data structure from such attacks, defensive methods such as fragmenting the dataset, either horizontally or even vertically to encrypt the high-weighted core attributes, may be implemented. Another threat possible in this phase is that of obtaining inappropriate results from the data analysis process [13]. Thus, it is important to follow the proper analytics process as well as to document it.

Lastly, in the knowledge creation phase, the knowledge created from the big data process usually comprises of sensitive information that should not be distributed publicly, especially to business rivals. Some threats to privacy, as well as security attacks, may intend to affect those who are related to the final result of this big data process. As a result, effective security policies and their proper implementation are required. Once everyone realizes the extent of damage possible due to security lapses in big data, they will hopefully begin to adhere to established standard procedures for access control. The development of efficient security awareness programs that could not just mitigate, but even prevent the occurrence of any threat, is needed.

5   Case Study of CIA—USA Documents Disclosed on WikiLeaks

Take for example, the WikiLeaks case [25]. WikiLeaks, in what is considered to be one of the greatest leaks, on March 7, 2017, released thousands of pages describing complex software tools as well as techniques used by the CIA to break into computers, smartphones, and even Internet-connected televisions. The documents provide a detailed catalog of highly specialized tools. They also include the instructions for hacking a wide range of common computer-oriented tools used for spying such as Skype, Wi-Fi networks, PDF documents, and even commonly used antivirus programs.

The Wrecking Crew program explains the process to crash a specific computer, while another program is used to steal passwords, by using the auto-complete function on Internet Explorer. Other similar programs called AngerQuake, CrunchyLimeSkies, McNugget, and ElderPiggy have been used. This document dump is the latest attack on the antisecrecy organization and of course a great blow to the CIA that uses its hacking capabilities to conduct espionage against foreign targets.

WikiLeaks remarked that this was just the first release of a larger set of confidential CIA material and includes more than 7800 Web pages with about 940 attachments, most of them partly redacted by WikiLeaks editors to avoid revealing the actual code for cyber-weapons. The entire archive of CIA material, however, consists of several hundred million lines of computer code, the group claimed. In one disclosure that may specifically trouble the tech world if confirmed, WikiLeaks said that the CIA and its allied intelligence services have managed to compromise both Apple and Android smartphones, allowing their officers to bypass the encryption on popular services such as Signal, Telegram, and even WhatsApp. WikiLeaks also confirmed that government agents can also collect audio as well as message traffic even before encryption is done.

The National Security Agency (NSA) documents given by Edward J. Snowden in 2013 to journalists do not include examples of how those tools had actually been used against foreign targets. Even though the liability for such a leak toward compromising national security was limited, the breach was highly embarrassing, specifically for an agency that depends on secrecy. Robert M. Chesney, a specialist in national security law at the University of Texas at Austin, likened the CIA episode, to a group calling themselves as the Shadow Brokers disclosing the set of hacking tools being used by the National Security Agency, last year. There was no public confirmation of the authenticity of the documents, which were produced by the CIA's Center for Cyber Intelligence and are mostly dated from 2013 to 2016. The authenticity of the documents has been confirmed by one government official, while another former intelligence officer said that few of the code names for the CIA programs, one of the organization charts as well as the description of a CIA hacking base, appeared to be genuine.

The CIA was surprised by this kind of a document dump. In some regard, the CIA documents confirmed and provided details on a few capabilities that have been suspected in technical circles for a long time. It is being contemplated by people who know a lot about security and hacking that the CIA was investing in enhancing these capabilities as, otherwise, other parties like China, Iran, Russia, and several private agents would actively examine and exercise the possibilities.

This episode would surely raise concerns in the USA and other nations regarding the trustworthiness of technology where cyber-security can impact human life and public safety. There is no evidence to suggest whether the CIA hacking tools have been in fact used against Americans themselves, although documents suggest the government has knowingly allowed such vulnerabilities in phones and other electronic devices to make spying easier, though these could be effectively used by hackers too. Therefore, patching these security holes immediately would be the best way to make everyone's digital life safer.

In the business world, the so-called Panama Papers and several other large-volume leaks have laid bare the details of secret offshore companies used by wealthy and corrupt people to hide their assets. Both government and corporate leaks have been made possible by the ease of downloading, storing, and transferring millions of documents in seconds or minutes, a sea change from the use of slow photocopying for some earlier leaks, including the Pentagon Papers in 1971.

6 Big Data Security Analytics Tools

Big data security analytics subsequently will be as common as malware detection and vulnerability scans. This is owing to the fact that these big data platforms permit enterprises to collect data from several, diverse data sources, collaborate that data in almost real time, explore patterns, and identify malicious activities, and also observe, report, and perform forensic investigations. The tools used for analyzing big data security are in turn capable of scrutinizing and processing huge volumes of disparate data types. Although most organizations may not require all of the features of big data security analysis tools, these are extremely efficient in providing a comprehensive solution for big data security. The leading tools providing security analytics for big data include solutions from IBM, RSA, Splunk, LogRhythm, Fortscale, Hexis, and Cybereason. Each of these solutions can be evaluated against five essential parameters that would determine the extent of its utility:

i. Unified data management

Any product for big data security analytics is based on the principle of unified data management. This is because the data management platform typically stores and also queries the data all through the enterprise. Moreover, it must also try and adjust the data management features with its cost and scalability features. The most commonly used analytics platforms usually prefer Hadoop, as it is a widely used platform for managing big data and its related ecosystems. For instance, Fortscale, uses a Cloudera distribution of Hadoop, which permits Fortscale to linearly scale up, as there are new nodes being added to the cluster.

A distributed data management system is used by IBM's QRadar. This enables the data storage to be scaled horizontally. Although it is rare, since most applications would need access to only local data, there are possibilities that the data store may be distributed, for instance, in case of forensic analysis. As a result, their corresponding security information management systems, or SIEM, would also need to be distributed. Such a big data SIEM makes use of data nodes to scale up to petabytes of data, rather than using storage area networks (SANs), resulting in substantial economies in costs as well as complexity. Similarly, RSA Security Analytics too uses an architecture that is able to scale linearly, as it is both federated as well as distributed. Events need to be prioritized so as to enhance the efficiency of the tool. Hawkeye Analytics Platform (Hawkeye AP) is built on a data warehouse platform for security event data. It is usually characterized with providing low-level and scalable data management (i.e., it is able store large amounts of data in multiple files across servers). Further, it is also crucial as it comprises of tools to query the data in a structured way. The Hawkeye AP, presented by Hexis Cyber Solutions, interestingly stores data in columnar form, as compared to traditional row-based methods. This, in turn, ensures not just optimized performance, but also tamperproof storage of big data.

ii. Support for multiple data types, including log, vulnerability, and flow

Variety, as we have seen, is one of the basic attributes of big data. This in turn poses considerable challenge to the integration of data, and that too securely. The security analytics tool by RSA employs a modular architecture to collaborate various data types, also allowing data from other sources to be added subsequently. Multiple kinds of data often dictate the need for various security tools. For instance, IBM's QRadar has a component called vulnerability manager that is able to combine data from several vulnerability scanners and enhance the data with situation-relevant details [11]. Incident Forensics is a special module to analyze security-based incidents through flow data and packet captures across the network. This detection tool comprises of a search engine which can effectively process even terabytes of network data. Another popular example of a platform with wide support for almost all types of data (machine data, security activities, system and audit logs, flow data, and application logs) is the Security Intelligence Platform by LogRhythm. This platform generates data regarding file veracity, process validity, network communication efficiency as well as user activity, by analyzing the big data store. The Enterprise Security package by Splunk lets analysts not just search for data but also depicts visual correlations among them to discover and collect data about malicious events.

iii. Scalable data ingestion

Big data security analytics can analyze large volumes of data comprising of a broad range of data types. For this, security products for big data analytics must consume data from servers, terminals, nodes, networks, and any other infrastructural components that vibrantly change states. The major risk of such data ingestion is, however, that the analytics processing cannot match pace with the rate of incoming data. Splunk is quite well recognized for its wide capabilities for data ingestion. This platform also allows for custom connections, in addition to the usual ones. Data here is maintained in a schema-less mode and is indexed at the time of ingestion itself to allow multiple data types and also responding rapidly to queries. IBM QRadar scales up to geographically distributed systems from simpler, single-appliance deployments. It is well designed to fulfill the demands of big enterprises. It can process millions of events in a second for real-world applications. Augmentation is another important point to be considered here. It is the process of supplementing the recorded event with additional information, specifying the context in which it has been collected. RSA security analytics, for instance, qualifies the network data, as it is being analyzed, by adding details of these network sessions, or maybe threat indicators, or any other details, which could help analysts to comprehend the wider picture enclosing the low-level security data. Moreover, the data collection is also a key concern. The time taken by the system to collect the data effectively sets a lower bound on the rate of detecting any security events. The positioning of data collection locations also determines the extent and the type of data being collected. Another platform, named Cybereason, uses sensors that execute in the user space of terminal operating systems, thereby allowing data to be collected without causing any adverse effect on either user experience or the kernel-level functions. These sensors can keep collecting data even if the devices are not linked with the enterprise network.

iv. Information security-specific analytic tools

Big data security analytic tools are required to scale up to be able to meet the amount of data produced by the enterprise. Likewise, the analysts must be able to execute queries on the event data at appropriate levels of abstraction keeping the view of an information security position. Fortscale uses machine learning as well as statistical analysis, commonly known as data science methods, to be able to adapt to the changes occurring in the security environment. As a result, the analysis can be based on the real data than just on predefined rules. Machine learning algorithms can identify changes in the network behavior, doing away with the need of any human intervention to modify the predefined set of rules.

   Security analytics largely depends on the intelligence regarding malicious activities. The RSA live service dictates data processing, as well as correlation rules to the deployments of RSA security analytics. New rules could thus be used to analyze the new data that is arriving in real time, as well as the historical data that is stored on the RSA security analytics system. Data science methods are often used to improve the quality of analysis. Analytics workflow by LogRhythm includes the processing, machine analysis, and forensic analytics stages. Processing transforms the data in many ways to raise the chance for detection of useful patterns from raw data. The processing comprises of data classification, time normalization, risk contextualization, and metadata tagging.

v. Compliance reporting, alerting and monitoring

Compliance reporting is an essential component for almost all enterprises today. The reporting rules that are to be included along with big data security platforms would thus need to satisfy the specific compliance requirements of the organization.

The risk manager is an add-on with IBM Security QRadar and provides tools to manage configurations of the network devices for the purpose of risk management and compliance. It has the following features including automated monitoring, compliance policies assessment, multi-vendor product audits, and threat modeling.

Fortscale, as mentioned earlier, utilizes machine learning to regularly assess changes to the baseline activities in order to detect anomalous events. The system can raise alerts and also provide information regarding the context of these events. RSA security analytics provides almost 90 templates to fulfill reporting needs of regulations including SOX, HIPAA, and PCI DSS with minimum efforts needed from the end users. The Cybereason platform has been specifically designed to identify any malicious activity. The platform comprises of an investigation console that coordinates the information and also visualizes the attack timelines and the users and devices affected. Continuous monitoring via dashboards is the strategy followed by Splunk Enterprise Security. The metrics include key security, as well as, performance indicators, along with trend pointers. Prioritized workflows are another strong point for this platform, which also supports the tracking of highly privileged users while reporting on any unauthorized attempts to access any critical applications. The Hawkeye AP package is stocked with 400 reports that can be customized to particular requirements. It provides the option to create custom reports, as it uses the relational technology and also supports the ANSI standard for SQL, and also the JDBC and ODBC drivers. Lastly, the LogRhythm platform consists of prioritized alarms, standardized reports, and a real-time reporting dashboard. It also comprises additional forensic analysis tools such as case management tools, evidence lockers, and incident tracking metrics.

7 Open Source Security Tools for Big Data

As the world is progressing toward IoT, the need for Web as well as infrastructure security has become a prime concern. We need to make our devices, systems, and the networks secure. Industry innovators such as Google, Facebook, and Netflix have looked into this concern and engage in the development of security tools with the open source fraternity. The rapid transition in the network infrastructure, from being closed to now scaling entire enterprises, would in turn increase the possibility of threats or attacks by virus, rootkits, malware, spyware, adware, and so on. Such security threats could in fact result in numerous disruptions, from denial-of-service (DoS) attacks to DNS poisoning, identity theft, etc., across the Internet. The frequency of security breaches occurring on the Web prompts organizations, and even professionals, to take appropriate precaution against such attacks. Several open source tools [5] are available in order to counter such threats and to protect our big data stores from risk. Following are ten open source tools widely used as big data security solutions in the industry.

1. OSQuery

OSQuery has been developed by Facebook. It is a simple tool for Linux and Mac OS X infrastructure. The important features of this tool include hardware changes, monitoring files and network traffic and process creation. This tool allows for easy access to data and also logs system information according to the queries posed. It further allows users to code automation scripts, to apply executive big data security intelligence, as well as to discover novel ways for the enterprise to upgrade servers.

2. Security Onion

Security Onion is a Linux-based solution for intrusion detection systems (IDS), log management, and network security monitoring (NSM). It is basically an intrusion detection system and is extremely simple to set up for an organization. Security Onion comprises of three major functions: full packet capture, network-based IDS (NIDS), and host-based IDS (HIDS) for detecting intrusion and several powerful tools to provide security and analytics for big data [23].

3. Skyline

Skyline provides features to uncover anomalies in the big data infrastructure. Operating in real time, it is built to facilitate offline monitoring of several thousands of metrics. It is generally used to monitor systems where huge quantities of data arise from high-resolution time series. Once any anomalous metric is detected, it floats up the entire time series to the Web app, so that the anomaly can be noticed and rectified.

4. Google Rapid Response

Google Rapid Response (GRR) is an approach where Google aims to examine incident responses instantaneously, even though from a remote site [7]. GRR typically comprises of an agent, or client, to be deployed to the target system, as well as a server infrastructure, which could communicate with and also manage the agent. It is available for most popular OS including Linux, Windows, and Mac OS X. Its major features include live, remote, memory analysis by the use of open source memory packages for Linux, Windows, and Mac OS X, and also a memory analysis framework, called Rekall.

5. OSSEC

OSSEC is an open source host-based IDS (HIDS) with excellent features such as file integrity checking, log analysis, policy monitoring, real-time alerting, rootkit detection, and dynamic responses. It executes on most popular operating systems, such as Linux, Windows, Mac OS, AIX, HP-UX, and Solaris [17].

6. Scumblr and Sketchy

Scumblr is a Web-based application, which allows its users to conduct regular searches and take appropriate actions based on the results generated. It also performs searches by using plug-ins, or APIs, called search providers. Every search provider recognizes the method to perform searches through a particular site or API, such as Google, Twitter, Bing. Moreover, searches could also be pre-configured within Scumblr itself, on the basis of the options offered by the search provider.

7. RAPPOR

RAPPOR, developed by Google, is another interesting privacy mechanism that allows the analysis of big data, such as wide demographic statistics, to make inferences about their populations, while preserving privacy of the individual users.

8. OpenVAS

OpenVAS is a powerful and comprehensive vulnerability management solution. It provides a robust framework, comprising several tools and services to incorporate powerful vulnerability scanning, and in turn, vulnerability management [19].

9. OpenSSH

OpenSSH, as the name suggests, is a free version of SSH (secure shell) association tools, which the technical users of the Internet can rely on. The users of telnet, ftp, or rlogin may not be aware that their passwords are transmitted in an unencrypted form, directly, across the Internet. OpenSSH therefore encrypts all the traffic (even passwords) so as to effectively eliminate attacks such as connection hijacking, eavesdropping. Moreover, OpenSSH also provides secure tunneling capabilities and offers various authentication methods, in addition to supporting all versions of the SSH protocol [18].

10. MIDAS

MIDAS is a framework to develop a Mac intrusion detection analysis system, which is based on the collaborative work and discussions by the Facebook and Etsy security teams. The repository offers a modular framework as well as a number of helper utilities, apart from an example module that can be used to detect alterations to usual OS X persistence mechanisms.

8 Summary

This chapter discussed the impact of big data and the need for securing it. We began the analysis by outlining the vulnerabilities faced by the existing big data architectures and discussed several competent security techniques for big data in the context of government and corporate sectors. We discussed in detail the different factors that make the current big data architecture vulnerable. Further, we discussed the possible techniques by which the government agencies, as well as corporate, could counter this issue. We presented results of a survey by MeriTalk that concludes that big data analytics is not a priority even though it would considerably improve data security. We also talked about the Anthem system that was attacked to steal vital information from medical records. Additionally, we described several best practices for big data security, which suggest the need for granular access control, as well as advanced encryption of big data. Compliance and security monitoring need to be real time, especially at end points. It is suggested that regular auditing is the best precautionary method though. Keeping the security and privacy aspects in view, the threats and attacks against big data have been discussed at length. We have analyzed the big data lifecycle threat model to elaborate on the types risks each phase is susceptible to. We presented a case study that showcased the possibility and impact of illegal access to big data stores of the CIA. The WikiLeaks case clearly demonstrated the vulnerability of big data, implying the need for big data security. We then described several big data analytics tools and explained in detail their desired characteristics. We also analyzed several open source security tools for big data that would provide an affordable and efficient solution to secure the big data stores.

References

1. Alshboul, Y., Wang, Y., & Nepali, R. K. (2015). Big data lifecycle: Threats and security model. In 2015, Twenty-first Americas Conference on Information Systems, Puerto Rico 2015.
2. Bisk. (2017). "What is big data?" Business intelligence by Villanova University. Retrieved on May 22, 2017.
3. Boyd, D., & Crawford, K. (2011). Six provocations for big data. In *Social Science Research Network: A Decade in Internet Time: Symposium on the Dynamics of the Internet and Society.* https://doi.org/10.2139/ssrn.1926431.2011.
4. Community cleverness required. (2008). *Nature, 455*(7209), 1. https://doi.org/10.1038/455001a. PMID 18769385.2008.
5. Dan, S. (2016) Comparing the top big data security analytics tools. At http://searchsecurity.techtarget.com/feature/Comparing-the-top-big-data-security-analytics-tools. Accessed on May 16, 2017.
6. Dev, H., Sen, T., Basak, M., & Ali, M. E. (2012). An approach to protect the privacy of cloud data from data mining based attacks. In *Proceeding of High Performance Computing, Networking Storage and Analysis, IEEE*, November, (pp. 1106–1115). https://doi.org/10.1109/SC.Companion.2012.133.
7. GRR Rapid Response at https://github.com/google/grr. Accessed on May 22, 2017.
8. Grimes, S. (2017). Big data: Avoid 'Wanna V' Confusion. InformationWeek. Retrieved May 25, 2017.
9. Hilbert, M., López, P. (2011). The World's Technological Capacity to Store, Communicate, and Compute Information". *Science, 332*(6025), 60–65. https://doi.org/10.1126/science.1200970. PMID 21310967.
10. Jensen, M. (2013). Challenges of privacy protection in big data analytics. In *Proceeding of the International Congress on Big Data, IEEE*, June, (pp. 235–238). https://doi.org/10.1109/BigData.Congress.2013.39.
11. Jitendra, C. (2014). Top 5 big data vulnerability classes. At http://www.cisoplatform.com/profiles/blogs/top-5-big-data-vulnerability-classes-1. Accessed on 12, 2017.
12. Kantarcioglu, M. (2017). Securing 'big' data. At http://www.utdallas.edu/~muratk/research-summary.pdf. Accessed on May 18, 2017.
13. Kim, S.-H., Eom, J.-H., & Chung, T.-M. (2013). Big data security hardening methodology using attributes relationship. In 2013 International Conference on Information Science and Applications (ICISA), IEEE, June, (pp. 1–2). https://doi.org/10.1109/ICISA.2013.6579427.
14. Mac Intrusion Detection Analysis System (MIDAS). Available at https://github.com/etsy/MIDAS. Accessed on May 26.
15. Mayer-Schönberger, V., & Cukier, K. (2013). *Big data: A revolution that will transform how we live, work and think.* London: John Murray.
16. MeriTalk. (2015). Survey analysis, How Government IT Can Counter Security Threats By Analyzing Big Data. At https://www.splunk.com/content/dam/splunk2/pdfs/white-papers/how-government-it-can-counter-security-threats-by-analyzing-big-data.pdf.
17. OSSEC—Open Source HIDS Security at http://www.ossec.net/ Accessed on May 22, 2017.
18. OpenSSH. (2017). At http://www.openssh.com/. Accessed on May 24, 2017.
19. OpenVAS. (2017). At http://openvas.org/. Accessed on May 25, 2017.
20. Oracle and FSN. (2017). Mastering big data: CFO strategies to transform insight into opportunity. December 2012.
21. Reichman, O. J., Jones, M. B., Schildhauer, M. P. (2011). Challenges and opportunities of open data in ecology. *Science. 331*(6018), 703–705. https://doi.org/10.1126/science.1197962. PMID 21311007.2011.
22. Rob, M. (2017). 10 best practices for securing big data. At http://in.pcmag.com/feature/107583/10-best-practices-for-securing-big-data. Accessed on May 12, 2017.
23. Security at http://blog.securityonion.net/p/securityonion.html. Accessed on May 27, 2017.

24. Segaran, T., Hammerbacher, J. (2009). *Beautiful data: The stories behind elegant data solutions* (p. 257). O'Reilly Media. ISBN 978-0-596-15711-1.2009.
25. Shane, S., Rosenberg, M., & Lehren, A. W. (2017). WikiLeaks releases trove of alleged C.I.A. hacking documents. *New York Times* March 7, 2017.
26. Skyline anomaly detection system at https://github.com/etsy/skyline. Accessed on May 28, 2017.
27. The Economist Newspaper. (2010, February 25). *Data, data everywhere.* Accessed on May 28, 2017.
28. What is big data?—Bringing big data to the enterprise. www.ibm.com. Retrieved May 20, 2017.
29. Wu, C., & Guo, Y. (2013). Enhanced user data privacy with pay-by-data model. In *Proceeding of the International Conference of Big Data*, IEEE, Ieee, October, pp. 53–57. https://doi.org/10.1109/BigData.2013.6691688.
30. Xu, L., Jiang, C., Wang, J., Yuan, J., & Ren, Y. (2014). Information security in big data: Privacy and data mining. *The Journal for Rapid Open Access Publishing, 2*, 1149–1176. https://doi.org/10.1109/ACCESS.2014.2362522.

Methods to Investigate Concept Drift in Big Data Streams

Nidhi, Veenu Mangat, Vishal Gupta and Renu Vig

Abstract The explosion of information from various social networking sites, Web clickstream, information retrieval, customers' records, users' reviews, business transactions, network event logs, etc. Results in generating a continuous deluge of data at different rates, called streaming data. Organizing, indexing, analyzing, or mining hidden knowledge from such a data deluge becomes a critical functionality for a broad range of content analysis tasks that includes emerging topic detection, interesting content identification, user interest profiling, and real-time Web search. But managing such 'Big Data' becomes even more challenging when streaming data is taken for analyzing and producing results in real time. The streaming data may include numeric, categorical, or mixed value. Most of the current research has been done on numeric data streams by exploiting the statistical properties of the numeric data. But now categorical/textual data streams have also gained researchers' interest due to the high availability of data in textual format on the Internet. Applying classification for managing data streams is an unrealistic approach as not every incoming data has a class label. So, in such a case, for managing unlabeled data streams, a clustering technique is applied. One property that can affect the results of any clustering algorithm is concept drift. Therefore, detecting and managing concept drift over a period imposes a great challenge to better cluster analysis. This chapter provides an in-depth critique of various algorithms that have been introduced to handle concept drift in a real environment. A framework for examining concept drift in big data streams is also proposed.

Nidhi (✉) · V. Mangat · V. Gupta · R. Vig
UIET, Panjab University, Chandigarh, India
e-mail: nidhi789@pu.ac.in

V. Mangat
e-mail: vmangat@pu.ac.in

V. Gupta
e-mail: vishal@pu.ac.in

R. Vig
e-mail: renuvig@hotmail.com

© Springer Nature Singapore Pte Ltd. 2018
S. Margret Anouncia and U. K. Wiil (eds.), *Knowledge Computing and Its Applications*, https://doi.org/10.1007/978-981-10-6680-1_3

Keywords Big data · Social networks · Social media · Clustering
Concept drifting

1 Introduction

The primary task of social networking sites like Facebook, Twitter, Snapchat, Instagram is to communicate and interact with a vast audience. Nowadays such sites play a significant role in disseminating information ranging from the entertainment industry, new medical science achievements, brand promotion, awareness among people, etc. Initially, this role was played by blogging sites, but reading and creating such lengthy blogs is a time-consuming task. Therefore, micro-blogging was introduced. Micro-blogging describes content in a concise and a meaningful way. Twitter is the most notable example of a micro-blogging site and is one of the greatest revolutions in social media. Here the user can post a single message up to 140 characters only, and these messages are well known as tweets [1]. According to the Twitter Web site (http://www.twitter.com), the broad coverage of this social network is confirmed by having 255 million monthly active users that post 500 million tweets per day [2]. An enormous amount of data that is being generated and shared across these micro-blogging sites serves as an excellent source of big data streams for analysis. Data shared on such sites is not confined to only textual data but also consists of photos, videos, gifs, URLs, etc., as content. Clustering is a widely used technique to organize such unlabeled data into homogenous groups. But implementation of clustering techniques for streaming data is very different from those for static data (i.e., data that does not change in the clustering process). The reasons for this are twofold. Firstly, it is hard to store an entire data stream once as data is arriving continuously, and secondly, due to the massive volume of streaming data, scanning data multiple times as it comes is not possible. Another major issue with streaming data is that the concepts behind the data evolve with time; therefore, discovering hidden concepts in data streams with time imposes an enormous challenge to cluster analysis [3, 4]. Therefore, identification and handling concept drift in 'Big Data' streams is a current area of interest [5]. For this, a learning system is required to detect and analyze the concept drift while producing results in a real time, i.e., results need to be temporally relevant and timely [3, 6]. In future, this work will enable linking concepts semantically to analyze topic convergence or divergence [7–9] and predicting coevolving events, if any, in the data streams. The work will also help in refining the results of clustering algorithm by reducing the clusters' count either by merging the clusters if there is any semantic relationship between clusters, or by declaring the clusters as outliers. The applications of this study are in the detection of events, product recommendation systems, campaign promotion, customer segmentation, etc. Many clustering algorithms that have been researched so far are mostly implemented for numeric data by exploiting the mathematical properties, e.g., by calculating the distance between data objects to form a cluster. Therefore, an effective clustering algorithm is required for textual streaming data while keeping semantic aspect in mind as very less amount of work has been done in this direction as compared to clustering static data and numeric data streams.

1.1 Social Media Analysis

Social media technologies are growing at a rapid rate, they are now considered to be a mainstream communication tool for much of the global population. Social media is a broad and continually evolving term but refers to Internet-based platforms, which enable users to connect, interact, and share information, ideas, and other content. They have become increasingly popular and numerous, with an estimated 1.5 billion users according to recent statistics (widely used examples include Facebook, Twitter, Snapchat, Instagram, and LinkedIn) with millions of users incorporating them into their daily routines. Recently, social networking sites have also been found to be utilized by the government for agenda setting, policy making, and to communicate new initiatives [4]. There are billions of active users on the social networking sites that are generating dynamic data in an enormous amount at different rates.

Twitter is one of the most well-known social media platforms, being characterized by providing a micro-blogging service where users can post text-based messages of up to 140 characters, known as tweets, mimicking the SMS (Short Message Service) messages [5, 6].

Twitter Entities: Entities provide metadata and additional contextual information about content posted on Twitter. Entities are always linked to the content they describe. In API v1.1, entities are returned wherever Tweets are found in the API [7] (Table 1).

1.2 Applications of Social Media Analysis

1. **Event Detection**: As content changes frequently with every second as new data arrives, it becomes necessary to detect the event of the data and to analyze topic convergence or divergence if any [8–10]. It also helps in predicting coevolving events.

Table 1 Twitter entities

Entity name	Description
Hashtag	For indexing information (#name)
User_Mention	For indexing users (@name)
Media	Consists photos, animated gifs, videos, etc.
URLs	URLs present in the tweets

2. **Recommendation Systems**: Such a system provides suggestions for users to buy a particular product by analyzing the user's profile, user's history, and information sharing [11].
3. **Campaign Promotion**: It deals with influencing people's behaviors/opinions/ decisions about a particular thing or concept. It is thus important to discover such campaigns, their promoter accounts, and how the campaigns are organized and executed as it can uncover the dynamics of Internet marketing [12].
4. **Customer Segmentation**: By examining users' behavior, liking/disliking, communities they follow, customer values, products/stocks they purchased, etc., customers can be segregated into different groups [13, 14].
5. **Awareness Programme**: This enables promotion of health care facilities among public [15].

1.3 Introduction to Streaming Data

The rate, at which data is being generated nowadays with offline and online users, is increasing exponentially. Data that is being generated continuously, temporally ordered, fast changing, and massive in size is called streaming data. Real-time examples of streaming data from different fields are following:

1. **Business/E-commerce**: Online transactions of the users, stock market; online advertisements; data generated from mobile apps.
2. **Social Media**: Textual data from user's posts on Twitter, Facebook, etc.; sharing images or videos on social media; multimedia data.
3. **Medical Science**: Doctors make use of large amounts of time-sensitive data for serving patients, including results of lab tests, pathology reports, X-rays, and digital imaging.
4. **Telecommunication and Networks**: It includes data from sensors or cameras for traffic analysis; clickstream data; network monitoring of packets or security; GPS data; data from satellites; server logs.

1.4 General Architecture of Clustering Streaming Data

Most of the algorithms for processing streaming data work on the assumption that the class labels of the arriving data are available. However, this assumption is very impractical, especially in the case of social media analysis where the data or content depends heavily on millions of users. There can be some association or relation between the contents depending upon users' interestingness, or there can be no association. In such a case, labeling each incoming data is very expensive and time consuming; it also requires a lot of skilled labors to do so.

Therefore, for processing and analyzing streaming data without class labels, clustering techniques are used. But the approach for clustering streaming data is different from clustering static data (i.e., data that does not change in the clustering process) as the entire data stream is not available at a time. Therefore, clustering of streaming data includes two phases: offline and online processing of streaming data [16].

Online Phase: In this component, initial processing of streaming data is done in chunks, which produces micro-clusters by storing summaries/clusters of each processed data chunk. It also includes incremental computation and maintenance of the micro-clusters to keep the micro-clusters updated.

Offline Phase: This component produces macro-clusters by analyzing the micro-clusters on various users' constraints for example number of clusters, a query related to a particular time or topic, etc.

1.5 Introduction to Concept Drift

In a given time window, each data sample in the data stream is assigned to an individual cluster that further accounts for a concept. As new data records arrive, two cases can be possible: (1) the existing clusters either update themselves to accommodate new arriving data (if there is any similarity between the two concepts); or, (2) a new cluster is created (if two concepts are completely different than each other). In the second case, the difference in the concepts of existing and incoming data is what is known as concept drift.

1.5.1 Types of Concept Drift

1. Sudden (ABRUPT) Concept Drift: Sudden concept drift occurs when there is a dramatic change in the concepts. As shown in Fig. 1, at time *t*1, there are only two concepts "A" and "B," as time *t*2 occurs, only "B" and "C" concepts are prevailing. Concept "A" is entirely replaced by concept "C".

Fig. 1 Sudden concept drift

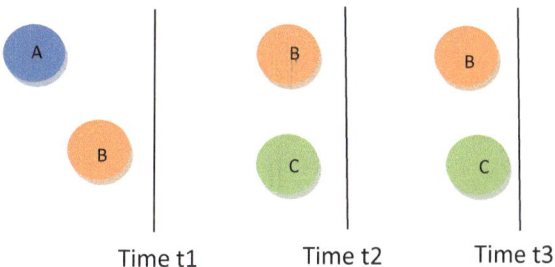

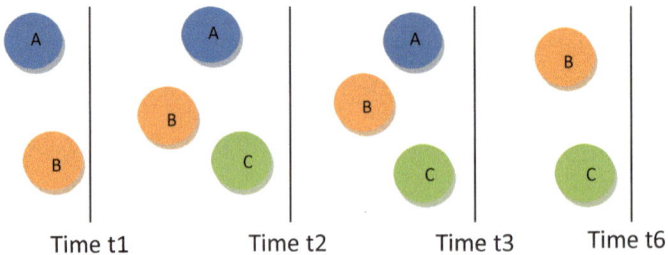

Fig. 2 Incremental concept drift

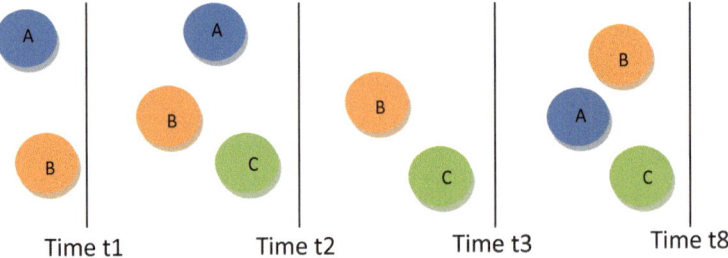

Fig. 3 Gradual concept drift

2. Gradual Concept Drift: In this concept drift, change in a concept occurs either gradually or incrementally over time. In incremental concept drift, concept "A" vanished from the site at time t6 and never occured again (as shown in Fig. 2), whereas in gradual concept drift, concept "A" is switching back and forth over time. As shown in Fig. 3, concept "A" is not present from time t3 to t7 but recurs at time t8.

But the major difficulty in dealing with concepts is identifying outliers in clusters as one may misinterpret outlier as concept drift by analyzing adjacent windows. At that time, it becomes necessary to distinguish between a concept drift and an outlier by analyzing all the succeeding and preceding windows.

1.5.2 Need to Detect Concept Drift in Streaming Data

Concept drift is the most undesirable yet prevalent property of streaming data as data streams are very unpredictable. Due to concept drift, the performance of mining techniques such as classification or clustering degrades as chances of misclassification increase. Therefore, it becomes necessary to identify such drifts in the data to get efficient results with accuracy.

2 Literature Review

In this chapter, the literature review is organized into three broad categories:

1. Study of clustering algorithms on categorical data streams
2. Study of detecting concept drift using unsupervised learning approach
3. Study of research done on social media analysis regarding concept drift.

2.1 Clustering Categorical Streaming Data

Undoubtedly, many clustering algorithms studied so far are mostly implemented for numerical data by exploiting various mathematical properties for calculating the distance between data objects to form a cluster. But, in a real environment, data consists of either categorical data or both numeric and textual data, e.g., Web clickstream data, information retrieval data, patients' records, customers' records. The field of clustering categorical streaming data has received very less attention as compared to applying clustering on static data and numerical data streams. In the following table, the conventional clustering algorithms for categorical/textual streaming data are mentioned with their key features. The latest research on clustering categorical data streams is also reviewed in this section (Table 2).

In [18], Wu proposed an algorithm for detecting outliers in categorical data streams. First, closed frequent patterns are discovered in each sliding window using HCI-MTree. Then, using weighted closed frequent pattern outlier factor

Table 2 Conventional clustering algorithms for categorical/textual streaming data [17]

Algorithms	Features
K-Modes; Fuzzy K-Modes	• Extended version of k-Means • Based on frequency method
CACTUS	• Consist of three phases: Summary (to compute candidate clusters); validation and clustering (to get actual clusters by validating the clusters formed)
ROCK	• Adaption of agglomerative hierarchical clustering • Each tuple is assigned separately to individual clusters; then clusters are merged by closeness between the tuples which is calculated by the sum of links between pairs of tuples
Squeezer	• Tuples are scanned one by one and are assigned/rejected to/by clusters on the basis of similarity
COOLCAT	• Entropy-based algorithm; clusters of similar entropy • Efficient for streaming data as it is based on incremental approach
CLOPE [8]	• Based on increasing height-to-width ratio of the cluster histogram • Efficient for large, high-dimensional transactional data, but requires little domain knowledge about dataset to control the count on clusters

(WCFPOF), outliers are detected which are stored in query indexed structure (QIS). To handle concept drift, the older categorical data are periodically replaced by newer ones. The accuracy achieved by the algorithm is equaled to Apriori algorithm.

Sora et al. [19] introduced a practical approach for clustering categorical data streams called FLoMSqueezer, an enhanced version of Squeezer algorithm (a clustering algorithm for categorical dataset). Squeezer provides high-quality clusters, but its performance degrades if the size of the histogram increases. The histogram is used to capture the distribution of values in the dataset. Therefore, by controlling the size of the histogram, the same method can be applied to categorical data streams, resulting in a new approach called FLoMSqueezer. The results show that proposed algorithm outperforms Squeezer algorithm in average clustering error and execution time parameters. The proposed algorithm also produces quality clusters with limited memory in data stream environment.

To reduce the dimensionality of the categorical dataset, entropy-based relevance index is computed for each attribute in the cluster to calculate its relevancy in the cluster [20].

Qin et al. [21] proposed a new information theory-based algorithm for clustering categorical data by computing mean gain ratio (MGR) of attributes and then selecting equivalence class on the chosen clustering attributes using entropy of the clusters. The proposed algorithm is compared with other information theory-based algorithms such as MMR (a rough set hierarchal algorithm), K-ANMI (K-means and mutual Information) and COOLCAT (Entropy based) on accuracy, adjusted Rand index (ARI) performance metrics. The results show that MMR achieves highest accuracy (0.931 for zoo dataset) and higher ARI (0.96 for zoo and cancer dataset) than other algorithms. The average running time of MGR is 6.67 s which is 21, 11.09, and 133.159 s for MMR, COOLCAT, and K-ANMI, respectively. In future work, MMR algorithm can be combined with K-ANMI for improving the performance of clusters.

Lenco et al. [22] introduced change detection algorithm for evolving categorical data streams in which summaries are extracted from the batches, and then Chebyshev's Inequality is used for detecting a change in those batches. The proposed algorithm is compared with [23] in which concept drift is detected by computing the difference between the two concepts using rough set theory; this may also help in measuring the velocity at which change is happening in the concepts. The proposed algorithm achieves an average accuracy of 75% for different batch sizes, i.e., 50, 100, 500, and 100. Lenco et al. [23] algorithm can detect the changes which were not identified. A high point of this approach is its ability to exploit historical information for decision making by adjusting the size of history window.

Li et al. [24] presented an integrated model for clustering categorical data streams that help in data labeling, change detection, and cluster evolving analysis. In this, three dissimilarity measures are proposed that are based on incremental entropy (for computing dissimilarity between point-cluster and cluster-cluster) and sample standard deviation (for computing dissimilarity between two cluster distributions). The algorithm is made to run on different sliding window sizes to analyze its efficiency on various datasets: zoo, soybean, dermatology, and DNA. The experimental results demonstrate that the labeling accuracy of the proposed algorithm is higher than [25–27] (other data labeling methods). For measuring concept detection performance, precision, and recall are calculated as follows:

$$\text{Precision} = \frac{no\_of\_concept\_drifting\_windows\_correctly\_detected}{total\_no\_of\_concept\_drift\_windows\_detected}$$

$$\text{Recall} = \frac{no\_of\_concept\_drifting\_windows\_correctly\_detected}{actual\_no\_of\_concept\_drift\_windows\_present}$$

The change detection algorithm of [24] is superior to [25], and somehow if the performance degrades then, it is due to the use of k-modes algorithm in the proposed work.

Talistu et al. [28] suggested a new approach to cluster data streams in a distributed manner which reduces the burden of a centralized processor to process the whole data streams by itself. In this approach, first local online summaries are generated at each local location using hierarchical clustering technique; then these summaries are distributed to every other node present in the system to get a global view of the data using gossip protocol. This collection and summarization of the data streams may help in analyzing the patterns over time and also in tracking cluster evolution, if any. After distributing local online summaries, nodes may get many duplicate summaries which can give a false interpretation of the data streams. Therefore, to avoid such false results, the merge-and-reduce algorithm is applied to remove the replicated summaries. This algorithm can also be used at hierarchical clustering phase to get the smaller sets of representative data points. For performance analysis, the average squared error (ASE) is computed for evaluating the quality of the clusters and scalability is analyzed by calculating the effect on cluster quality as numbers of nodes in the system increases. Gossip protocol does not guarantee delivering local online summaries among nodes; this could be a limitation when some nodes in the system increase plus the merge-and-reduce algorithm is not efficient in detecting and deleting the duplicate summaries at each node. The experimental results demonstrate that the value of ASE increases and scalability performance of the system decreases with the increase in the number of nodes.

Xhafa et al. [29] discussed one of the issues of processing big data streams, i.e., streaming consistency. Streaming consistency is the desirable property for some of the applications where the time of occurring events plays a significant role in generating results; in that scenario, it becomes necessary to preserve the order of the events in the output as in the input streams. In the paper, Yahoo!S4 is implemented for processing real-time data streams generating from FlightRadar24. The implementation showed that the average time for updating the flight status is 4 s which is a very practical result but still several inconsistencies are observed in the streaming due to the heterogeneous nature of the clusters.

A novel approach based on consensus or trade-off is introduced by Erdi and Gil [30] to aggregate atomic similarity semantic measures using fuzzy logic (consensus) and then applying a high degree of trade-off to get the overall score. To aggregate the similarity measures between two concepts, aggregation function (if-else rule) is developed. The significant result of 0.85 is achieved by the proposed method than other standard measures. A second experiment is conducted to measure the degree of similarity between the terms by exploiting the search logs of Google search engine using following measures: midrange, quadratic mean, arithmetic mean, maximum, minimum median, and normalized Google distance. In this case also, the proposed algorithm outperforms other measures by achieving 0.64 accuracy.

Rehioui et al. [31] provided an improved version of density-based technique (DENCLUE-IM) for clustering big data streams. The hill climbing step of standard DENCLUE had high execution time and produced poor clustering results. Therefore, in the improved version, this step is modified from calculating density attractor at each step to finding an equivalent variable to density attractor that will represent all points, hence result in reducing the computations and increasing the clustering efficiency. To evaluate the effectiveness of the proposed method on the following datasets: page blocks, spambase and cloud services, Dunn index (intra-cluster dissimilarity), Davies–Bouldin index (inter-cluster dissimilarity), execution time, and clustering accuracy are taken as performance measures. Following results are obtained by the proposed algorithm (Table 3).

For cloud services dataset, clustering accuracy cannot be computed as labeling data is not available. The proposed algorithm provides acceptable results as compared to DENCLUE but remarkable results are achieved regarding execution time. The proposed algorithm is taking 12 times less execution time than DENCLUE for page block dataset; also for classifying cloud services, DENCLUE takes approx. 32 h where DENCLUE-IM takes only 28 min.

Laohakiat et al. [32] developed a clustering algorithm for high-dimensional data by incorporating dimension reduction step in the clustering framework. The algorithm performs dimension reduction at the arrival of each datum by finding local subspace using unsupervised linear discriminant analysis (LDA) method where classes are replaced by clusters and each class mean is replaced by the center of each cluster. Normalized mutual information matrix (NMI), Rand index (RI), adjusted Rand index, (AR) and Hubert's index (HI) are taken as performance indices. KDD CUP 99 and forest cover type datasets are used to examine the performance of the algorithm in comparison with other streaming algorithms named DenStream, HDDStream, and HPStream. The window size of 10,000 data instances

Table 3 Results of DENCLUE-IM [31]

	Page blocks	Spambase	Cloud services
Dunn index	0.693	0.831	0.899
Davies–Bouldin index	0.412	1.041	1.262
Execution time (in minutes)	5.479	27.54	1657.508
Clustering accuracy	0.911	0.701	–

and 2000 data instances is taken for KDD CUP 99 dataset and forest cover type dataset, respectively. The experimental results show that for KDD CUP 99, 0.86, 0.87, 0.97, and 0.94 values are obtained for NMI, RI, AR, and HI, respectively, which is much higher than the performance indices values of other clustering algorithms taken for comparison. Similarly, for forest cover type dataset, proposed algorithm yields better performance than DenStream, HDDStream, and HPStream. Even the runtime of the algorithm is relatively lower than other clustering algorithms. For future work, this work can also be extended for categorical clustering attributes.

The literature review of clustering categorical data streams indicates that almost all the clustering algorithms exploit only one property to cluster the categorical data streams, i.e., computing the distance between the contents of two consecutive windows of the data stream to measure the similarity between two adjacent windows. No algorithm has utilized or discussed the semantic relation between the contents of windows to cluster the evolving data streams. The semantic relations, if considered, can help in reducing clusters' count in the result, hence reducing the execution time, computational cost, memory usage and may lead to better prediction of the content of next incoming data instances (Table 4).

2.2 Detecting Concept Drift in Categorical/Textual Data Streams

To handle concept drift in any data stream with efficiency, classification is the most common approach. But the primary requirement of the classification task is class labels which are not always available in the streaming data. And assigning class labels manually to the incoming data streams is very time consuming and requires a lot of skilled workforce. Therefore, in this section, different clustering or semi-supervised techniques are studied to detect concept drift depending on whether the researcher has knowledge about the domain of the incoming data or is completely unaware of their contents.

Barddal et al. [33] focused on a different kind of concept drift called feature drift occurring in the streaming data. Feature drifts are relevant to the learning algorithm as they result in dimensionality reduction and the decision can also be made based on these features. The experiment is conducted on spam corpus dataset, and the performance of the proposed method is evaluated regarding classifiers' accuracy, the runtime of the algorithm, and memory usage. The proposed landmark-based drift detection algorithm runs on different chunks created from the streaming data, these chunks are analyzed to determine relevant feature subsets present in chunks using discriminative factor, and then finally classifier is trained using these subsets. The results reveal that the proposed method produces interesting features subset with fast computation and less memory usage as compared to conventional algorithms such as information gain, correlation, and gain ratio.

Table 4 Summary of clustering categorical/textual data streams

Author	Primary task	Methodology	Dataset	Performance measure	Window size	Results
QunHui and ShiLong [18]	Detecting outliers from categorical data streams	Discovering closed frequent patterns in sliding window then measuring weighted factor	KDD CUP 99 network intrusion detection data set	Precision, running time of algorithm	50 K	Algorithm is sensitive to data size; precision achieved equals to Apriori algorithm
Sora et al. [19]	To overcome the scalability problem of Squeezer clustering algorithm	Sampling is introduced into the update histogram module	Mushroom Dataset (22 Atributes, 8124 tuples)	Execution time; accuracy	Varying window size	Proposed algorithm outperforms Squeezer algorithm in efficiency and execution time
Joel and Abel [20]	Clustering categorical data using entropy-based k-modes (EBK) algorithm	Measure the relevancy of each attribute in the dataset to reduce dimensionality by entropy-based relevance index	Vote (435 tuples, 17 attributes, two classes); mushroom (8124 tuples, 23 attributes, two classes); breast cancer (286 tuples, ten attributes, two classes); Soyabean (683 tuples, 36 attributes, 19 classes); genetic promoters (106 tuples, 58 attributes, two classes)	Accuracy, f-measure. Adjusted Rand index	No window size; static data	The accuracy results of EBK ranges from 70–88% for all the datasets. Even in f-measure and adjusted Rand Index metrics, the proposed algorithm outperforms the standard K-modes algorithms and its versions
Qin et al. [21]	Clustering categorical data	Using mean gain ratio and entropy	Zoo, congressional votes (Votes), Wisconsin breast cancer (breast cancer), mushroom, balance scale, car evaluation, chess, Hayes-Roth, and nursery	Accuracy, adjusted Rand index (ARI)	No sliding window; static data	Running time less than other three algorithms: MMR (a rough set hierarchal algorithm), K-ANMI (K-means and Mutual Information) and COOLCAT (Entropy based); MMR achieves highest 0.96 ARI

(continued)

Table 4 (continued)

Author	Primary task	Methodology	Dataset	Performance measure	Window size	Results
Lenco et al. [22]	Detecting change in evolving categorical data streams	Extracting summaries and applying Chebyshev's Inequality	Electricity, Forest, Airlines, KDD99	Accuracy; percentage of change detected	Batch size: 50, 100, 500, 1000	Achieves average accuracy of 75%; also able to detect changes, not detected by [23]
Li et al. [24]	Data labeling, change detection, analyzing evolving clusters	Incremental entropy and standard deviation	Soybean, zoo, dermatology and DNA	Precision, recall (for measuring concept detection performance) and accuracy, time (for evaluating clustering efficiency)	Varying sliding window sizes	Superior to [25–27] in detecting concept drift and generating clusters
Talistu et al. [28]	Clustering data streams	Distributed processing, gossip protocol merge-and-reduce algorithm to manage summaries in online phase	2-D dataset consisting 1,920,000 data objects	Average squared error, scalability	Distributed approach	ASE increases and scalability performance decreases with increase in the number of nodes
Xhafa et al. [29]	Streaming consistency	Clustering approach	FlightRadar24	Average time to update flight status	Clusters	Average time of 4 s is achieved for updating flight status

Fast evolutionary algorithm for clustering (FEAC) of data streams is proposed by Silva et al. [34] that aims at relaxing the assumption of prior knowledge of a number of clusters to be used in the clustering algorithm. In the suggested method, the number of clusters is estimated from the first incoming data using silhouette width criterion, then these clusters are incrementally updated, discarding the previous clusters as new data arrives; or change is reflected in any cluster by Page-Hinkley method. KDD CUP 99 intrusion detection dataset, forest cover type, and localization data for person activity are taken for experimental results in chunks of 2000 data instances. The algorithm can capture the changes present in both intra-class and inter-class characteristics by providing best trade-off between accuracy and runtime. In future, this work can be extended to predict cluster changes.

Liu and Zio [35] suggested a new algorithm based on feature selection vector to detect recurring drift in the streaming data. The major drawback with the existing algorithms is the replacement of the old patterns with new ones in the updating procedure. As a result, when that same old pattern reoccurs, the model needs to relearn again and hence computational time increases. To overcome this drawback, online ensemble based on feature vector (OE-FV) is introduced to store all the past patterns and updating the submodels by their weights. The algorithm adaptively updates the ensemble with the arrival of new data and thus can learn new patterns on time. Unlike, other window-based or chunk-based approach where the model has to wait for sufficient amount of data to be available. The results prove that the proposed algorithm decreases the count of changes patterns, leading to a reduction in computational burden.

Bai et al. [36] introduced a new clustering algorithm for categorical data streams and a detection index for drifting concept. Here, the clusters are merged by computing the difference between the old clusters and new ones. If the difference is greater than the threshold value, then a new cluster is formed. Otherwise, clusters are merged. At the end of this iterative step, the final clusters are analyzed to capture the evolution in the trends with time. For detecting drifts, following two facts are considered between the last and the new cluster: (1) distribution variation (for computing cluster representative) and (2) certainty variation (to capture change direction). The performance indices used to evaluate the proposed algorithm are accuracy, effectiveness in detecting drifts, and scalability. The datasets used for testing are letters, DNA, nursery, and KDD CUP 99. The results indicate that the proposed algorithm produces valid clusters irrespective of the data distributions in comparison with Chen et al. [37] and Cao et al. [38] algorithm. The average precision, recall, and accuracy value of the proposed algorithm for all the three datasets are above 85%. To evaluate the results of detective index, precision, recall, and Euclidean distance are computed. Even, in this case, the proposed algorithm produced less number of clusters while detecting drifts in the data streams. The computational time of the proposed algorithm is much higher than other two algorithms because, in other two algorithms, clustering problem is seen as data labeling step that needs a single iteration whereas the proposed algorithm considers the clustering problem as an iterative learning, requiring many iterations for better results.

For detecting both local and global changes in the transactional data streams, Koh [39] presented an algorithm where transactions falling in a given window are represented as a tree. Spearman correlation is computed to check the homogeneity between the current and previous window. If the distance calculated is greater than the threshold value then global drift is detected. In a similar fashion, when the difference in support of any particular item in the current window and the previous window comes out to be greater than the threshold value, then local drift is detected. True detections, false alarms, and detection delay are computed to measure the performance of the algorithm on following datasets: Kosarak, Airlines, and BMS-POS. The results show that the proposed algorithm gives high true positive rate and very less false alarm rate for determining structural changes in the dataset, even with noise.

A novel ensemble model called dynamic clustering forest that employs clustering trees (CTs) for clustering textual streams was proposed in [40]. In training phase, first clustering trees (CTs) of incoming chunks of textual streams are built, then discriminative CTs are selected by comparing the accuracy weight of each CT with defined threshold value. Next, for classifying test instances, the creditability of each discriminative CT is computed, and finally, with the help of both the weights, i.e., accuracy and credibility weights, the test set is classified. Creditability is calculated by computing the similarity between the test instance and centroid of each CT. To evaluate the performance, the proposed algorithm is compared with accuracy weighed ensemble (AWE) and accuracy updated ensemble (AUE) algorithms. For the spam stream dataset, DCF achieves around 1–5% improvement in classification accuracy. DCF running time is also low indicating its applicability to real-world datasets.

Haque et al. [41] introduced a semi-supervised algorithm based on existing SAND algorithm (k-NN type) for detecting concept drift and concept evolution in data streams by computing classifier confidence score, i.e., association and purity. The confidence score is also used to select limited labeled data instances for labeling the current chunk. A difference in the confidence scores of two windows represents the occurrence of a concept drift. The change detection algorithm is invoked only when the value of confidence score is less than the threshold value, resulting in a reduction in execution time. The algorithm is tested on three real-time datasets: forest cover (FC), power supply (PS), and physical activity monitoring (PAM). The performance metrics used are misclassification error (error %), novel class instances misclassified as existing class (M-new), existing class instances misclassified as new classes (F-new), and F-score. The error % of the proposed algorithm is 3.55, 3.85, and 0.01 for FC, PS, and PAM, respectively. For FC dataset, the M-new, F-new, and F2 achieved by the proposed algorithm are 13.55, 2.13, and 0.71, respectively.

An unsupervised, distribution independent, online incremental-based algorithm is introduced for detecting concept drift in unlabeled data streams in Sethi et al. [42]. The critical blindspot cardinality (CBC) is analyzed for detecting drifts in the data streams. For each of the incoming data samples, CBC is computed that indicates whether the input sample belongs to a CBC region or not. If the drift is detected, then experts are asked to label the samples to retrain the classifier. The average accuracy achieved by the proposed methodology when tested on four real-world datasets is 92.05%. The algorithm can reduce false alarms by capturing only those changes that can affect the performance of the classifiers.

By exploiting the temporal relationship among data, four unsupervised concept drift detection algorithms are introduced in [43], namely (1) stable clustering concept drift detection (SCCDD), (2) cross recurrence concept drift detection (CRCDD), (3) SCCDD-decomposition, and (4) CRCDD-decomposition. Third and fourth algorithms are the extended versions of first and second algorithms, respectively, in which preprocessing of the data is done before applying the algorithm. In the first experiment, dendrograms of two successive windows are created and then analyzed using Gromov–Hausdorff distance to detect concept drift. In the second experiment, a matrix of recurrence is created from the data of two successive windows, and then this matrix is analyzed to detect recurrent patterns by computing the longest diagonal in the matrix. The results of each proposed algorithm are compared with conventional concept drift detection methods, i.e., Page-Hinkley test and cumulative sum. The following metrics are used to measure the performance: (i) missed detection rate (MDR); (ii) mean time to detection (MTD); (iii) mean time between false alarms (MTFA). Following are the experimental results of the algorithms (Table 5).

The results show that SCCDD performs remarkably better than other algorithms for detecting concept drifts in the data streams, even in the presence of noise.

Lughofer and Mouchaweh [44] work on merging and splitting clusters by computing Mahalanobis distance between the clusters. If the distance is greater than the threshold value, then a new cluster is created. Otherwise, parameters of existing clusters are updated. The algorithm works efficiently for high-dimensional data but has the limitation of over-clustering for complex clusters with arbitrary shape and inability to handle concept drifts.

A mechanism is proposed by Yang and Fong [45] to handle concept drifts by introducing the optimized node-splitting method in the learning tree. Hoeffding bound with weighted Naïve Bayes classifier (to handle biased data) is used to construct an incremental tree for incoming data streams. The experimental results show that the proposed mechanism requires less memory as the final decision tree is compact in size.

Table 5 Results of paper [43]	MDR	MTD	MTFA
PHT	1.000	0.000	0.989
CUSUM	0.850	0.002	0.213
SCCDD	0.250	0.006	0.319
CRCDD	0.050	0.018	1.000
SCCDDd	0.300	0.001	0.548
CRCDDd	0.100	0.009	0.600

Wu et al. [46] proposed a semi-supervised classification algorithm for unlabeled data streams. First, an incremental decision tree is generated using arriving data streams. In parallel, unlabeled data instances are labeled using K-modes algorithm and concept drift is detected by measuring the deviation between previous and current concept clusters. The proposed algorithm outperforms other semi-supervised algorithms CVFDT and BagBest regarding classification accuracy by 30%. The algorithm is also flexible to a degree of noise in the presence of the major unlabeled data.

Approaches for detecting concept drift are very diverse as can be concluded from this section. Every researcher is exploring a new methodology to handle concept drift according to their dataset either by finding the similarity between the concepts semantically or computing the distance between two windows. Hence, proving that there is still a wide scope of developing an efficient algorithm for not only analyzing concept drift but also describing the reason for occurrence of the concept drift.

2.3 Research Done for Detecting Concept Drift on Social Media

In social media analysis, two cases can be possible (1) an outlier in a particular window can become a trend in all the succeeding windows; (2) a trend in a current window can be evolved or vanished in succeeding windows. These two cases can help in producing results that either predict popular topics by analyzing current messages or contents of social media or analyze topics that may revive or evolve itself over the time. For obtaining such results, concept drift needs to be detected and handled efficiently. This section includes the literature review of work done on social media data for detecting concept drift.

In [47], re-tweets count of a particular message is taken as the measure to capture the popularity on Twitter data. But this approach ignores the temporal correlation between the messages.

A formal popular multimedia detection algorithm in social media is described in [48] by tackling two classification problems. First, is predicting whether a particular message M of time T will be re-shared by the users in time $T + 1$, this is called *re-share classification*. And second is multiclass classification problem, called *popularity score classification*; in this, the prediction is made by calculating the popularity score of social multimedia. The features used in the prediction task are (i) information diffusion feature (this includes the information of a set of users who post a particular message consisting of multimedia M in time t; (ii) multimedia meta-information (extracting meta-information of multimedia shared in time t from the source). Euclidean and overlap metric is used to measure the distance between numerical values and categorical values, respectively. The proposed algorithm performs remarkably better than SVM and J48 in terms of accuracy and F-score for the classification task.

In order to predict future networks, two approaches are proposed in [49]: (1) unweighted approach and (2) weighted approach. In unweighted approach, if a pair of nodes is connected to each other in the current window, then there will be a higher probability of connecting common nodes in the future networks. But this probability decreases with the increase in the neighbors of the network. The results show that the prediction performance of the algorithm for the static network is much better than evolving networks. In weighted approach, weights are assigned to the nodes by analyzing the dominance of human behavior, the decrease in the dominance will also reduce the weights allocated to the nodes. This approach also performs well for the static network.

In this paper [50], a novel method is presented to extract and visualize events from social media streams. The concept map is created where nodes are represented as sender and words found in the tweets post. To enhance the concept map, events are time labeled, and semantic similarity is computed between the nodes using Lin similarity. The method can extract the concept drift occurs in the events.

Miller et al. [51] implemented a combination of StreamKM++ and DenStream clustering algorithm for detecting spam in Twitter streams and also studied the effect of StreamKM++ and DenStream clustering algorithms individually on Twitter streams. The results prove that StreamKM++ achieves 99% recall and a 6.4% false positive rate and DenStream produces 99% recall and a 2.8% false positive rate, whereas the proposed algorithm, i.e., a combination of these algorithms reaches 100% recall and a 2.2% false positive rate.

To identify topic-specific post in Twitter streams, adjusted information gain (AIG) index is proposed and compared with other existing term scoring indexes such as IDF, Dice, Jaccard, and TF in [52]. The results show that 8–40% of the area under ROC curve that signifies true positive rates versus false positive rates when AIG is used as term scoring index.

Malik et al. [53] introduce two models: (i) to group related tweets into automatically generated topics (ii) an interactive visualization tool called TopicFlow to display similar topics in a specific time and identifying topic emergence or convergence or divergence over a given period. The proposed algorithm works as follows: (1) partition incoming tweets into bins, (2) run an unsupervised clustering algorithm called latent Dirichlet allocation (LDA) on each bin, (3) align topic into four cases: emerging, ending, continuing, and standalone, and (4) displaying results in visualization tool. The evaluation shows that TopicFlow is very useful in capturing the frequency of a particular topic over the given period and also provide details on demand to users through hovering over the extracted topic.

Context free method or variable length Markov model [54] is a model to predict the next symbol in a stream by observing the set in the preceding symbols, instead of relying on distance measure methods. This method is also effective for mixed data types.

A distributed and incremental temporal model for extracting topic from massive micro-blogs data streams called bursty event detection (BEE+) is proposed in [55], and the same is implemented on spark engine for real-world applications. BEE+ preserves the latent semantic indices by processing the post-stream incrementally to track the topic drifting of events over time. The algorithm is compared with TwitterMonitor (TM) to detect bursty events. Weibo dataset of 6,360,125 posts is taken for the experiment. The results show that for top-10 results, TM and BEE+ achieve 50 and 70% precision, respectively. TM also not able to group the keywords based on the temporal information of events whereas BEE+ does and hence can detect maximum concept drift in the streams.

For solving infrequent and semantic gap problem in the data streams, an algorithm is developed in which DBpedia is used as a knowledge base to link the data semantically. And for resolving word disambiguation problem, for each tweet, a graph is constructed based on the graph centrality theory. The proposed algorithm is compared with spotlight, a publicly available NEL tool. The proposed algorithm and spotlight achieve 60 and 51% F-score, respectively [56].

An improvised version of SFPM (Soft Frequent Pattern Mining) to recognize real-time social events in Twitter streams is introduced in [57]. In this paper, the new term selection algorithm is also developed that selects not only relevant keywords from the current window but also identifies the emerging terms from the current window to handle concept drift. Experiments are conducted on the posts related to FIFA World Cup 2014. The proposed framework outperforms in all the performance indices such as topic recall, keyword precision, and keyword recall, in comparison with standard SFPM version.

In paper [58], DBpedia is used as a knowledge base to link twitter entities based on graph centrality with the assumption that entities present in the topic-specific tweets are related to each other. The results show that the proposed algorithm achieves 69.5% F-score in comparison to TAGME model that attains only 46.8% F-score. In a similar way, to classify tweets related to specific events, semantic relations between the Twitter entities are exploited by again using DBpedia Ontology as a knowledge base. The algorithm [59] works in three main steps: (i) entity detection from the tweets; (ii) extracting relevant features related to the specific event; (iii) calculating entity score and class similarity to avoid word ambiguities.

3 Proposed Framework for Detecting Concept Drift on Social Media

From the literature review, following two significant research gaps are identified:

- More efficient clustering algorithm is required for analyzing concept drift on big data streams by determining the strength of association between data streams and correlating them.
- There is a need to explore other means of trusted sources of data and various content-based features to find the actual context of data, to identify ambiguities or synonyms in the data streams.

3.1 Proposed Framework

The proposed framework for detecting and handling concept drift in streaming data is divided into the following phases:

Phase 1: Data Collection
Collecting data from various online sources such as Twitter, Web sites, news articles.

Phase 2: Data Clustering and Labeling

2.1   Dividing data streams into windows.
2.2   Applying appropriate clustering algorithm on a window to group the data into clusters.
2.3   Labeling clusters to get knowledge about the concept hidden in that cluster.
2.4   Steps 2.2 and 2.3 are repeated for each window of the streaming data.

Phase 3: Detection of Concept Drift
Clusters of two adjacent windows are then analyzed for the following tasks: (i) to identify concept drift; (ii) to find any relationship between the concepts of two windows; (iii) to analyze coevolving events, if any.

For analyzing clusters, graph edge weight technique can be implementing by using content-based features of the streaming data, entities of social networking sites, and different data resources such as Web sites, new articles for making a decision.

Phase 4: Evaluation Phase
The performance of the algorithm can be evaluated by various performance metrics such as accuracy, precision, recall and execution time, classification, or clustering error.

4   Conclusion

The primary problem faced while clustering streaming data is the concept drift. The state-of-the-art clustering techniques and concept drift detection techniques mainly work on exploiting statistical features of data that often produce too many clusters in a final output. However, the clusters' count can be reduced by merging clusters that are semantically similar.

This paper highlights the recent research on clustering categorical data streams and concept drift detection techniques. From the literature review, it can be concluded that clustering algorithms which handle concept drift are often lacking in some major performance parameters such as computational cost, execution/running time of the algorithm, the quick response to the query, varied data density. Even, for exploiting semantic similarity between the clusters, the assurance of any knowledge-based resources such as Web sites, new articles or proper utilization of social networking sites entities is also required to validate the results.

Hence, for analyzing and mining any hidden knowledge from the streaming data efficiently and effectively, it becomes necessary to develop a system that can tackle the problem of concept drift using conceptual features. Several applications of such a system can be: identifying new emerging trends, top news, thematic categorization, top-influencers, queries suggestion, etc.

References

1. Zhang, B., Qin, S., Wang, W., Wang, D., & Xue, L. (2016). Data stream clustering based on fuzzy C-mean algorithm and entropy theory. *Journal of Signal Processing, 126,* 111–116.
2. Lifna, C., & Vijaylakshmi, M. (2015). Identifying concept drifts in Twitter streams. In *International Conference on Advanced Computing Technologies and Applications* (pp. 86–94).
3. Xhafa, F., Naranjo, V., Barolli, L., & Takizawa, M. (2015). On Streaming Consistency of Big Data Stream Processing in Heterogenous Clusters. In *18th IEEE International Conference on Network-Based Information Systems* (pp. 476–482).
4. Schnitzler, K., Davies, N., Ross, F., & Harris, R. (2016). Using Twitter™ to drive research impact: A discussion of strategies, opportunities and challenges. *International Journal of Nursing Studies, 59,* 15–26.
5. Costa, J., Silva, C., Antunes, M., & Ribeiro, B. (2014). Concept Drift Awareness in Twitter Streams. In *13th IEEE International Conference on Machine Learning and Applications* (pp. 294–299).
6. Wang, Y., Liu, J., Huang, Y., & Feng, X. (2016). Using Hashtag graph-based topic model to connect semantically-related words without co-occurrence in microblogs. *IEEE Transactions on Knowledge and Data Engineering, 28*(7), 1919–1933.
7. Entities—Twitter Developers. (2017). In Dev.twitter.com. https://dev.twitter.com/overview/api/entities. Accessed May 8, 2017.
8. Eskandari, S., & Javidi, M. (2016). Online streaming feature selection using rough sets. *International Journal of Approximate Reasoning, 69,* 35–57.
9. Li, J., Tai, Z., Zhang, R., Yu, W., & Liu, L. (2014). Online bursty event detection from microblog. In *IEEE/ACM 7th International Conference on Utility and Cloud Computing* (pp. 865–870).
10. Adedoyin-Olowe, M., Gaber, M., Dancausa, C., Stahl, F., & Gomes, J. (2016). A rule dynamics approach to event detection in Twitter with its application to sports and politics. *Expert Systems with Applications, 55,* 351–360.
11. Villanueva, D., González-Carrasco, I., López-Cuadrado, J., & Lado, N. (2016). SMORE: Towards a semantic modeling for knowledge representation on social media. *Science of Computer Programming, 121,* 16–33.
12. Li, H. (2014). Detecting campaign promoters on Twitter using Markov random fields. In *IEEE International Conference on Data Mining* (pp. 290–299).
13. Kuo, R., Mei, C., Zulvia, F., & Tsai, C. (2016). An application of a meta-heuristic algorithm-based clustering ensemble method to APP customer segmentation. *Neurocomputing, 205,* 116–129.
14. Wang, B., Miao, Y., Zhao, H., Jin, J., & Chen, Y. (2016). A biclustering-based method for market segmentation using customer pain points. *Journal of Engineering Applications of Artificial Intelligence, 47,* 101–109.

15. Giannitsioti, E., Athanasia, S., Plachouras, D., Kanellaki, S., Bobota, F., Tzepetzi, G., et al. (2016). Impact of patients' professional and educational status on perception of an antibiotic policy campaign: A pilot study at a university hospital. *Journal of Global Antimicrobial Resistance, 6,* 123–127.

16. Han, J., Kamber, M., & Pei, J. (2011). *Data mining* (3rd ed.). Amsterdam: Elsevier/Morgan Kaufmann.

17. He, Z., Xu, X., & Deng, S. (2011). Clustering categorical data streams. *Journal of Computational Methods in Sciences and Engineering, 11*(4), 185–192.

18. Wu, Q., & Ma, S. (2011). Detecting outliers in sliding window over categorical data streams. In *Eighth International Conference on Fuzzy Systems and Knowledge Discovery* (pp. 1663–1667).

19. Sora, M., Roy, S., & Singh, I. (2011). FLoMSqueezer: An effective approach for clustering categorical data stream. *International Journal of Computer Science Issues, 8*(6), 1.

20. Carbonera, J., & Abel, M. (2014). An entropy-based subspace clustering algorithm for categorical data. In *IEEE 26th International Conference on Tools with Artificial Intelligence* (pp. 272–277).

21. Qin, H., Ma, X., Herawan, T., & Zain, J. (2014). MGR: An information theory based hierarchical divisive clustering algorithm for categorical data. *Knowledge-Based Systems, 67,* 401–411. https://doi.org/10.1016/j.knosys.2014.03.013.

22. Lenco, D., Bifet, A., Pfahringer, B., & Poncelet, P. (2014). Change detection in categorical evolving data streams. In *29th Annual ACM Symposium on Applied Computing* (pp. 792–797).

23. Cao, F., & Huang, J. Z. (2013). A concept-drifting detection algorithm for categorical evolving data. In J. Pei, V. S. Tseng, L. Cao, H. Motoda & G. Xu (Eds.), *Advances in knowledge discovery and data mining. PAKDD* 2013. *Lecture notes in computer science* (vol. 7819). Berlin, Heidelberg: Springer.

24. Li, Y., Li, D., Wang, S., & Zhai, Y. (2014). Incremental entropy-based clustering on categorical data streams with concept drift. *Knowledge-Based Systems, 59,* 33–47. https://doi. org/10.1016/j.knosys.2014.02.004.

25. Chen, H. L., Chen, M. S., & Lin, S. C. (2009). Catching the trend: A framework for clustering concept-drifting categorical data. *IEEE Transactions on Knowledge Data Engineering, 21*(5), 652–665.

26. Cao, F., Liang, J., Bai, L., Zhao, X., & Dang, C. (2010). A framework for clustering categoricaltime-evolving data. *IEEE Transactions on Fuzzy System, 18*(5), 872–882.

27. Cao, F., & Liang, J. (2011). A data labeling method for clustering categorical data. *Expert Systems with Applications, 38,* 2381–2385. https://doi.org/10.1016/j.eswa.2010.08.026.

28. Talistu, M., Moh, T. S., & Moh, M. (2015). Gossip-based spectral clustering of distributed data streams. In *International Conference on High Performance Computing and Simulation* (pp. 325–333). https://doi.org/10.1109/HPCSim.2015.7237058.

29. Xhafa, F., Naranjo, V., Barolli, L., & Takizawa, M. (2015). On streaming consistency of big data stream processing in heterogenous clusters. In *18th International Conference on Network-Based Information Systems* (pp. 476–482).

30. Martinez-Gil, J. (2016). CoTO: A novel approach for fuzzy aggregation of semantic similarity measures. *Cognitive Systems Research, 40,* 8–17. https://doi.org/10.1016/j.cogsys.2016.01. 001.

31. Rehioui, H., Idrissi, A., Abourezq, M., & Zegrari F. (2016). DENCLUE-IM: A new approach for big data clustering. In *7th International Conference on Ambient Systems, Networks and Technologies* (pp. 560–567).

32. Laohakiat, S., Phimoltares, S., & Lursinsap, C. (2017). A clustering algorithm for stream data with LDA-based unsupervised localized dimension reduction. *Information Sciences, 381,* 104–123. https://doi.org/10.1016/j.ins.2016.11.018.

33. Barddal, J., Gomes, H., Enembreck, F., & Pfahringer, B. (2017). A survey on feature drift adaptation: Definition, benchmark, challenges and future directions. *Journal of Systems and Software, 127,* 278–294. https://doi.org/10.1016/j.jss.2016.07.005.

34. Andrade Silva, J., Hruschka, E., & Gama, J. (2017). An evolutionary algorithm for clustering data streams with a variable number of clusters. *Expert Systems with Applications, 67,* 228–238. https://doi.org/10.1016/j.eswa.2016.09.020.
35. Liu, J., & Zio, E. (2016). A SVR-based ensemble approach for drifting data streams with recurring patterns. *Applied Soft Computing, 47,* 553–564. https://doi.org/10.1016/j.asoc.2016.06.030.
36. Bai, L., Cheng, X., Liang, J., & Shen, H. (2016). An optimization model for clustering categorical data streams with drifting concepts. *IEEE Transactions on Knowledge and Data Engineering, 28,* 2871–2883. https://doi.org/10.1109/tkde.2016.2594068.
37. Chen, H.-L., Chen, M.-S., & Lin, S.-C. (2009). Catching the trend: A framework for clustering concept-drifting categorical data. *IEEE Transactions on Knowledge and Data Engineering, 21,* 652–665. https://doi.org/10.1109/tkde.2008.192.
38. Cao, F., Liang, J., Bai, L., Zhao, X., & Dang, C. (2010). A framework for clustering categorical time-evolving data. *IEEE Transactions on Fuzzy Systems, 18,* 872–882. https://doi.org/10.1109/tfuzz.2010.2050891.
39. Koh, Y. S. (2016). CD-TDS: Change detection in transactional data streams for frequent pattern mining. In *International Joint Conference on Neural Networks* (pp. 1554–1561). https://doi.org/10.1109/IJCNN.2016.7727383.
40. Song, G., Ye, Y., Zhang, H., Xu, X., Lau, R., & Liu, F. (2016). Dynamic clustering forest: An ensemble framework to efficiently classify textual data stream with concept drift. *Information Sciences, 357,* 125–143. https://doi.org/10.1016/j.ins.2016.03.043.
41. Haque, A., Khan, L., Baron, M., Thuraisingham, B., & Aggarwal, C. (2016). Efficient handling of concept drift and concept evolution over Stream Data. In *IEEE 32nd International Conference on Data Engineering* (pp. 481–492).
42. Sethi, T. S., Kantardzic, M., & Arabmakki, E. (2016). Monitoring classification blindspots to detect drifts from unlabeled data. In *IEEE 17th International Conference on Information Reuse and Integration* (pp. 142–151).
43. da Costa, F., Rios, R., & de Mello, R. (2016). Using dynamical systems tools to detect concept drift in data streams. *Expert Systems with Applications, 60,* 39–50. https://doi.org/10.1016/j.eswa.2016.04.026.
44. Lughofer, E., & Mouchaweh, M. S. (2015). Autonomous data stream clustering implementing split-and-merge concepts—Towards a plug-and-play approach. *Information Sciences, 304,* 54–79. https://doi.org/10.1016/j.ins.2015.01.010.
45. Yang, H., & Fong, S. (2015). Countering the concept-drift problems in big data by an incrementally optimized stream mining model. *Journal of Systems and Software, 102,* 158–166. https://doi.org/10.1016/j.jss.2014.07.010.
46. Wu, X., Li, P., & Hu, X. (2012). Learning from concept drifting data streams with unlabeled data. *Neurocomputing, 92,* 145–155. https://doi.org/10.1016/j.neucom.2011.08.041.
47. Hong, L., Dan, O., & Davison, B. D. (2011). Predicting popular messages in Twitter. In *ACM International Conference on World Wide Web*(WWW).
48. Li, C., Shan, M., Jheng, S., & Chou, K. (2016). Exploiting concept drift to predict popularity of social multimedia in microblogs. *Information Sciences, 339,* 310–331. https://doi.org/10.1016/j.ins.2016.01.009.
49. Shang, K., Yan, W., & Small, M. (2016). Evolving networks—Using past structure to predict the future. *Physica A: Statistical Mechanics and its Applications, 455,* 120–135. https://doi.org/10.1016/j.physa.2016.02.067.
50. Lipizzi, C., Dessavre, D., Iandoli, L., & Marquez, J. (2016). Social media conversation monitoring: Visualize information contents of Twitter messages using conversational metrics. *Procedia Computer Science, 80,* 2216–2220. https://doi.org/10.1016/j.procs.2016.05.384.
51. Miller, Z., Dickinson, B., Deitrick, W., Hu, W., & Wang, A. (2014). Twitter spammer detection using data stream clustering. *Information Sciences, 260,* 64–73. https://doi.org/10.1016/j.ins.2013.11.016.

52. Karunasekera, S., Harwood, A., Samarawickrama, S., Ramamohanrao, K., & Robins, G. (2014). Topic-specific post identification in microblog streams. In *IEEE International Conference on Big Data* (pp. 7–13). https://doi.org/10.1109/BigData.2014.7004416.

53. Malik, S., Smith, A., Hawes, T., Papadatos, P., Li, J., Dunne, C., et al. (2013). TopicFlow: Visualizing topic alignment of Twitter data over time. In *IEEE/ACM International Conference on Advances in Social Networks Analysis and Mining* (pp. 720–726). https://doi.org/10.1145/2492517.2492639.

54. Jiang, W., & Brice, P. (2009). Data stream clustering and modeling using context-trees. In *6th IEEE International Conference on Service Systems and Service Management* (pp. 932–937).

55. Li, Wen J., Tai, Z., Zhang, R., & Yu, W. (2015). Bursty event detection from microblog: A distributed and incremental approach. *Concurrency and Computation: Practice and Experience, 28*(11), 3115–3130.

56. Kalloubi, F., Nfaoui, E. H., & Beqqali, O. El. (2014). Named entity linking in microblog posts using graph-based centrality scoring. In *9th International Conference on Intelligent Systems: Theories and Application* (pp. 501–506). https://doi.org/10.1109/SITA.2014.6847286.

57. Gaglio, S., Re, G., & Morana, M. (2015). Real-time detection of Twitter social events from the user's perspective. In *IEEE International Conference on Communications (ICC)* (pp. 1207–1212).

58. Kalloubi, F., Nfaoui, E., & Beqqali, O. (2014). Graph-based tweet entity linking using DBpedia. In *IEEE/ACS 11th International Conference on Computer Systems and Applications (AICCSA)* (pp. 501–506). https://doi.org/10.1109/AICCSA.2014.7073240.

59. Kumar, N., & Muruganantham, D. (2016). Disambiguating the Twitter stream entities and enhancing the search operation using DBpedia ontology. *International Journal of Information Technology and Web Engineering, 11*(2), 51–62. https://doi.org/10.4018/IJITWE. 2016040104.

Semantic Interpretation of Tweets: A Contextual Knowledge-Based Approach for Tweet Analysis

Nazura Javed and Muralidhara B. L.

Abstract Tweets are cryptic and often laced with insinuation. Hence, interpretation of tweets cannot be done in isolation. Human beings can interpret the tweets because they possess the requisite *Contextual Knowledge*. This knowledge enables them to understand the context of tweets and interpret the text. Emulating interpretation ability in machines requires the machine to acquire this contextual knowledge. Tweets pertaining to political and societal issues contain domain-specific terms. Interpretation of such tweets solely on the basis of sentiment orientation of words produces incorrect sentiment tags. Polarity of terms is based on the topic of reference. Thus, an understanding of the pertinent domain terms and their associated sentiment is essential to guide the sentiment mining process. A resource of relevant domain-specific contextual terms and associated sentiments can help to achieve an enhanced sentiment mining performance. With the objective of equipping the machine with the contextual knowledge to facilitate semantic interpretation, we tap the Web resources, process them and structure them as Contextual Knowledge Structures (CKS). We then leverage the CKS to *enable* a semantic interpretation of tweets. We construct a CKS-based training set to train the Naïve Bayes classifier and classify the tweets. We further transform the CKS into sentiment training set (STS) and use it for detecting sentiment polarity tags for tweets. CKS provide the necessary background knowledge pertaining to issues, events, and the related domain-specific terms, thus facilitating semantic sentiment mining. All our experiments are conducted in the context of political/public policy, trending topic, and event-related tweets with an objective of obtaining a pulse of the political climate in India. Our CKS-based classifier exhibits an accuracy of 94.23% in mapping the tweets to the political topic. The distance-based CKS-Sentiment mining algorithm exhibits a consistent performance with an accuracy of 70.90%.

N. Javed (✉) · Muralidhara B. L.
Department of Computer Science and Applications, Bangalore University,
Bengaluru, India
e-mail: nazuraj@gmail.com

Muralidhara B. L.
e-mail: murali@bub.ernet.in

© Springer Nature Singapore Pte Ltd. 2018
S. Margret Anouncia and U. K. Wiil (eds.), *Knowledge Computing and Its Applications*, https://doi.org/10.1007/978-981-10-6680-1_4

The relevance of this contribution is: (a) a novel method which leverages the Web content to derive an optimum training set for tweet analysis, (b) a high degree of Accuracy, Precision, and Recall in tweet classification and sentiment mining with a small CKS-based training set, (c) a topic-adaptive model which can adapt to any domain or topic and exhibit improved tweet analysis performance.

Keywords Social media analysis · Tweet classification · Contextual Knowledge Structures · Sentiment mining

1 Introduction

Social media has emerged as the de facto medium of dissemination, expression and deliberation in the recent years. Users resort to social media to post status updates, to voice their opinions and express their sentiments. Hence social media mining has assumed tremendous relevance today. Twitter is a microblogging service that allows people to communicate with short 140 character messages. It is increasingly used by society to express views, sentiments, concerns, and debate issues. It has gained significance in the political context with political leaders and entities leveraging it to disseminate, promote, support, and debate causes. Analyzing tweets provides an insight into societal trends and scenarios. Hence, techniques and frameworks for improved tweet analysis are being investigated.

Tweet analysis has garnered much interest of researchers because of challenges involved in analyzing them. Unlike the other documents, articles, and blog contents, tweets are extremely noisy. The tweet text is unstructured with no adherence to grammatical syntax. It contains slangs, incorrectly spelled or phonetically spelled words, abbreviations, and conjoined words. Colloquial terms are liberally used. Moreover, most of the tweets do not contain a hashtag. Hence, determining the subject or topic of the tweet itself is a challenge. Also, tweets are terse, cryptic, and sparse. They lack sufficient context, and hence, cannot be interpreted per se. Applying typical machine learning and mining techniques on them can be inconclusive.

The tweet text though limited to 140 characters can speak volumes, subject to the fact that it is interpreted with reference to the context. Mining and analyzing microblogs like Twitter has proved challenging because of the sparse content and assumption of preliminary or contextual knowledge. Hence, interpretation on face value or in isolation can be a misleading exercise. Human beings can interpret the tweets because they possess *World knowledge*. They can decipher the sparse and cryptic tweets because they are guided by the *Contextual knowledge*. Emulating this interpretation ability in machines requires the machine to acquire this background knowledge. We propose a novel methodology of extracting structured contextual knowledge for popular topics/events, and building knowledge structures using mining and computational linguistics techniques. We first discover the trending topics from the tweets, then tap the Web for the related, relevant content, and harness the same to build Context Knowledge Structures (CKS). CKS are built using text mining techniques and Computational linguistics and are relevant Subject-Predicate-Object triples that depict a specific topic or event. We leverage the CKS for analysis and interpretation of tweets. Interpretation includes semantic classification and sentiment mining from tweets. CKS provide a semantic dimension to tweet interpretation resulting in enhanced mining performance.

Contextual knowledge can be exploited for tweet classification. Most of the public tweets do not contain a hashtag or an explicit event/topic mention. However, they contain certain domain- or topic-specific terms which are indicative of the topic. We use the CKS to train the Naïve Bayes classifier and equip it with the requisite domain-specific/contextual knowledge. This training set, though small, does not suffer from sparseness. It *enables* the classification of tweets and maps them to the associated topics/events.

CKS enable subjective and background-based meaningful sentiment analysis. Two approaches are generally used for sentiment mining. The first approach is a lexicon-based unsupervised technique wherein sentiment polarity of a document is computed with the aid of sentiment mining tools on the basis of sentiment orientation of words. The second approach uses supervised machine learning techniques for deriving the sentiment classes. In this approach, a large training set of labeled tweets is used to train the classifier and achieve sentiment polarity detection [1]. Both these techniques have their own limitations. While the supervised techniques involve a tedious process of labeling and building a training set, it is found that the sentiment mining tools do not perform consistently [2]. They do not exhibit good results in detecting sentiments from the sparse, cryptic tweets which lack structure. Also, sentiment associated with a term is not static. It depends on the context in which the term is used. Hence, the tweet terms need to be interpreted with respect to the context. Our work in sentiment mining adds the required contextual dimension to sentiment analysis through means of CKS. A resource of relevant domain-specific contextual terms and the associated sentiments can help to achieve an enhanced sentiment mining performance. CKS terms with the associated sentiment tags are leveraged as a sentiment training set (STS) for enhanced sentiment polarity classification. This training set guides our distance-based CKS-Sentiment mining algorithm to learn and derive the applicable sentiment labels.

Our CKS-based Naïve Bayes classifier exhibited an accuracy of 94.23% in classifying the tweets. Our distance-based CKS-Sentiment mining algorithm exhibited a consistent performance with an Accuracy, Precision, Recall, and F-score of approximately 70%. The performance of this algorithm was compared with the popular sentiment mining tools. The results revealed an improvement in the performance by 10–20%.

The proposed approach has a twofold advantage. It shows that a small training set of knowledge structures can be effectively used for machine learning. Also, the CKS-based model adapts to any topic. Trending topics are dynamically discovered, associated CKS are built, and improved mining results are attained.

This chapter is further structured as follows: Sect. 2 describes the background and related works. Section 3 describes the procedure for CKS construction. CKS are channelized to achieve semantic classification and sentiment detection. Section 4 discusses the CKS-based Naïve Bayes classifier. In Sect. 5, we present the distance-based CKS-Sentiment mining technique. Section 6 describes the relevant experiments and the results thereof. This chapter finally summarizes the relevance and contribution of our technique and methodology.

2 Background and Related Work

The relevance and challenges of social media analysis are discussed in the works [3–5]. Social media analytics is concerned with developing and evaluating tools and frameworks in order to collect, monitor, analyze, summarize, and visualize social media data. However, social media analytics involves several challenges such as semantic inconsistency or inaccuracies, misinformation, lack of structure, size, and dynamic nature. Thus, tweets need to be normalized and converted into standard form of English in order to make them suitable for machine translation and natural language processing (NLP) [5]. In our work, we pre-process the tweets; we remove stop words and punctuations, convert the phonetically spelled words into correct English words, split the conjoined words into standard form, and separate the hashtag, @tags, and embedded URLs. This makes them more amenable for processing.

The different approaches used for classification and mining of tweets are discussed in [6, 7]. Machine learning, lexicon-based approaches, and natural language processing (NLP) techniques are primarily used for the analysis of tweets. The role of the Web sources and the approaches and techniques for leveraging them to achieve semantic interpretation of text content are discussed in the following works. The relevance of Web resources is highlighted in [8]. N-gram summaries are found from the tweets using internal, external, and mined internal content [9]. This paper uses external sources like news articles and other Web content to enrich the tweet and facilitate a broader interpretation. The relevance of the semantic models in interpretation and inference is investigated [10]. Semantic models built can be used for machine intelligence and can become a basis for inference for the semantic Web. In our work too, we leverage the Web resources and build semantic structures in the form of CKS. These CKS are then used for inferring topics or classifying the sparse tweets and performing sentiment analysis.

The different approaches and techniques that leverage the additional resources for a more optimized result are discussed in the following works. It is difficult to interpret some tweets in isolation. But if the context of this tweet is available in the

Knowledge base (KB), then inference and interpretation are possible. Social genome, a large real-time social knowledge base, was built. It was built, using Wikipedia, a set of other data sources, and social media data [11]. Metadata and external resources can help guide the interpretation of tweets. Enhanced classification and mining can be achieved by using these resources. Metadata like location, date, day of the week, and time of the day is used to discover relationships between the tweet sentiment and day, time, etc., to aid analysis [12]. The twitter messages are classified into different topics [13]. Here, features are extracted from the words in tweets, from the URLs mentioned in tweets, and words from the user profile. All these features are thought as a Bag-Of-Words (BOW) for achieving classification.

The problems and challenges in classification and mining of short texts like tweets are highlighted in the following works. Short texts do not provide sufficient word occurrences, and hence, traditional classification methods such as "Bag-Of-Words" fail to exhibit good results. To solve this problem, the short text is appended with a small set of domain-specific features extracted from the author's profile and text [13]. The tweet is enhanced with context [15]. The context of the tweet is obtained by mapping onto the most similar Wikipedia pages. This context adds to the sparse tweet and thus helps to achieve a more accurate classification. A distantly supervised approach for classifying tweets is used in [16]. The tweets contain a plethora of distinctive named entity types. Almost all of these are infrequent, and hence, even a large sample of annotated tweets is not a good sample for training the classifier. Hence, a distantly supervised approach using large dictionaries of entities gathered from freebase is used for obtaining the entity's context. Enriching the tweets by appending them with discriminatory terms from the contemporary Web corpus facilitates and enhances classification [17]. The sparse issue is addressed by grouping short texts to form long text clusters and using the clusters to train the classifier [18]. A system that automatically classifies Twitter messages into a set of predefined categories is proposed in [13]. This system takes into account external features such as words from linked URLs, mentioned user profiles, and Wikipedia articles for classification.

Lexicon-based unsupervised techniques for sentiment mining use sentiment mining tools that compute sentiment scores on the basis of sentiment lexicons. The sentiment orientation of words in the lexicon is used to compute the sentiment polarity of the content. Lexicon-based unsupervised approach is used for sentiment mining in [19–22]. As against this, the machine learning approach involves the use of a labeled training set. Here, the sentiment detection problem is treated as a classification problem. A training set is used to train the classifiers like Naïve Bayes, SVM, and neural networks to learn the model and derive the sentiment output. The machine learning approach for sentiment determination is followed in [23–27]. Though both the above techniques are suitable for mining sentiments from documents, articles, etc., they may not be efficient for detecting sentiments from tweets. The challenges of Twitter sentiment analysis are discussed in [28]. The issues like data sparseness and sarcasm are responsible for incorrect classification of tweets [29]. Topic-based search for tweets cannot produce good results because the tweets are sparse. Hence, tweets are enriched by exploiting the links mentioned in the tweets [30]. Hashtag in the tweets can be leveraged to determine the sentiment. If the hashtag is made up of sentiment words, these words can guide the determination of sentiment polarity [31].

Lexicon-and ontology-based unsupervised techniques of sentiment mining are discussed in the following works. Since the sentiment of a term is not static but is dependent on the context in which it is used, the co-occurrence pattern of given term with the other terms is considered in order to obtain the semantics of the given term [32]. A Twitter-specific lexicon for a single subject is built and exploited for achieving improved classification accuracy [33]. Semantics are added as additional features to the training set for a semantic sentiment analysis. Semantics are added in the form of synsets and SentiWordNet features for achieving an enhanced result [34]. A hybrid approach, which combines machine learning approach with a domain-specific lexicon for sentiment analysis, is used in [35]. Domain ontology is leveraged to calculate sentiment scores. The ontology is built using the tweets retrieved by performing a topic-based tweet search [36]. Our CKS are functionally comparable to domain ontology or domain lexicon. However, unlike the previous works, our CKS are not constructed from the sparse tweets. They are built by scraping relevant and related Web content, and hence, they are more comprehensive.

A comparative study of the different sentiment analysis methods is done [2]. This paper presents an evaluation and comparison of twentyfour popular sentence-level sentiment analysis tools using gold standard datasets. Some tools detect three polarity classes—positive, negative, and neutral while others detect only the positive and negative classes. In the context of social networks, the best method for 3-class experiments was Umigon, followed by LIWC15 and VADER. AFINN builds and uses a Twitter-based sentiment lexicon including Internet slangs and obscene words. Hence, LIWC15, VADER, and AFINN tools were selected in the category of 3-class sentiment for comparing and evaluating the performance of our CKS-Sentiment mining algorithm. In the class of 2-class experiment, we evaluated our results using Sentiment140.

On the basis of our study, we can say that the recent works in Twitter analysis solve the problem of sparseness by enriching the tweets by appending them with context words, metadata, or making use of knowledge bases (KB) like Wikipedia and ontology. Our work too addresses the sparseness problem. However, we do not enrich the tweets; instead we equip the machine with contextual knowledge in the form of CKS. Our objective is to arm the machine with knowledge of related terms, entities, and events in the form of CKS and use the same to guide tweet classification.

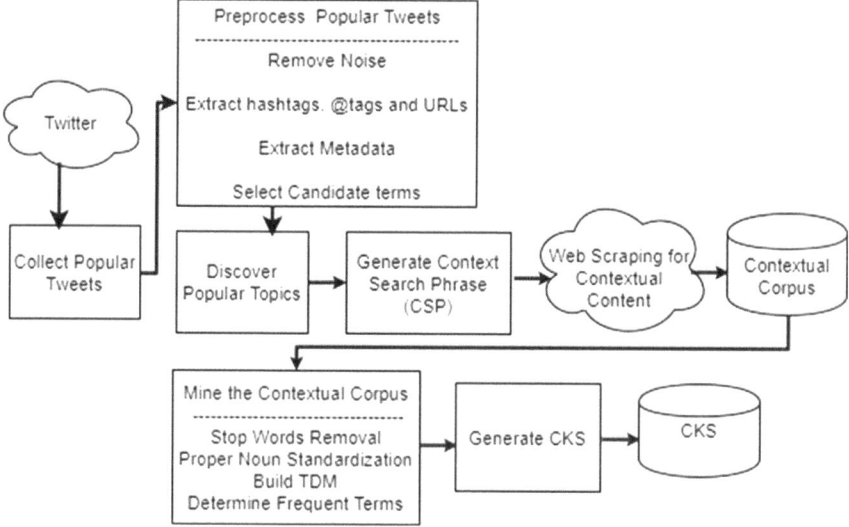

Fig. 1 CKS generation for the popular/trending topics

3 Constructing the Contextual Knowledge Structures (CKS)

CKS are developed with the objective of equipping machine with the relevant background knowledge about the trending topics. We build CKS for the trending topics and channelize them for semantic interpretation of tweets. Figure 1 depicts the steps in CKS generation.

The following subsections contain a description of the processing steps.

3.1 Capturing Popular Tweets

The political leaders with more than a million followers and belonging to different political parties of India are considered as popular leaders. The twitter messages tweeted by them are termed as "Popular tweets." Political entities or leaders tweet about the current and relevant events, issues, or policies. Hence, parsing and processing these tweets yield popular, trending topics or events.

3.2 Pre-process Popular Tweets

The popular tweets are subjected to cleaning and pre-processing. We correct the spellings of the incorrectly/phonetically spelled words, replace the colloquial terms, split conjoined words and thus clean the tweets. We extract metadata like username, location, retweet count, followers count, etc., for further analysis [37].

3.3 Discovering Popular Topics (PT) and Generating Context Search Phrase (CSP)

Popular or trending topics are discovered from the popular tweets by constructing a document-term matrix (DTM). Frequent terms and the count of the hashtagged terms collectively determine the popular topics.

The PTs may be specific terms like "Budget2016" or relatively general terms like "JNU." In order to build a semantic contextual corpus, it is necessary to scrap out only the related Web contents. Search for general terms like "JNU" may yield irrelevant Web content. Hence, we generate a Context Search Phrase (CSP) by appending the PT with the month, the year of tweet, and the country. Additionally, we also append the suggestion given by Google Trends and construct a CSP. Thus, CSP = PT $\cup$ month and year of tweets $\cup$ Suggestion by Google Trends.

3.4 Web Scraping

CSP is used as a search string, and the relevant, related documents, news articles, and blogs are searched from the Web. The contents from the retrieved URLs are scraped. This yields multiple documents of context data for each PT. These documents are compiled and organized into a contextual corpus.

3.5 Mining the Contextual Corpus

We remove punctuations, stop words, special symbols, and Unicode characters from the contextual corpus. We perform standardization of proper nouns. For example, Prime Minister Narendra Modi, PM Mr. Modi, etc., are the different ways in which Prime Minister Narendra Modi is represented. These proper nouns are translated into a standard form like "NarendraModi." After pre-processing, we generate a term-document matrix (TDM). We filter out the infrequent terms for reducing the dimensionality. This is done using the feature ranking method. The top ranked terms are selected using the Pareto analysis[1] principle. The filter vector is constructed using these frequent terms. Thus, one filter vector is derived for each PT in the contextual corpus. This filter vector is used to construct CKS.

3.6 Generate CKS

The filter vector contains the frequent terms for a given PT. We select from the contextual corpus only the sentences that contain the filter vector terms. This yields a subset of sentences which are relevant and important for a specific topic. POS tagging, dependency parsing, and chunking[2] enable us to extract meaningful verb phrases from the contextual corpus. These verb phrases are subjected to further processing so as to a) eliminate duplicates and b) find semantically inclusive or superset phrases and perform elimination, for example, "provide" "employment guarantee" "under MGNREGA" is considered a superset of "provide" "employment" "under MGNREGA". Hence, the latter triple is eliminated. The verb phrases are restructured as <Subject> <Predicate/Verb> <Object> triples and are termed as CKS. Table 1 illustrates the extracted CKS.

4 Semantic Classification with CKS

Users generally tweet about trending topics. As per our study, 85% of the user tweets do not contain a hashtag. Most of the user tweets do not contain sufficient content or context. Hence, classifying them on the basis of the content becomes difficult. For example., in the tweet—"*jnusu president kanhaiyakumar, umarkhalid and anirbanbhattacharya granted regular bail by a Delhi court in a sedition,*" there is no hashtag/mention of the related "JNU February 9 2016"[3] incident. Hence, classifying this tweet and mapping it to the above incident are only possible if contextual information is learnt by the machine. Our CKS approach powers the machine with the requisite knowledge to drive the semantic classification. Figure 2 shows the different processing steps for classification of tweets.

---

[1]Pareto Analysis is a statistical technique in decision-making used for the selection of a limited number of task/features that produce significant overall effect. It uses the Pareto Principle (also known as the 80/20 rule) the idea that by doing 20% of the work you can generate 80% of the benefit of doing the entire job.

[2]Chunking is the process of grouping various words which have Part-Of-Speech (POS) tags into phrases like Noun phrases, Verb phrases etc.

[3]On February 9, 2016, students of Jawaharlal Nehru University (JNU) held a protest on their campus against the capital punishment meted out to the 2001 Indian Parliament attack convict Afzal Guru.

Table 1 Verb phrases generated using MontyLingua tool and the extracted CKS

Relevant sentences from corpus	Verb phrases	CKS
JNU is caught in a row over an event against the hanging of Parliament attack convict AfzalGuru during which antinational slogans were allegedly raised	("catch" "JNU" "in row" "over event") ("hang" " " "of Parliament attack convict AfzalGuru") ("raise" "antinational" "slogan")	"JNU " "catch " "in row" "over event" " " "hang" "of Parliament attack convict AfzalGuru" "antinational slogan" "raise"
In the Union Budget for 2016–17, the finance minister Arun Jaitley announced an allocation of Rs 38,500 crore for providing employment under MGNREGA	("announce" "finance minister Arun Jaitley" "allocation" "of Rs 38,500 crore") ("provide" "employment" "under MGNREGA")	"finance minister Arun Jaitley" "announce" "allocation of Rs 38,500 crore" "employment" "provide" "under MGNREGA"

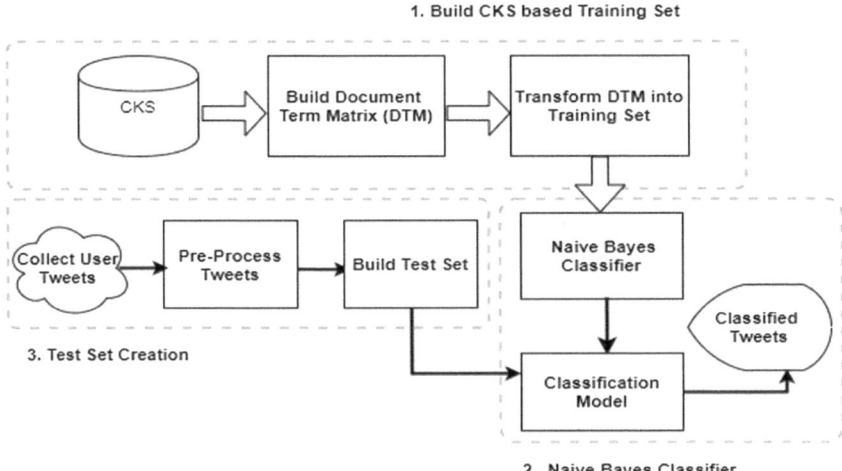

Fig. 2 Semantic classification with CKS training set

4.1   CKS-Based Training Set Construction

CKS constructed in the previous section are converted into a training set. Each triple in a CKS set is considered as a separate document, and a DTM is constructed. DTM terms are annotated with the appropriate class labels and transformed into a training set (Table 2). There is one DTM for each topic. All the DTMs are combined to form a CKS training set.

Table 2 DTM transformed into CKS training set

Term	$\sum tf$ (sum of term frequencies)	Class label
$Term_1$	$\sum tf_1$ (weight$_1$) for $Term_1$	Class label
$Term_2$	$\sum tf_2$ (weight$_2$) for $Term_2$	Class label
...		
$Term_n$	$\sum tf_n$ (weight$_n$) for $Term_n$	Class label

4.2 Training the Classifier

The Naïve Bayes (NB) classifier is a suitable machine learning algorithm for text classification because it considers both presence and frequency of N-grams [38]. The NB algorithm is appropriate and can be applied to our methodology because our model is based on both presence and frequency of the context terms in CKS. The following section contains a brief discussion of the NB classifier.

4.2.1 The Naïve Bayes Classifier

The training set (Table 2) is used to train the NB classifier. Each tweet tuple is represented by an n dimensional attribute vector $X = (x_1, x_2, \ldots, x_n)$. There are m prospective topic classes $C_1, C_2, \ldots, C_m$. The objective of Naïve Bayes classification is to derive the maximum a posteriori and accordingly map the tuple to the associated class.

Maximum a posteriori probability, i.e., the maximal $P(C_i|X)$ can be derived from Bayes' theorem as follows: $P(C_i|X) = \frac{P(X|C_i)P(C_i)}{P(X)}$.

Since $P(X)$ is constant for all classes, only $P(X|C_i)P(C_i)$ needs to be maximized. Thus, given the tuples with $X = (x_1, x_2, \ldots, x_n)$, we compute

$$P(X|C_i) = \prod_{k=1}^{n} P(x_k|C_i)$$

4.3 Test Set Creation

In order to test our CKS-based Naïve Bayes classifier, we build a test set by capturing the streaming tweets using the Twitter streaming API. Only English tweets are collected. We perform a comprehensive cleaning and pre-processing process in order to clean and normalize the tweets [37].

4.4 The CKS-Based NB Classifier

Applying the NB theorem in our work, we compute the value of $P(X|C_i)$ for each class $C_1, C_2, \ldots, C_m$. We compute $P(X|C_i)$ by first computing $P(x_1|C_i), P(x_2|C_i) \ldots P(x_k|C_i)$ individually and then applying:

$$P(X|C_i) = \prod_{k=1}^{n} P(x_k|C_i)$$

For example, by taking x_1 as "kanhaiyakumar" and C_1 as "JNU FEB9," $P(x_1|C_i)$ is computed as follows:

$$P(x_1|C_1) = \frac{\sum tf_{\text{kanhaiyakumar for classJNUFEB9}}}{\sum tf_{\text{JNUFEB9}}}$$

$P(C_1)$ = Priori Probability of class "JNU FEB9" is computed as follows:

$$P(C_1) = \frac{\sum tf_{\text{for class JNU FEB9}}}{\sum tf_{\text{for classes } C_1 \ldots C_m}}$$

We compute $P(X|C_i))$ as described above for each class $C_1, C_2, \ldots, C_m$. We select the label C_i such that $P(X|C_i)P(C_i)$ is maximized.

The advantages of the CKS-based classification are as follows:

1. Supervised learning algorithms require a large training set. Since the tweets are sparse, even a large set does not provide sufficient term occurrences. Generating a training set by manually labeling tweets is a herculean task. In our work, we leverage CKS for training the classifier and demonstrate how a small set of CKS can achieve high classification accuracy.

2. This approach does not suffer from the overfit problem caused by the overlap of training and test set. Since the CKS are extracted independently, they have no bearing or relation with the test set.

3. The proposed methodology does not use labeled tweets as training examples. Rather, an optimized set of CKS are used to equip the machine with the requisite knowledge for guiding and facilitating classification.

4. We address the problem of topic sensitivity of a classifier. Classifiers trained for a particular domain or topic cannot adapt to a different domain and hence are topic-sensitive. In our work, we follow a dynamic approach. We discover the trending or popular topics and automatically develop the associated CKS. Thus, the CKS can be generated for any trending topic and be used for training the classification model. We can thus say that this approach is flexible and has the ability to adapt itself to any domain or topic.

5 Enhanced Sentiment Detection with CKS-Sentiment Mining Algorithm

Sentiment polarity determination should be subjective or contextual to the subject. Sentiment mining that does not consider the domain terms, their relevance, and the associated sentiment fails to produce correct sentiment tags, for example, the tweet "ATMs dry, queues everywhere" with respect to the topic "Demonetization[4], November 2016, India" should be assigned higher negative score than the score normally assigned by mining tools. Similarly, the phrase "stone pelting has stopped" is associated with terrorism and implies reduction in the acts of terrorism in Kashmir. It should be assigned a high positive score. Hence, we can say that there is a need to modify the sentiment polarity or sentiment score based on the topic of reference. In our work, CKS provide the necessary reference for a given topic/event and thus facilitate semantic sentiment detection. We associate the sentiment with the CKS terms and develop a STS which helps us to achieve an improved result in tweet sentiment mining.

This work demonstrates the efficacy of our distance-based CKS-Sentiment mining technique and evaluates its performance by comparing it with the outputs of state-of-art sentiment mining tools. The processing steps in sentiment detection are depicted by Fig. 3. The three main stages in achieving improved sentiment mining results are described hereunder.

1. *Construct CKS-based Sentiment Training Set (STS)*

CKS facilitate correct interpretation of domain-specific terms and thus enable a meaningful, semantic sentiment analysis. We need to organize, enhance, label and thus transform the CKS into an effective STS. We follow the steps depicted in Fig. 4 to perform the transformation. The adjectives, verbs, nouns, adverbs, and negation terms are extracted by parsing the CKS. These terms are enriched with their synonyms/synsets using WordNet. This enriched CKS provides a semantic dimension to our distance-based sentiment mining algorithm.

The terms and synsets are assigned sentiment tags as stated and illustrated in Table 3. The enriched and labeled CKS are transformed into STS.

Thus, STS = CKS ∪ Synsets ∪ Sentiment Tags

2. *Building a Test Corpus for Sentiment Mining*

We capture the real-time tweets and subject them to a comprehensive cleaning process. We use our CKS-based NB classification algorithm (see Sect. 4.4) to map the tweets to the associated topic/event. The tweets that cannot be mapped to a topic are classified as "Other." We filter out these tweets and thus obtain a clean and

---

[4]On 8 November 2016, the Government of India announced the demonetization of all ₹500 (US 7.80) and ₹1,000 (US 16) banknotes of the Mahatma Gandhi Series.

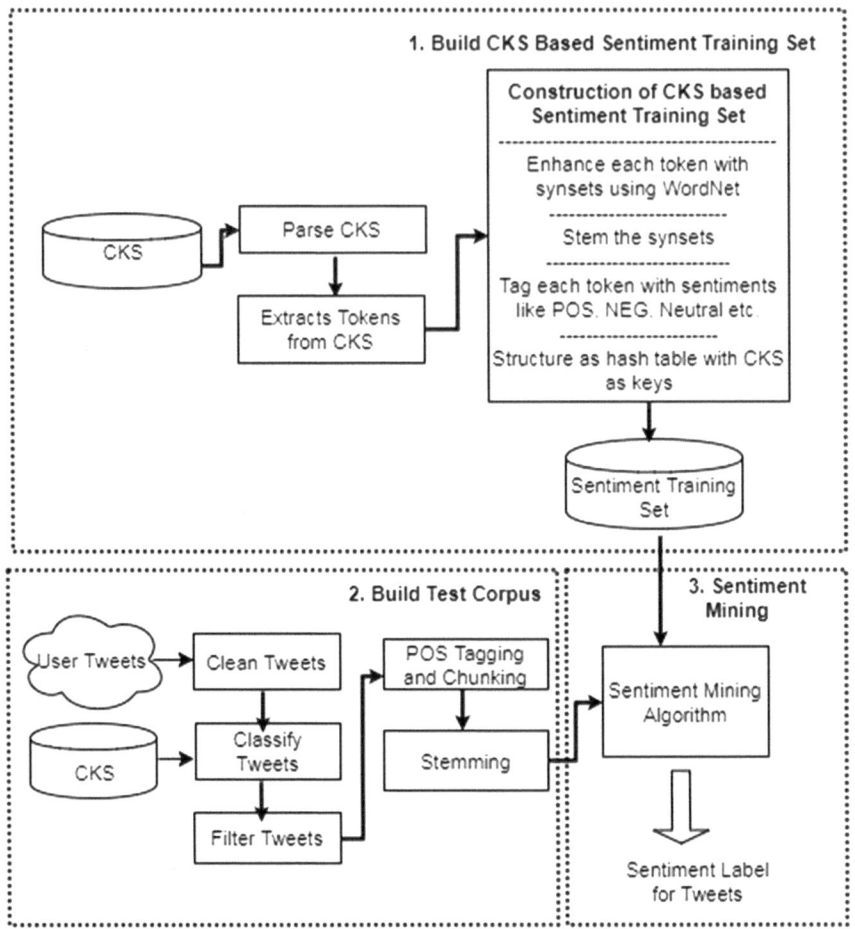

Fig. 3 CKS-based sentiment mining

classified test corpus of user tweets. The tweets in the test corpus are further chunked so that the relevant noun and verb phrases can be extracted and dimensionality can be reduced. This chunked tweet corpus is used as test set for sentiment mining.

3. *The Sentiment Mining Process*

We propose a distance-based CKS-Sentiment mining algorithm which uses the STS for training and derives semantic sentiment labels. Similarity metric measures the similarity or dissimilarity between two data objects. Euclidean distance between two points measures the numerical difference for each corresponding attributes of point p and point q. Then, it combines the square of differences in each dimension

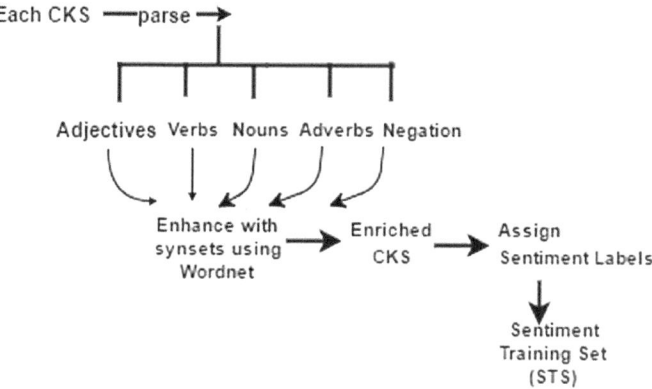

Fig. 4 CKS transformed into sentiment training set

Table 3 Sentiment tags, meanings, and examples

Sentiment Tags	Meaning	Examples
POS	Positive polarity	Digital, applaud, transparent, empower, optimistic
NEG	Negative polarity	Blackmoney, counterfeit, terror, disruption, launder, queue, dry
NEU	Neutral	Cash, money, currency, payment
DOMNOUN	Domain-specific nouns	Demonetization, NarendraModi, GDP, ATM
POSMOD	Modifier which modifies/ enhances the sentiment	Accept, support, enhance, promote, strong
NEGMOD	Modifier which modifies/ negates the sentiment	Curb, end, no, not, refuse, strike

into an overall distance. It can be used in higher-dimensional space and is specified as follows:

$$d(p,q) = \sum_{i=1}^{n} \sqrt{(p_i - q_i)^2}$$

Our algorithm derives the sentiment by comparing the tweet tokens with the STS terms. This algorithm computes similarity measure S as follows:

$$S(p,q) = \sqrt{(p_1 - q_1)^2 + (p_2 - q_2)^2 + \ldots (p_n - q_n)^2} = \sum_{i=1}^{n} \sqrt{(p_i - q_i)^2}$$

wherein

p_1, p_2, p_3 ...: tokens from user tweet

q_1, q_2, q_3 ...: STS terms

n: number of tokens in user tweet

$p_i - q_i = \sim 0$, i.e., 1: if $p_i \in$ Key of STS

$p_i - q_i$ = similarity score by Wu and Palmer similarity metric[5]: if $p_i \in$ Synsets

$p_i - q_i$ = sentiment score as returned by SentiWordNet[6]: if $p_i \notin$ STS

Table 4 shows the STS organized as a hash table with the CKS terms as the keys. The tweet is tokenized, and POS tagging is done for the tokens. The tweet token is compared to the STS terms in order to determine the associated sentiment. Each tweet token is assigned sentiment orientation according to the sentiment tag specified in the STS. However, if the tweet token is preceded by a negative modifier/negation (NEGMOD), the sentiment orientation is reversed or negated. If a tweet token has a neutral (NEU) tag or a domain noun (DOMNOUN) tag in the STS, the sentiment orientation for that tweet token is determined on the basis of the preceding tokens. If the preceding token is a positive modifier (POSMOD), then the sentiment orientation is modified to positive. If the preceding token is a negative modifier (NEGMOD), then the sentiment orientation is modified to negative. However, modification of the sentiment is done subject to the fact that the distance between the modifiers and the tweet token is within the threshold limits (Θ). The tweet tokens (nouns, adjectives, verbs) which are not included in the STS (because they are not domain-specific words) are assigned the sentiment orientation and score based on the polarity and score returned by SentiWordNet lexicon. Since SentiWordNet returns scores in the range of [0,1] for each of the sentiments (positive, negative and objective), only the sentiment tag and score associated with highest score is considered.

Our distance-based CKS-Sentiment mining algorithm assigns sentiment polarity to each tweet. It assigns sentiment tags like "POS," "NEG," and "NEU" based on the computed polarity scores. Our algorithm exhibits a consistent performance with an accuracy of 70.90%. This algorithm has the following limitations: (i) It fails to detect correct sentiment when the tweet is laced with irony; (ii) sentiment detection may be incorrect when the tweet is framed as a question, for example, "What do you think Modi? You have eliminated corruption?"—this tweet is not correctly tagged by the algorithm. Moreover, polarity detected may be incorrect when the tweet is absolutely ungrammatical. However, this is not considered as a limitation of the algorithm because in such cases of semantic ambiguity even human beings fail to correctly identify the correct polarity.

---

[5]The Wu and Palmer [39] similarity metric is used to measure the depth of the given concepts in the Word Net taxonomy, the least common subsumer (LCS) depth and combines these figures into a similarity score.

[6]SentiWordNet is a lexical resource for opinion mining. SentiWordNet assigns to each synset of WordNet three sentiment scores: positivity, negativity and objectivity. It gives scores in the range [0,1] for each of the sentiments i.e. positive_score + negative_score + objective_score = 1.

Table 4 STS organized as a hash table

Key	Values	
	Sentiment tag	Array list of synsets
Curb	NEGMOD	Stop, end, anti, impair, curtail, …
Support	POSMOD	Endorse, join, promote, …
Praise	POS	Appreciate, applaud, laud, …
Blackmoney	NEG	Unaccounted_money, black_money
Counterfeit	NEG	Fake, imitation, forge, …
Cash	NEU	Money, currency, rupee, dollar, …
Demonetization	DOMNOUN	Demonetize

6 CKS-Based Enhanced Tweet Interpretation: Experiments and Results

Our experiments demonstrate the effectiveness of CKS-based enhanced tweet interpretation. Section 6.1 describes the experiments conducted for CKS-based classification, and Sect. 6.2 describes the sentiment mining experiments.

6.1 Experiment—CKS-Based Semantic Classification

Our experiments demonstrate the effectiveness of our novel methodology which leverages the extracted CKS to function as a training set and achieve a meaningful classification. The experiment involves the stages enumerated hereunder.

1. Determination of CSP for Popular Topics and CKS Construction

We used Twitter search API to capture the tweets of important Indian politicians. In the experiment carried out by us, we discovered five popular topics on the basis of 482 tweets of popular political leaders collected during the period from February 2016 to April 2016. "JNU" and "MGNREGA" were two such PTs. After determining the PTs, we constructed the CSP. For example, the PT "JNU" was appended with "February 2016, India, antinational slogans" to form a CSP. Using the CSP, we searched the Web for the related content using Microsoft Bing API. Content was extracted by Web scraping content from these URL sites. After reducing dimensionality using the procedure stated in Sect. 3.6, the content collected was compiled into a contextual corpus.

This corpus was subjected to POS tagging. POS tagging and dependency parsing were done using MontyLingua [40]. Triples in the form of Subject-Predicate/Verb-Object were extracted for each sentence of the corpus. For our experiment, we used two set of CKS triples—(i) pertaining to the "JNU Feb 9 2016" event and

(ii) "MGNREGA" (Mahatma Gandhi National Rural Employment Guarantee Act) and translated the same into a training set.

2. Test Set Construction and Classifier

To test the validity of our approach, we collected 31,307 tweets randomly during the period from February 2016 to April 2016. We cleaned, pre-processed the tweets, eliminated the retweets, and finally had a test corpus of 17,260 cleaned tweets. We classified the tweets in the test corpus using our Naïve Bayes classification algorithm coded in python. The Naïve Bayes classifier demonstrated an accuracy of 94.23%. Table 5 shows the confusion matrix constructed for evaluating the performance of the classifier. The computation of the Accuracy, Precision, Recall, and F-score is based on this confusion matrix. In order to reduce the false positives, we had fixed a minimum threshold for all classes. This threshold was fixed on the basis of results of the validation tests in repeated runs of classification algorithm. Only the tweets, whose computed posterior probability crossed the fixed threshold, were classified as belonging to that class. The tweets which could not pass the threshold test for the above classes were classified as "Other." Thus, the tweets which could not be associated with the events/topics with *certainty* were assigned the "Other" class.

We compared the performance of our CKS-based classifier with the BOW-based classifier. The BOW classifier was trained using a corpus of manually labeled tweets. We used the same test corpus comprising of 17,260 pre-processed tweets and carried out six-fold cross validation testing using the R mining tool.

The confusion matrix containing the classification results of the BOW approach is shown in Table 6. The CKS-based classifier demonstrates an accuracy of 94.23% while the BOW classifier exhibits an accuracy of only 46.75%. Table 7 compares the performance of CKS- and BOW-based classification. Figure 5 shows a graphical comparison of the models. The tables and charts reveal that the CKS classifier exhibits consistent performance with high Accuracy, Precision, Recall and F-score. The low accuracy of the BOW classifier can be attributed to the sparseness of tweet which results in insufficient training. However, the BOW classifier exhibits a high Recall score for the class "MGNREGA". This is so because, the "MGNREGA" tweets though less in number (1510), are not sparse. Hence the BOW classifier, when trained/tested using six-fold cross validation testing exhibits good performance in recognizing the True Positives for "MGNREGA" class.

6.2   Experiment—Enhanced Sentiment Mining with Distance-Based CKS-Sentiment Mining Algorithm

The efficacy of our distance-based CKS-Sentiment mining methodology is substantiated by the following experiments.

Table 5 Confusion matrix—classes "JNU Feb9," MGNREGA, and other (CKS classification)

Actual	Predicted		
	JNU FEB9	MGNREGA	Other
JNU FEB9 (2740)	2580	0	160
MGNREGA (1510)	0	1510	0
Other (13010)	700	135	12,175

Table 6 Confusion matrix class—"JNU Feb9," "MGNREGA," and "Other" (BOW-based classification)

Actual	Predicted		
	JNU FEB9	MGNREGA	Other
JNU FEB9 (2740)	2500	240	0
MGNREGA (1510)	0	1510	0
Other (13010)	6075	2875	4060

Table 7 Comparison of performance of CKS and BOW

	Accuracy %		Precision %		Recall %		F-score %	
	BOW	CKS	BOW	CKS	BOW	CKS	BOW	CKS
JNU	46.75	94.23	38.36	94.24	91.24	94.16	54.01	94.20
MGNREGA	46.75	94.23	41.65	93.68	100.0	100.0	58.80	96.73
Other	46.75	94.23	94.35	96.23	31.20	93.58	46.90	94.89

Fig. 5 Graphical comparison of the performance of CKS and BOW classification

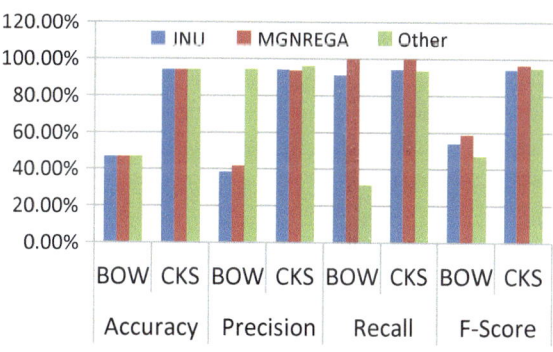

"Demonetization in India" was found to be the trending topic in November 2016. Web scraping was done to collect the background content for this topic. In our experiment, we collected articles and news items on "Demonetization, November 2016, India." We applied the computational linguistic techniques using MontyLingua and transformed the contextual content into CKS.

User tweets were collected during the period from November 9, 2016 to December 15, 2016 in order to analyze the public sentiments with respect to the trending topic "Demonetization November 2016, India". We collected 28,785 tweets using Twitter streaming API. After cleaning the tweets and removing the duplicate tweets/retweets, we had a test corpus of 15,435 clean tweets. The test corpus was classified using our Naïve Bayes classifier. This was done to ensure that the tweets in the corpus were actually pertaining to demonetization. Our classification experiment which exhibited an accuracy of approximately 94% was used to filter out tweets that could not be associated to demonetization with certainty. At the end of filtration through classification, we had a clean corpus of 15,208 classified tweets. We used POS tagging and chunking using context-free-grammar (CFG) and extracted the relevant parts from the tweets. This was done using NLTK in python. This yielded a test corpus of chunked tweets with a dimensionality reduction of approximately 16%.

The CKS were transformed into STS as described in Sect. 5. The sentiment mining algorithm which was coded in Python assigned polarity tags like "POS," "NEG," and "NEU" to the tweets in the test corpus.

The results of our algorithm were compared with the results of the popular state-of-art sentiment mining tools like LIWC 2015 [41], VADER [42], AFINN [43], and Sentiment140 [44]. Table 8 tabulates the performance of these tools and compares the same with our CKS-Sentiment mining algorithm. While Fig. 6 shows the graph which compares the Accuracy, Fig. 7 compares the Precision, Recall, and F-score. CKS-Sentiment miner shows a consistent performance in detecting the three sentiment classes. CKS-Sentiment miner exhibits 10–20% improved performance in terms of Accuracy, Precision, Recall and F-score as compared to the mining tools. This improved performance is on account of the correct subjective interpretation *enabled* by the STS. The sentiment classification exhibited by the sentiment mining tools is not satisfactory and can be attributed to the following reasons: (i) Tweets in the test set are ungrammatical. (ii) Tweets are sparse and contain many domain specific terms. These domain specific terms like blackmoney, counterfeit etc. are not typical sentiment terms and hence they are not correctly tagged/scored by the tools that rely on lexicons. Hence many tweets get classified with the Neutral tag.

7 Conclusion and Future Scope

Contextual knowledge equips machine with the requisite background knowledge. This knowledge of domain-specific terms functions as a training set and yields an enhanced mining output. This work dynamically builds the requisite knowledge structures (CKS) by tapping the Web sources. These CKS are leveraged for achieving improved semantic interpretation of tweets.

Our study reveals that approximately 85% of the user tweets do not contain a hashtag. Hence, a tweet per se cannot be mapped to a topic. Mapping a tweet to a topic or tweet classification is essential for further analysis like sentiment mining. Our CKS-based classifier performs consistently and shows high accuracies.

Table 8 Comparison of CKS-Sentiment mining and other sentiment mining tools

	Recall %			Precision %			F-score %		
	POS	NEG	NEU	POS	NEG	NEU	POS	NEG	NEU
CKS	72.18	69.62	71.23	70.40	71.65	70.72	71.28	70.62	70.97
LIWC	61.25	49.01	61.56	55.10	55.10	54.25	58.01	51.88	57.67
VADER	53.85	33.67	72.08	52.32	52.32	42.31	53.08	40.97	53.33
AFINN	56.05	63.11	41.11	52.43	52.43	60.08	54.18	57.27	48.82
Sentiment-140	15.55	27.83	82.47	54.36	54.36	22.57	24.18	36.81	35.44

Fig. 6 Comparison of accuracy of CKS—sentiment algorithm and other tools

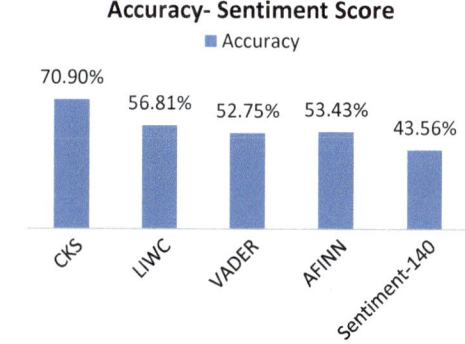

Fig. 7 Comparison of the performance of CKS—sentiment algorithm and other tools

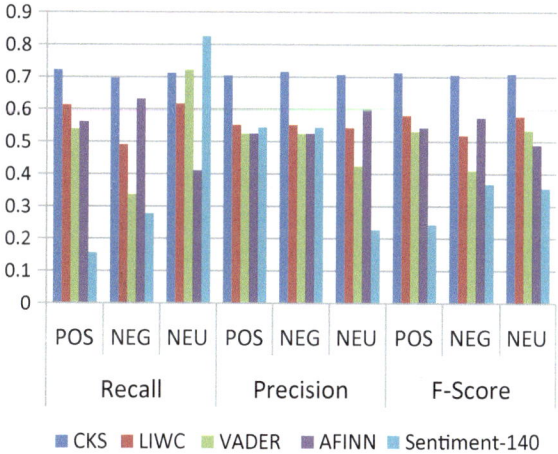

CKS-based training set is comprehensive and meaningful and hence is able to achieve a much more meaningful classification. Unlike the BOW classifier which is insufficiently trained (learns only a few terms) because of the sparseness of the tweets, our classifier is not biased by *the presence of few specific terms*. Domain terms, events, and entities in the CKS together guide the classification output. The tweets that lack relevant terms and cannot be mapped to a specific class label are filtered out as noise. This helps in reducing computing overheads for further processing/sentiment mining of insignificant tweets.

CKS lend a semantic dimension to machine learning. While sentiment mining tools compute sentiment on the basis of occurrence of sentiment words, they fail to consider subjective-/domain-specific angle to sentiment. Building reference in the form of CKS and leveraging the same for training can enhance the performance of mining algorithms. Our set of experiments in sentiment mining describe a framework for performing a domain-specific sentiment analysis wherein sentiment is discovered and guided by contextual knowledge.

The CKS-based approach does not take into consideration emoticons, irony in tweets, and tweets posed as questions. A CKS-based algorithm which detects irony and questions and accordingly determines/modifies the sentiment orientation needs to be explored and is our next milestone.

References

1. Pang, B., & Lee, L. (2008). Opinion mining and sentiment analysis. *Foundations and Trends in Information Retrieval, 2*(1–2), 1–135.
2. Ribeiro, F. N., Araújo, M., Gonçalves, P., Gonçalves, M. A., & Benevenuto, F. (2016). Sentibench-a benchmark comparison of state-of-the-practice sentiment analysis methods. *EPJ Data Science, 5*(1), 1–29.
3. Zeng, D., Chen, H., Lusch, R., & Li, S. H. (2010). Social media analytics and intelligence. *IEEE Intelligent Systems, 25*(6), 13–16.
4. Li, C., Sun, A., Weng, J., & He, Q. (2015). Tweet segmentation and its application to named entity recognition. *IEEE Transactions on Knowledge and Data Engineering, 27*(2), 558–570.
5. Kaufmann, M., & Kalita, J. (2010, January). Syntactic normalization of twitter messages. In *International Conference on Natural Language Processing, Kharagpur, India*.
6. Volkova, S., Bachrach, Y., Armstrong, M., & Sharma, V. (2015, January). Inferring latent user properties from texts published in social media. In *AAAI* (pp. 4296–4297).
7. Burnap, P., Rana, O. F., Avis, N., Williams, M., Housley, W., Edwards, A., et al. (2015). Detecting tension in online communities with computational Twitter analysis. *Technological Forecasting and Social Change, 95*, 96–108.
8. Gomadam, K., Yeh, P. Z., Verma, K., & Miller, J. A. (2012, June). Data enrichment using web APIs. In *2012 IEEE First International Conference on Services Economics (SE)* (pp. 46–53). IEEE.
9. Jadhav, A. S., Purohit, H., Kapanipathi, P., Anantharam, P., Ranabahu, A. H., Nguyen, V., et al. (2010). Twitris 2.0: Semantically empowered system for understanding perceptions from social data.
10. Villanueva, D., González-Carrasco, I., López-Cuadrado, J. L., & Lado, N. (2016). SMORE: Towards a semantic modeling for knowledge representation on social media. *Science of Computer Programming, 121*, 16–33.
11. Chai, X., Deshpande, O., Garera, N., Gattani, A., Lam, W., Lamba, D. S., ... et al. (2013). Social media analytics: The Kosmix story. *IEEE Data Engineering Bulletin, 36*(3), 4–12.
12. Vosoughi, S., Zhou, H., & Roy, D. (2016). Enhanced twitter sentiment classification using contextual information. arXiv preprint arXiv:1605.05195.
13. Theodotou, A., & Stassopoulou, A. (2015, November). A system for automatic classification of twitter messages into categories. In *International and Interdisciplinary Conference on Modeling and Using Context* (pp. 532–537). Cham: Springer.

14. Derczynski, L., Maynard, D., Rizzo, G., van Erp, M., Gorrell, G., Troncy, R., et al. (2015). Analysis of named entity recognition and linking for tweets. *Information Processing & Management, 51*(2), 32–49.
15. Liu, S., Cheng, X., Li, F., & Li, F. (2015). TASC: Topic-adaptive sentiment classification on dynamic tweets. *IEEE Transactions on Knowledge and Data Engineering, 27*(6), 1696–1709.
16. Ritter, A., Clark, S., & Etzioni, O. (2011, July). Named entity recognition in tweets: An experimental study. In *Proceedings of the Conference on Empirical Methods in Natural Language Processing* (pp. 1524–1534). Association for Computational Linguistics.
17. McDonald, G., Deveaud, R., McCreadie, R., Macdonald, C., & Ounis, I. (2015). Tweet enrichment for effective dimensions classification in online reputation management.
18. He, Y., Yang, C. S., Yu, L. C., Lai, K. R., & Liu, W. (2015, December). Sentiment classification of short texts based on semantic clustering. In *2015 International Conference on Orange Technologies (ICOT)* (pp. 54–57). IEEE.
19. Hatzivassiloglou, V., & Wiebe, J. M. (2000, July). Effects of adjective orientation and gradability on sentence subjectivity. In *Proceedings of the 18th Conference on Computational linguistics—Volume 1* (pp. 299–305). Association for Computational Linguistics.
20. Hu, M., & Liu, B. (2004, August). Mining and summarizing customer reviews. In *Proceedings of the Tenth ACM SIGKDD International Conference on Knowledge Discovery and Data Mining* (pp. 168–177). ACM.
21. Taboada, M., & Grieve, J. (2004, March). Analyzing appraisal automatically. In *Proceedings of AAAI Spring Symposium on Exploring Attitude and Affect in Text, Stanford University, CA* (AAAI Technical Re# port SS# 04# 07) (pp. 158–161). AAAI Press.
22. Read, J., & Carroll, J. (2009, November). Weakly supervised techniques for domain-independent sentiment classification. In *Proceedings of the 1st International CIKM Workshop on Topic-Sentiment Analysis for Mass Opinion* (pp. 45–52). ACM.
23. Pang, B., Lee, L., & Vaithyanathan, S. (2002, July). Thumbs up? Sentiment classification using machine learning techniques. In *Proceedings of the ACL-02 Conference on Empirical Methods in Natural Language Processing* (Vol. 10, pp. 79–86). Association for Computational Linguistics.
24. Pang, B., & Lee, L. (2004, July). A sentimental education: Sentiment analysis using subjectivity summarization based on minimum cuts. In *Proceedings of the 42nd Annual Meeting on Association for Computational Linguistics* (p. 271). Association for Computational Linguistics.
25. Boiy, E., Hens, P., Deschacht, K., & Moens, M. F. (2007, June). Automatic sentiment analysis in on-line text. In *ELPUB* (pp. 349–360).
26. Zhao, J., Liu, K., & Wang, G. (2008, October). Adding redundant features for CRFs-based sentence sentiment classification. In *Proceedings of the Conference on Empirical Methods in Natural Language Processing* (pp. 117–126). Association for Computational Linguistics.
27. Narayanan, R., Liu, B., & Choudhary, A. (2009, August). Sentiment analysis of conditional sentences. In *Proceedings of the 2009 Conference on Empirical Methods in Natural Language Processing: Volume 1* (pp. 180–189). Association for Computational Linguistics.
28. Hassan, A., Abbasi, A., & Zeng, D. (2013, September). Twitter sentiment analysis: A bootstrap ensemble framework. In *2013 International Conference on Social Computing (SocialCom)* (pp. 357–364). IEEE.
29. Khan, F. H., Bashir, S., & Qamar, U. (2014). TOM: Twitter opinion mining framework using hybrid classification scheme. *Decision Support Systems, 57,* 245–257.
30. Abel, F., Celik, I., Houben, G. J., & Siehndel, P. (2011). Leveraging the semantics of tweets for adaptive faceted search on twitter. *The Semantic Web–ISWC 2011,* 1–17.
31. Simeon, C., & Hilderman, R. (2015, October). Evaluating the effectiveness of hashtags as predictors of the sentiment of tweets. In *International Conference on Discovery Science* (pp. 251–265). Cham: Springer.
32. Saif, H., He, Y., Fernandez, M., & Alani, H. (2016). Contextual semantics for sentiment analysis of Twitter. *Information Processing and Management, 52*(1), 5–19.

33. Ghiassi, M., Skinner, J., & Zimbra, D. (2013). Twitter brand sentiment analysis: A hybrid system using n-gram analysis and dynamic artificial neural network. *Expert Systems with Applications, 40*(16), 6266–6282.
34. Saif, H., He, Y., & Alani, H. (2012). Semantic sentiment analysis of twitter. *The Semantic Web–ISWC 2012*, 508–524.
35. Bahrainian, S. A., & Dengel, A. (2013, December). Sentiment analysis and summarization of twitter data. In *16th International Conference on Computational Science and Engineering (CSE), 2013 IEEE* (pp. 227–234). IEEE.
36. Kontopoulos, E., Berberidis, C., Dergiades, T., & Bassiliades, N. (2013). Ontology-based sentiment analysis of twitter posts. *Expert Systems with Applications, 40*(10), 4065–4074.
37. Javed, N., & Muralidhara, B. L. (2015). Automating corpora generation with semantic cleaning and tagging of tweets for multi-dimensional social media analytics. *International Journal of Computer Applications, 127*(12), 11–16.
38. Han, J., Pei, J., & Kamber, M. (2011). *Data mining: Concepts and techniques*. Amsterdam: Elsevier.
39. Wu, Z., & Palmer, M. (1994). Verbs semantics and lexical selection. In *Proceedings of the 32nd annual meeting on Association for Computational Linguistics* (pp. 133–138). Association for Computational Linguistics, (1994, June).
40. Liu, H. (2004). MontyLingua: An end-to-end natural language processor with common sense.
41. Pennebaker, J. W., Booth, R. J., Boyd, R. L., & Francis, M. E. (2015). *Linguistic inquiry and word count: LIWC2015*. Austin, TX: Pennebaker Conglomerates.
42. Hutto, C. J., & Gilbert, E. (2014, May). Vader: A parsimonious rule-based model for sentiment analysis of social media text. In *Eighth International AAAI Conference on Weblogs and Social Media*.
43. Nielsen, F. Å. (2011). A new ANEW: Evaluation of a word list for sentiment analysis in microblogs. arXiv preprint arXiv:1103.2903.
44. Go, A., Bhayani, R., & Huang, L. (2009). Twitter sentiment classification using distant supervision. *CS224N Project Report, Stanford, 1*(2009), 12.

Workload Assessment for a Sustainable Manufacturing Paradigm Using Social Network Analysis Method

V. K. Manupati, M. Anthony Xavior, Akshay Chandra
and Muneeb Ahsan

Abstract In this paper, focus is on the sustainable manufacturing systems functionalities, i.e., process planning and scheduling for effective and efficient performance of the system to achieve the desired objectives. The desired objectives for this research work have been considered according to the above-mentioned situation. Hence, makespan, throughput time, and energy consumption were identified as the most appropriate performance measures in line with the context of the problem. Mathematical model will be formulated for the performance measures by considering the realistic constraints. Unpredictable events such as machine breakdown or scheduled maintenance are most common in any manufacturing unit. Workload assignment with these disruptions is a challenge, and therefore, a new methodology has been developed for effective and efficient solutions. In this paper, a new social network analysis-based method is being proposed to identify the key machines that should not be disturbed due to its contribution toward achieving the best system's performance. Moreover, an illustrative example along with three different configurations will be presented to demonstrate the feasibility of the proposed approach. For execution, a Flexsim-based simulation approach will be followed, and with different instances, the proposed methodology can be executed. The validation of the proposed approach and its effectiveness will be evaluated through comparison with different instances, and finally the efficiency of the proposed approach will be confirmed with the results.

Keywords Manufacturing systems · Social network analysis · Workload assessment · Simulation

V. K. Manupati · M. A. Xavior (✉) · A. Chandra · M. Ahsan
School of Mechanical Engineering, VIT University, Vellore, India
e-mail: manthonyxavior@vit.ac.in

V. K. Manupati
e-mail: manupativijay@gmail.com

© Springer Nature Singapore Pte Ltd. 2018
S. Margret Anouncia and U. K. Wiil (eds.), *Knowledge Computing and Its Applications*, https://doi.org/10.1007/978-981-10-6680-1_5

1 Introduction

Recent manufacturing company's requirements are inclined toward sustainability due to several issues such as environmental concerns, strict government policies, inflated energy costs, and diminishing of non-renewable resources. Therefore, the traditional shop floor activities need to be changed according to the new requirements in particular to fulfill the triple bottom-line aspects. Here, the configurations such as parallel and hybrid in nature are proposed wherein most of the manufacturing systems in real environments can use them. While processing jobs of the machines, different strategies for successful execution of workload on the said configurations can be implemented. Plethora of literature is available for a single unit system particularly in case of degradation modeling. Barabâsi et al. [1] had presented about the various modeling efforts developed for the systematic investigation on the evolution of social networks. But, limited research could be found on sustainable manufacturing paradigm for identifying the degradation rate of the multiunit system with social network analysis which is the current area of interest. Hao et al. [2, 3] presented a methodology for controlling the residual life distribution of parallel unit systems through workload adjustment. Simultaneous signal separation and prognostics of identical multi-component systems have been suggested by Bonacich [4]. Wasserman and Faust [5] have mentioned about the various methods and applications of social network analysis. Social network method can be used for FlexSim job shop scheduling problem by investigating the reconfigurable effect on makespan of the product as discussed in Reddy et al. [6]. Varela et al. [7] analyzed the importance of using social network analysis techniques to analyze important relations between entities in a manufacturing environment, such as jobs and resources in the context of industrial plant layout analysis. The capacity or ability of a system to work under an enhanced workload in order to maintain the same efficiency has been suggested by Manupati et al. [8]. Borgatti and Li [9] presented an overview of social network analysis including the specific concepts and the generic explanatory mechanisms which are used to relate the network variables to the preferred outcome. Reddy et al. [10] made an attempt to utilize social network analysis to evaluate the identified key machines in flexible and dynamic job shop scheduling to analyze its impact on the manufacturing system performance. It was reported that the social network analysis method helped to identify the key machines dynamically rather than randomly which enhanced the performance measure. Manupati et al. [11] used social network analysis method to identify the key machines of the dynamic job shop scheduling problem from which the effectiveness and stability of scheduling, i.e., makespan and starting time deviations of the computational complex NP-hard problem have been solved with hybrid algorithm.

In this paper, out of many different types of sustainable manufacturing process available in the literature, the key machine was obtained using the social network analysis method and the FlexSim simulation software. Due to environmental concerns and manufacturing uncertainty, it was further assumed that a machine can handle only one job at a time. The objective in this research work is to determine the key machine that is the maximum amount of workload given to a particular machine in a unit time. To achieve the above objective, jobs were made to follow an effective process plan with two benchmark instances, namely BM-1, BM-2 along with the PM were BM-1 and BM-2 followed equal and random distribution of workload, respectively, and PM was our proposed methodology. Whereas in case of FlexSim simulation, these benchmark instances along with the proposed methodology were followed for both parallel and hybrid configuration. FlexSim simulation software was able to identify the machine which used to work for maximum amount of time along with its processing efficiency. The validation of the approach's effectiveness has been shown by the simulation results on two different configurations along with the proposed methodology having different benchmark instances. The later sections review the detailed description of the problem description, proposed framework, and experimentation.

2 Problem Description

A sustainable manufacturing environment with sixty multiple jobs that are assigned equally to six machines was considered. Each machine corresponds to different set of operations and a set of alternative process plans. Upon the arrival of these sixty jobs, they are distributed among the machines by following two benchmark instances that are BM-1 and BM-2. The proposed methodology is also implemented by following a process plan which was framed so as to increase the productivity of the machines. As a part of the objective in this paper, makespan has been considered as one such performance measure to conduct social network analysis (SNA). In addition to SNA method, a FlexSim simulation-based approach with parallel and hybrid configuration is considered for further justification. In a network manufactured environment, the machines with different capabilities are assigned with different process plans so as to perform various operations. The problem mentioned above makes several assumptions that are worth highlighting.

Assumptions: (1) The processing of the product on a machine is continued without interruption until; (2) Once the operation of a job is completed, it is immediately transferred to the succeeding machine for. (3) A machine can handle only one job at a time.

3 Framework of the Proposed Dynamic Workload Adjustment

The workload distribution in case of six jobs and six machines (6 × 6) configuration is applied to predict the key machine. The workload distribution is based on two benchmark instances: BM-1 (equal workload distribution), BM-2 (random workload distribution), and the one according to our PM (proposed methodology) is considered. Experimentation is carried out first using SNA-based approach, and then the result has been verified using FlexSim simulation software.

In SNA-based approach, results were established on the above-mentioned workload plans on a single machine arrangement based on the process plan given in Table 1. A matrix was formed based upon different process plans for BM-1, BM-2, and PM, respectively, in which, if a machine processes a particular job and a particular operation it is denoted as 1 and 0 if it does not process it. In equal distribution, i.e., for BM-1, each machine had a job and executed every operation, whereas for BM-2 and PM this was not the case. In distribution according to proposed methodology, process plan and machines corresponding to each operation were identified based on the principle that the machines which had the smallest processing time would get greater workload and hence degrade faster, but the overall makespan of the PM will be less than BM-1 and BM-2. Each process plan has a certain set of operations from which the machines are selected. In random distribution, i.e., BM-2, process plan and machines were selected at random from the given set of plans.

Table 1 Input data for the 6 × 6 problem

Job	PP	O1	O2	O3	O4	O5	O6
J1	PP1,1	{1,2} [6,5]	{3,4,5} [7,6,6]	{6} [8]			
	PP1,2	{1,3} [4,5]	{2,4} [4,5]	{3,5} [5,6]	{4,5,6} [5,5,4]		
J2	PP2,1	{2} [4]	{1,3} [2,3]	{2,4,6} [4,3,5]	{3,5} [2,4]	{2,4} [3,4]	{4,6} [3,5]
J3	PP3,1	{2,3} [5,6]	{1,4} [6,5]	{2,5} [5,6]	{3,6} [6,5]		
	PP3,2	{1} [9]	{3,4} [8,8]	{5} [9]			
	PP3,3	{2,3} [7,6]	{4} [7]	{3,5} [4,6]	{4,6} [5,5]	{2,4} [6,4]	
J4	PP4,1	{1,2} [7,8]	{3,4} [7,6]	{6} [9]			
	PP4,2	{1,3} [4,3]	{2} [4]	{3,4} [4,5]	{5,6} [3,5]		
J5	PP5,1	{1} [3]	{2,4} [4,5]	{3} [4]	{5} [3]	{4,6} [5,4]	
	PP5,2	{2,4} [5,6]	{5} [7]	{3,6} [9,8]			
J6	PP6,1	{1,2} [3,4]	{3,4} [4,3]	{2,5} [5,3]	{3} [4]	{4,5} [4,6]	{3,6} [5,4]
	PP6,2	{1,3} [4,4]	{2,3} [5,6]	{2,3} [6,7]	{6} [7]		
	PP6,3	{1,2,3} [3,5,8]	{4,5} [7,10]	{3,6} [9,9]			

In FlexSim-based approach, results were established based on the same work-load distributions, each on two different machine arrangements: parallel and hybrid. A total of 60 jobs were taken, with each different job having an input of 10. For both parallel and hybrid configurations, selection of machines corresponding to each operation for each type of workload distribution was based on the same table mentioned above using the same ideas as used in SNA-based approach.

4 Experimentation

In this section, the performance of the proposed dynamic workload adjustment methodology with two other benchmark instances is detailed. The computation was conducted on a *PC* with Intel Corei3-4005U with 107 GHz and 3 MB L3 cache running under Windows 10 professional operating system with 4 GB RAM. A case study was implemented in SNA approach that dealt with different manufacturing scenarios. Sixty jobs along with six machines and six operations were considered. These sixty jobs were made to follow two benchmark instances, namely BM-1, BM-2 along with the PM were BM-1 and BM-2 followed equal and random dis-tribution of workload, respectively, and PM was our proposed methodology. Hence, three sets of six by six data were framed using the above input parameters and were implemented using SNA for the different benchmark instances. On the analysis of six by six data, different features of individual nodes helped us to identify the key machine in each benchmark instance. It was identified that, in case of BM-1, machine 4 obtained higher degree, betweenness and closeness centrality. For BM-2, machine 3 obtained higher degree, betweenness and closeness centrality and in case of PM, it was machine 3 which obtained higher degree, betweenness and closeness centrality. The identified machines with higher centrality acted as key machines for the entire network, and their influence on the entire network was much higher.

In the FlexSim simulation experiment, the above three six by six data sets were used for parallel and hybrid configuration in order to verify the key machines that we have earlier obtained from SNA method. In FlexSim, the distribution of workload was performed using two benchmark instances that were BM-1, BM-2, and our proposed methodology PM. In case of BM-1, the workload was distributed equally among the six machines, while for BM-2, the distribution was random. For proposed methodology (PM), we had distributed the workload using a process plan so as to identify the key machine (the machine which performs the maximum work). These benchmark instances along with the proposed methodology were followed for both parallel and hybrid configuration. FlexSim simulation software was able to identify the machine which used to work for maximum amount of time along with its processing efficiency. Figure 1 shows the results obtained for equal distribution of workload using SNA-based approach. The three centrality measures for each were obtained. It shows that machine 4 has a degree centrality of 64.000, betweenness centrality of 91.026, and closeness centrality of 209.954 which is the

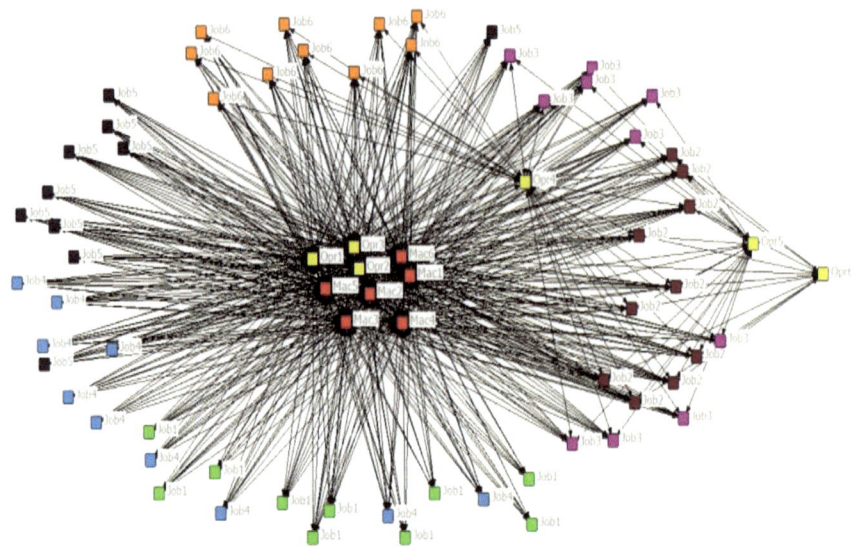

Machines	Degree Centrality	Betweenness Centrality	Closeness Centrality
Machine 1	61.000	87.654	189.765
Machine 2	62.000	88.750	209.811
Machine 3	63.000	89.873	189.813
Machine 4	**64.000**	**91.026**	**209.954**
Machine 5	63.000	89.873	204.845
Machine 6	63.000	89.873	254.978

Fig. 1 SNA-based result for equal distribution of workload showing graphical representation and table showing centrality measures

highest out of all machines. From the above results, it is clear that machine 4 is the key machine.

Figure 2 shows the arrangement of machines in FlexSim as parallel configuration for equal workload distribution and the corresponding processing times of machines. It shows that machine 2 has highest processing time of 330 h which is also the makespan of the configuration. It is also clear from the graph that machine 2 is the key machine as it works for most number of times having highest workload. But it is not in accordance to the results obtained from SNA as machine 4 is the key machine there. Figure 3a shows that machine 6 has highest processing time of 319 h which also is the makespan of the configuration. It is lesser than the makespan for equal workload distribution which shows random distribution is better. It is also clear from the graph that machine 6 is the key machine as it works for most number of times having highest workload. But it is not in accordance to the results obtained from SNA as machine 3 is the key machine there. Figure 3b shows that machine 4 has highest processing time of 150 h which is also the makespan of the configuration. This is least among all the configurations and the difference between their makespans is too high, which as expected shows that our

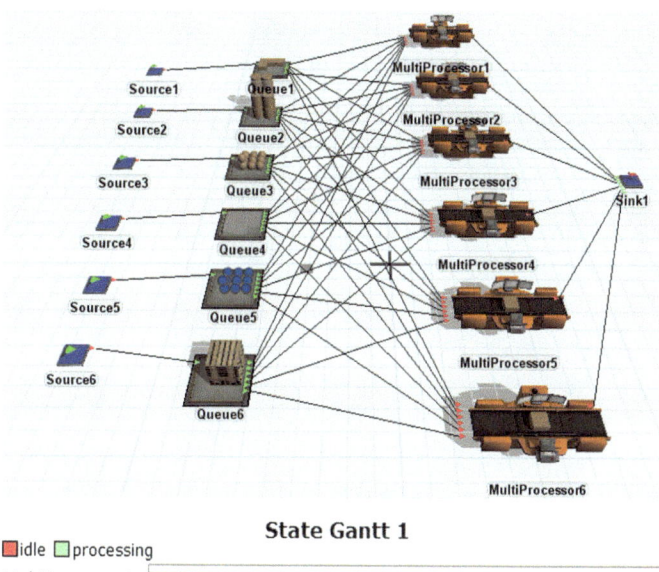

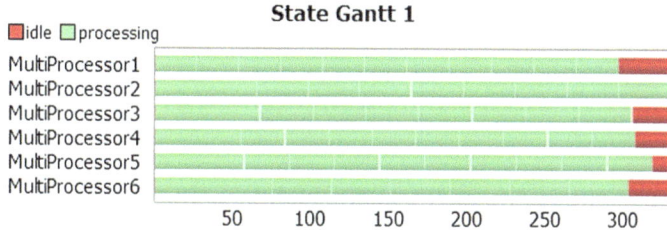

Fig. 2 Parallel configuration for equal workload distribution using FlexSim simulation with screenshot of the arrangement and Gantt chart showing total processing time of each machine

PM is the best plan among all the three. It is also clear from the graph that machine 4 is the key machine as it works for most number of times having highest workload. But it is not in accordance to the results obtained from SNA as machine 2 is the key machine there. Figure 3c shows that machine 4 has highest processing time of 330 h which is also the makespan of the configuration. It is also clear from the graph that machine 4 is the key machine as it works for most number of times having highest workload. Also it is in accordance to the results obtained from SNA. Although it has same makespan than that obtained from parallel configuration, it is better than the latter as it verifies the result from SNA.

Figure 4a the corresponding processing times of machines. It shows that machine 3 has highest processing time of 360 h which is also the makespan of the configuration. It is also clear from the graph that machine 3 is the key machine as it works for most number of times having highest workload. It is greater than the makespan for equal workload distribution which shows equal distribution is better among the two. Also, it is in accordance to the results obtained from SNA. Although it has higher makespan than that obtained from parallel configuration, it is better than the latter as it verifies the result from SNA. Figure 4b the corresponding

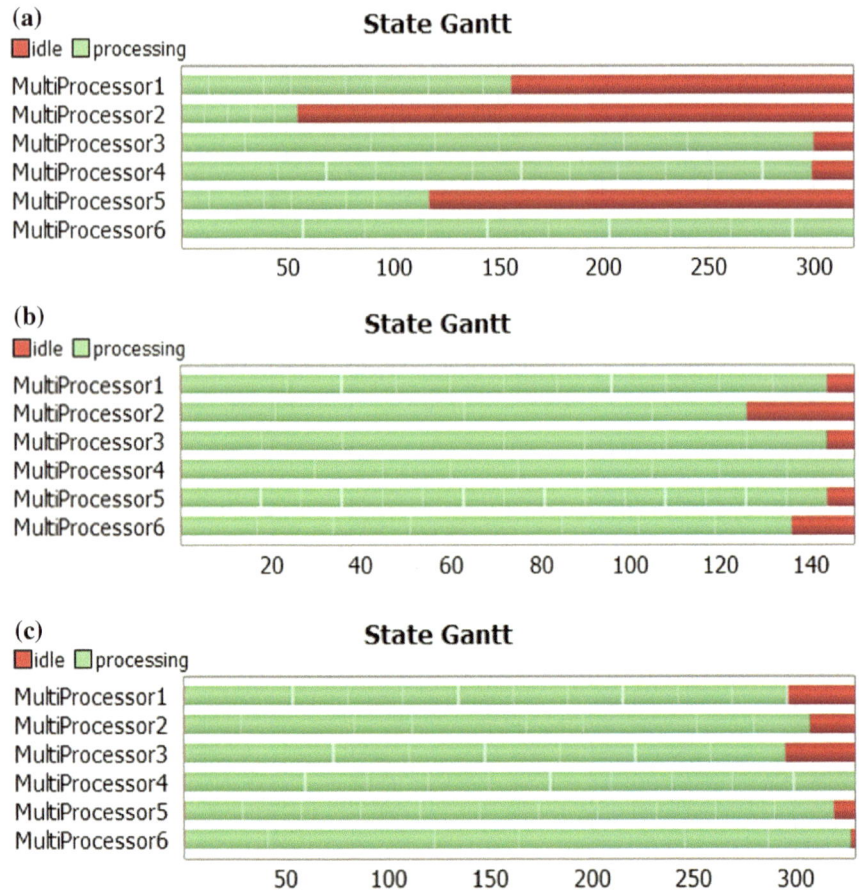

Fig. 3 a Parallel configuration Gantt chart showing total processing time of each machine. **b** Hybrid configuration Gantt chart showing total processing time of each machine. **c** Hybrid configuration for equal workload distribution and its total processing time of each machine

processing times of machines. It shows that machine 2 has highest processing time of 189 h which is also the makespan of the configuration. It is also clear from the graph that machine 2 is the key machine as it works for most number of times having highest workload. This is least among all the configurations, and the difference between their makespans is too high, which as expected shows that our PM is the best plan among all the three. Also, it is in accordance to the results obtained from SNA. Although it has higher makespan than that obtained from parallel configuration, it is better than the latter as it verifies the result from SNA.

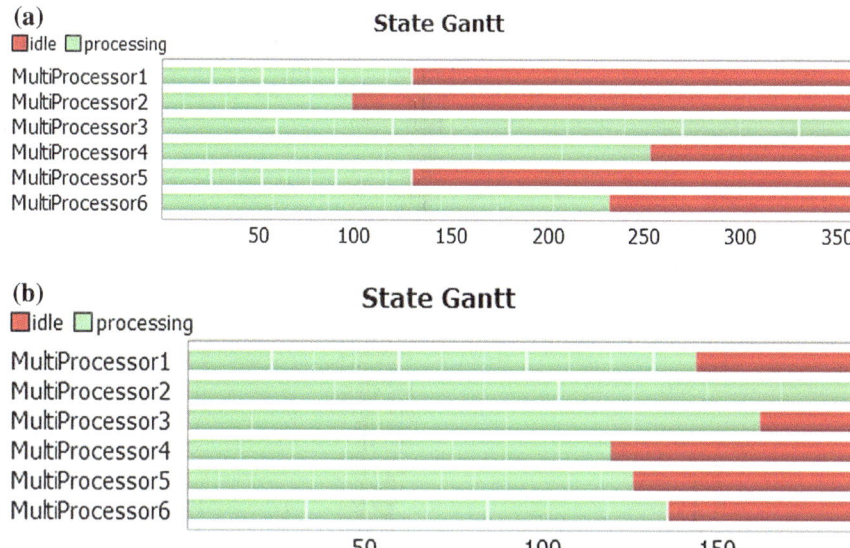

Fig. 4 **a** Hybrid configuration for random workload distribution and Gantt chart showing total processing time of each machine. **b** Hybrid configuration for workload distribution according to proposed methodology showing total processing time of each machine

5   Conclusion and Future Work

The focus of this paper is on sustainable manufacturing for workload assessment of multiple jobs. An effective process plan was implemented on social network analysis, and FlexSim simulation software was used for further analysis. To demonstrate the proposed method, a numerical study was conducted and a contrast between two different benchmark strategies was performed, the former with evenly distributed and the later with randomly distributed workload. In addition to the above two benchmark strategies, a new methodology was proposed which strictly followed the process plan. In case of FlexSim simulation, these benchmark instances followed two different configurations that are parallel and hybrid, respectively. The outcomes clarify that the proposed method is certainly better than these two benchmarks under different situations.

References

1. Barabâsi, A. L., Jeong, H., Néda, Z., Ravasz, E., Schubert, A., & Vicsek, T. (2002). Evolution of the social network of scientific collaborations. *Physica A: Statistical Mechanics and its Applications, 311*(3), 590–614.

2. Hao, L., Gebraeel, N., & Shi, J. (2015). Simultaneous signal separation and prognostics of multi-component systems: the case of identical components. *IIE Transactions, 47*(5), 487–504.
3. Hao, L., Liu, K., Gebraeel, N., & Shi, J. (2015). Controlling the residual life distribution of parallel unit systems through workload adjustment. *IEEE Transactions on Automation Science and Engineering*.
4. Bonacich, P. (2007). Some unique properties of eigenvector centrality. *Social Networks, 29* (4), 555–564.
5. Wasserman, S., & Faust, K. (1994). *Social network analysis: Methods and applications* (Vol. 8). Cambridge: Cambridge university press.
6. Reddy, M. S., Ratnam, C., Agrawal, R., Varela, M. L. R., Sharma, I., & Manupati, V. K. (2017). Investigation of reconfiguration effect on makespan with social network method for flexible job shop scheduling problem. *Computers & Industrial Engineering*.
7. Varela, M. L. R., Manupati, V. K., Manoj, K., Putnik, G. D., Araújo, A., & Madureira, A. M. (2016, December). Industrial plant layout analyzing based on SNA. In *Proceedings of International Conference on Intelligent Systems Design and Applications* (pp. 728–737). Berlin: Springer.
8. Manupati, V. K., Putnik, G., & Tiwari, M. K. (2015). Resource scalability in networked manufacturing system: Social network analysis social network analysis based approach. In *Handbook of Manufacturing Engineering and Technology* (pp. 3439–3450). London: Springer.
9. Borgatti, S. P., & Li, X. (2009). On social network analysis in a supply chain context. *Journal of Supply Chain Management, 45*(2), 5–22.
10. Reddy, M. B. S. S., Ratnam, C. H., Sharma, I. V., & Manupati, V. K. (2016). Social network analysis based evolutionary algorithmic approach to identify the influence of hubs on flexible scheduling problems. In *Proceedings of the 2016 International Conference on Industrial Engineering and Operations Management* (pp 794–800). Kuala Lumpur, Malaysia: IEOM Society International.
11. Manupati, V. K., Arudhra, N., Vigneshwar, P., Rajashekar, D., & Yaswanth, M. (2015). A multi-objective based evolutionary algorithm and social network analysis approach for dynamic job shop scheduling problem. *International Journal on Cybernetics and Informatics, 4*(2), 75–82.

Part II
Knowledge Computing for Large Organizations

Virtual Data Warehouse Model Employing Crypto–Math Modus Operandi and Intelligent Sensor Algorithm for Cosseted Transference and Output Augmentation

Rajdeep Chowdhury, Olive Roy, Soupayan Datta
and Saswata Dasgupta

Abstract A data warehouse is a storehouse which comprises all specifics and statistics of the institution. With the contemporary leaning of hacking beside the initiation of fresh hacking proficiency, the safekeeping for data warehouse has transpired to be an influential portion. The projected work exemplifies how an encryption technique is germane in the internal structure of a data warehouse. Apiece instant, any user endeavor for entrance to the information, the data in encrypted form is furnished to the user. Only an authenticated user would be able to obtain the original form of the data, because a well-defined decryption algorithm is installed at the authentic user's end only. The formulation of the spotless Intelligent Sensor Algorithm and the proposed cryptographic modus operandi ascertain that there would be considerable attenuation of admittance time, with observance for the cosseted transference, enhanced infringement avoidance and output augmentation. The applicable employment of the proposed virtual data warehouse model with improved safekeeping might be in its noteworthy utility in an assortment of institution where accrual of cosseted information is of extreme extent. The assortment of associations is inclusive of edifying institutions, business communities, medicinal associations, classified enterprises and there on.

Keywords Virtual Data Warehouse · Crypto–math · Intelligent Sensor
Encryptor · Decryptor · Pseudo Cryptographic Algorithm

R. Chowdhury (✉)
Department of Computer Application, JIS College of Engineering,
Block–A, Phase–III, Kalyani, Nadia 741235, West Bengal, India
e-mail: dujon18@yahoo.co.in

O. Roy · S. Datta · S. Dasgupta
Department of Computer Science and Engineering,
JIS College of Engineering, Block–A, Phase–III, Kalyani,
Nadia 741235, West Bengal, India

© Springer Nature Singapore Pte Ltd. 2018
S. Margret Anouncia and U. K. Wiil (eds.), *Knowledge Computing and Its Applications*, https://doi.org/10.1007/978-981-10-6680-1_6

1 Introduction

Data warehouse is an ingrained stockpile of an institution's electronically collective information [1–7]. Data warehouse is deliberated with the intention to aid all-inclusive coverage and adept scrutiny. Bestowing protection for the data warehouse is nearly a colossal susceptibility for apiece institution.

The projected effort is predominantly unprompted, as it bestows the devising of the ingenious construction confirming the intention of augmenting safety procedures and data warehouse output augmentation in the procedure [1–7].

The principal focal point is on how the projected effort could be functional on the anticipated representation. The projected effort portrays that apiece instant any user endeavor to ensure admittance to the information, it hurls a query to the Warehouse. Subsequent to a triumphant implementation of the query by the Warehouse Query Processor, the information in the encrypted structure would be retreated to the pertinent View Table of the Dimension Table [1–5].

In the formulated paper, it has been projected that apiece Dimension Table has an associated Intelligent Sensor Algorithm, which with intellect sense and decide whether the records appealed by the user is previously encrypted or not.

If the records are encrypted, then it transmits the records unwaveringly to the pertinent View Table, or else it transmits an appeal to the Encryptor for encrypting the records.

Here Encryptor is a hardware appliance which is fabricated into the Data Warehouse structural design. For the intact Data Warehouse, there is merely a solitary Encryptor and hence it might be demanding for the majority of the instance.

It is responsibility of the Intelligent Sensor Algorithm to ensure if the Encryptor is demanding or not earlier than conveying any appeal. Consequently, the Intelligent Sensor Algorithm utilizes an additional Encryption Request Algorithm.

2 Literature Survey

The segments focal point is on the interconnected effort obtainable in the identical type, hopeful in the compilation of the proposed paper [1–16].

There are fairly a few interconnected efforts on data warehouse and its safety procedures, which have previously been accepted, although the specific devise, realization and integration of the pioneering algorithm is impressive and exceptionally pleasing for modern researchers. Apiece instant the insight rouses in the wits concerning devise, realization and integration of safety means for data warehouse as well as structural point of reference, the widespread study of few papers need careful reveal and the crisp tips are avowed in minutiae for ease in mention [1–5].

In the paper titled "Design and Implementation of Proposed Drawer Model Based Data Warehouse Architecture Incorporating DNA Translation Cryptographic Algorithm for Security Enhancement," the DNA Translation cryptographic algorithm is integrated in the projected structural design in couple of distinctive tier of protection methodology, initially through the changeover from Operational Data Store to Data Vault and then through the changeover from Data Vault to Data Mart, in that way ascertain two-tier safety augmentation for the anticipated replica [3].

In the paper titled "Design and Implementation of Security Mechanism for Data Warehouse Performance Enhancement Using Two Tier User Authentication Techniques" the devise and execution of safety leads to data warehouse output augmentation with assimilation of the exclusively crafted twin tier user endorsement methodology.

At the primary layer, the user has to opt for his/her user classification. The classification variety bestows with a conceptual apparition of the information.

At the subsequent layer, a habitual crafted code would be transmitted to the meticulous user's mobile, who at present would attempt to ensure admittance to the warehouse and whilst that specific code is equipped by the user from his mobile, then merely the user would be competent to sight his/her desired information [4].

In the paper titled "A Data Warehouse Architectural Design Using Proposed Pseudo Mesh Schema" dimension tables are situated to be interconnected with each other, ascertaining enhancement in the time intricacy of a data warehouse. Fact tables and dimension tables have been engaged here, in their normalized type to reduce duplication. The numeral associations could be specifically evaluated employing n (n−1)/2, where n exemplifies the quantity of tables signified in the formation. The formation is actually a lithe one, as any enhance or refuse of solitary or additional database in the constitution would not impinge on the intact representation constitution of the data warehouse in concern [5].

In the paper titled "A Survey on Data Security in Data Warehousing: Issues, Challenges and Opportunities" the unearthing of all the precedent tasks that has been manoeuvred in the data warehouse and the safety concern for the requisite and execution. The effort deals with presently obtainable information safety technique ensuring focal point on precise concern and requisites, pertaining to their utilization in data warehousing ambiance [8].

In the paper titled "Towards Data Security in Affordable Data Warehouse," the data warehouse methodology is based on clustering and the star schema is fragmented over the nodes of the cluster. Dimension table is duplicated in all the nodes of the cluster and fact table is fragmented utilizing round robin or hash partitioning. Encryption technique is nix utilized in fact table owing to performance concern [9].

In the paper titled "An Integrated Conceptual Model for Temporal Data Warehouse Security," it has been anticipated that the preliminary incorporated abstract model for addressing temporal data warehouse safety requirements essential pattern. The model is the preliminary incorporated model that amalgamates ETL model with temporal data warehouse model [12].

In the paper titled "Towards Design, Analysis and Performance Enhancement of Data Warehouse by Implementation and Simulation of P2P Technology on Proposed Pseudo Mesh Architecture" Peer-to-Peer or P2P technology facilitates unwavering allocation of computer assets among couple or more computers in a computer arrangement. In this technology, every computer functions as mini server, where every computer could distribute its assets with additional computers competently.

Alike concept has been employed in the data warehouse structural devise employing projected pseudo mesh schema, where each and every dimension table in the structural design act as super peer and in effect typify a peer community, with assortment of peers. Every peer in identical peer community or from dissimilar peer community could have unwavering rapport in the midst of them utilizing view fact table [17].

Fact table is fragmented into copious clusters and fact data cannot be simply allied to the dimension data as they are encrypted [1–5, 8, 9].

In the paper titled "Study and Comparison of Indexing Models in Data Warehouse" the intent is to establish the effective indexing technique for data warehouse appliance based on relative edification of varied indexing modus operandi which are engaged to augment the processing instant of the queries.

The alike is ascertained on the origin of ostentatious essential and appliance of indexing, sustaining the variety of query, operational modus operandi of indexing, function on data, data treatment and functional outcome, cost benefit scrutiny of the indexing modus operandi and pros and cons [18].

3 Proposed Work

For ease in considerate, the flow chart of the projected effort has been deliberated beneath (Fig. 1):-

Previous to stating any kind of academic illumination, the projected effort requisites to be established diagrammatically, succeeded by extravagance in bitty conversation.

In Fig. 2, each and every Dimension Table comprises of an Intelligent Sensor Programming which is apparent by the short form IS in the box beside each Dimension Table.

The crucial underlying principle for Intelligent Sensor Programming is to establish whether the information that is currently being appealed by a user query is encrypted or nix. If the information is previously encrypted by the Encryptor, then it desires to be transmitted to its pertinent View Fact, or else the information requires sending to the Encryptor for encryption. For supervising over appositely, the Intelligent Sensor utilizes an algorithm called Decision Making Algorithm.

Encryptor and Decryptor are hardware devices which would be ingrained in ally with the Data Warehouse Structure. Furthermore, Encryptor and Decryptor comprise of the anticipated algorithm for encrypting and decrypting information. For entire Data Warehouse, each Encryptor and each Decryptor is utilized. Consequently, when the Intelligent Sensor of the Dimension Table wishes to fling any information, it might so emerge that Encryptor or Decryptor is full of activity. To triumph over such an obscure situation, Intelligent Sensor utilizes an additional algorithm termed Request Algorithm.

Fig. 1 Flow chart of the projected effort

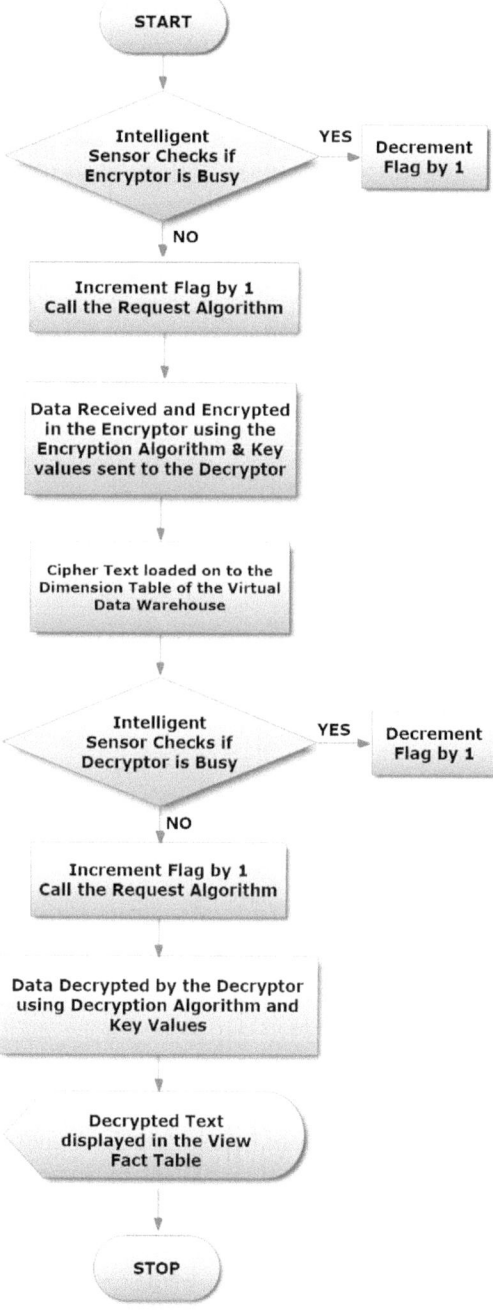

Request Algorithm

The concept of virtual database has been employed which comprises of all the encrypted values of the data present in the raw database. The raw database should be accessible only to the server administrator and should be made available offline. Upon transient triumphant encryption modus operandi, the data should be subsequently passed or laden on to the virtual database, which is an online database available to the clients for viewing.

However, the clients should not be able to craft any amendment, as they could not envisage the data which is present in its encrypted form and could merely be decrypted by the Decryptor present in the virtual database.

Encryptor would comprise of definite keys which would be selective for each and every user and for each user those keys would be summative with the Encryptor. Each Decryptor would comprise of definite keys too.

Apparently there would be 'n' users and the Encryptor would hence contain 'n' keys, signified by, K_1, K_2, K_3...K_n (K_ith key for ith user). Since a Decryptor is endowed with apiece user, it would only contain solitary key sufficient enough for decryption.

Initially Encryptor/Decryptor Semaphore is Zero.
After user verification, the query is considered.
Every time any record pierce the Encryptor,
Encryptor_Semaphore/Decryptor_Semaphore is assigned 1.

Step 1 Confirm admittance to Encryptor_Semaphore/Decryptor_Semaphore erratic status.
Step 2 If Encryptor_Semaphore/Decryptor_Semaphore is Zero, then eventually pierce the Encryptor.

Now Encryptor/Decryptor does the subsequent:

i) Set Encryptor_Semaphore/Decryptor_Semaphore to One.
ii) Encryption Function Call is requisite.

Else
Enter into Encryptor/Decryptor waiting queue and then lay Priority for apiece admittance in the waiting queue and linger till the advent of spin.

When a user is certified and legitimate to employ the data warehouse, an administrator would accrue an inimitable set of keys in the warehouse's Encryptor and would furnish the user with the Decryptor which would also contain the identical key values.

Now, the two algorithms used by the Intelligent Sensor and the Encryptor/Decryptor are exemplified below:

Intelligent Sensor Decision-Making Algorithm

Primarily flag is placed to 0.

Step 1 If flag is 0 then arrange to fling the appealed information from Dimension Table to Encryptor and ensure the subsequent:

i) Augment flag by 1.
ii) Encrypt Request Algorithm is initiated.

Else

Transmit the Dimension Table information to its apt View Table and Decrease flag by 1, so as to transpire again to 0.

The subsequent steps confer an enhanced observation about the distinguished transactions occupied in the proposed data warehouse:

i) Raw data from the dimension tables is laden on to the Encryptor upon its availability in accordance with the Intelligent Sensor Algorithm and the Request Algorithm.
ii) The value of keys 'm' and 'r' is computed using mentioned formulas and is passed into the Decryptor for apposite processing later.
iii) The ciphertext(s) obtained after encryption are reloaded into the dimensional tables of the virtual database.
iv) The cipher text(s) are then loaded on to the Decryptor, subsequent to its availability as per the Intelligent Sensor Decision Algorithm and the Request Algorithm.
v) Finally with the aid of the computed values already sent by the Encryptor, the ciphertext(s) are decrypted and laden on to the view fact table for viewing.

Encryption Algorithm used by the Encryptor

Step 1 Consider the data to be encrypted be 'Octorived71!=4prdm-p4ch4i9l8over7thandi [nto; he74e'.

Step 2 The length of the data denoted by d (L) is 48.

A set of characters is appended to the end of the data, to craft d (L) as a perfect square.

Since, d (L) is 48 here and its nearest perfect square is 49, 1character needs to be appended to it.

Let, that character be '0'. Hence d (L) = 49.

Choosing the would-be-appended character might appear to be tough, but could be resolved effortlessly.

If the data is Name, append only the requisite number of '0's. If the data is account number or balance or DOB, append 'X's to it, as no alphabet would emerge in these sorts of data. If the data is password, append some character that has been posed to be excluded while crafting the password (In general basis). The augmented data becomes Octorived71! =4prdmp4ch4i9l8over7thandi [nto; he74e0.

Step 3 The entire data is converted to its ASCII characters and placed row-wise into a square matrix.

In the example, the data becomes [79, 99, 116, 111, 114, 105, 118, 101, 100, 55, 49, 33, 61, 52, 112, 114, 100, 109, 112, 52, 99, 104, 52, 105, 57, 108, 56, 111, 118,

101, 114, 55, 116, 104, 97, 110, 100, 105, 91, 110, 116, 111, 59, 104, 101, 55, 52, 101, 48] after ASCII conversion.

A 7×7 matrix named S is created.

Step 4 A key is amassed for apiece user, and that key is enabled whenever a user is verified.

A fraction of the key is a set of numbers $K = \{k_i\}$, such that k_i is a positive integer and $k_i \leq 231$, $\forall$ i = 1(1) d (L).

It is proposed that K would usually be of length 64, since any data that the user desires to access would not usually be that long. Even if it is so, the data would be fragmented and then encrypted.

In this case, let the key be $K = \{3, 186, 14, 9, 20, 52, 23, 108, 145, 96, 109, 129,$ 40, 197, 35, 155, 166, 185, 42, 113, 52, 67, 157, 90, 171, 53, 140, 12, 159, 82, 70, 137, 67, 62, 17, 97, 196, 88, 52, 192, 139, 75, 79, 155, 107, 170, 47, 172, 41, ...\}

Let the ith element be denoted as k_i.

Key is randomized; if feasible, a relation could be developed for engendering the key.

The size of apiece digit in the key has not been mentioned. It is prudent to employ either 3-digit or 2-digit key set or else it might be intricate to split the key values from the key set, when sent to the Decryptor employing the pseudocryptographic algorithm, explained later.

Only 49 keys are requisite as the data is of length 49 also.

Hence, only the first 49 elements of the key set are considered and placed in another 7×7 matrix denoted by $K^{\backprime}$.

This key set would also be amassed in the user's Decryptor.

Note: If odd length is padded using the character '0' having a fixed ASCII value, its recurrence thereafter, if required numerous times, might make it predictable.

Hence, set of keys are to be sent to the Decryptor using a channel, as soon as they are engendered.

Step 5 Three very large prime numbers p, q and r are selected. Compute m = p*q*r.

Very large prime numbers make factorization of their product exceedingly intricate and hence finding out the Vs (mentioned later) would be even more intricate.

In this case, since a pseudocalculation is performed, the three prime numbers 11, 13 and 17 are considered.

The computed m is 2431.

The 'm' is exclusive and unique for apiece user and would be amassed in both the Encryptor and the Decryptor.

'm' value is to be set to the Decryptor using the channel (which could be declared to be simplex in nature).

Step 6 Now, a matrix V is originated, such that

$V = \{v_i: i = \{1(1) \ d \ (L) \ and \ v_i \equiv s_i^2 \ (mod \ m)\}$,
that is, $v_i - s_i^2 = m*r$.
For any particular user, 'r' would be fixed.
In this case, let $r = 3$.
'r' would also be amassed in both the Encryptor and the Decryptor.
In this case, V = [13534, 17094, 20749, 19614, 20289, 18318, 21217, 17494, 17293, 10318, 9694, 8382, 11014, 9997, 19837, 20289, 17293, 19174, 19837, 9997, 17094, 18109, 9997, 18318, 10542, 18957, 10429, 19614, 21217, 17494, 20289, 10318, 20749, 18109, 16702, 19393, 17293, 18318, 15574, 19393, 20749, 19614, 10774, 18109, 17494, 10318, 9997, 17494, 9597].

The value of 'r' is whether single digit or double digit is not mentioned, but suggested to be moderate, in accordance with size of m.

Note: The formula $v_i \equiv s_i^2 (mod \ m)$ is employed because even if an unauthorized individual intercepts the encrypted message, determining the modular square root would be very intricate when the factorization of m is unidentified.

Furthermore, 'm' has been set to multiplication of three large prime numbers; hence, computing the factors of m would be exceedingly intricate.

Step 7 A polynomial function $y = f(x)$ is adhered, such that $f^{-1}(x)$ is identified.

A polynomial function is ensured because it is invertible.
Let the function be $y = x - 8000$.

Step 8 Compute y_i, such that $y_i = f(v_i)$ and form the set

$$Y = \{y_i : i = 1(1) \ d \ (L)\}.$$

In this case, Y = [5534, 9094, 12749, 11614, 12289, 10318, 13217, 9494, 9293, 2318, 1694, 382, 3014, 1997, 11837, 12289, 9293, 11174, 11837, 1997, 9094, 10109, 1997, 10318, 2542, 10957, 2429, 11614, 13217, 9494, 12289, 2318, 12749, 10109, 8702, 11393, 9293, 10318, 7574, 11393, 12749, 11614, 2774, 10109, 9494, 2318, 1997, 9494, 1597].

Step 9 The following operation is performed:

Compute $r_i = y_i \% k_i$, and form the set $R = \{r_i\}$.
Find the quotient fraction denoted by q_i, when y_i is divided by k_i.
The set $Q = \{q_i\}$ would be sent to the user along with the encrypted file, as the set Q would be of utmost significance in the decryption phase.
In this case, R = [2, 166, 9, 4, 9, 22, 15, 98, 13, 14, 59, 124, 14, 27, 7, 44, 163, 74, 35, 76, 46, 59, 113, 58, 148, 39, 49, 10, 20, 64, 39, 126, 19, 3, 15, 44, 81, 22,

34, 65, 100, 64, 9, 34, 78, 108, 23, 34, 39] and Q = {1844, 48, 910, 1290, 614, 198, 574, 87, 64, 24, 15, 2, 75, 10, 338, 79, 55, 60, 281, 17, 174, 150, 12, 114, 14, 206, 17, 967, 83, 115, 175, 16, 190, 163, 511, 117, 47, 117, 145, 59, 91, 154, 35, 65, 88, 13, 42, 55}.

Step 10   Compute R` = R+32, that is, apiece element of the set R is added to 32. This is performed because characters and symbols start from 33 in ASCII.

In this case, R` = {34, 198, 41, 36, 41, 54, 47, 130, 45, 46, 91, 156, 46, 59, 39, 76, 195, 106, 67, 108, 78, 91, 145, 90, 180, 71, 81, 42, 52, 96, 71, 158, 51, 35, 47, 76, 113, 54, 66, 97, 132, 96, 41, 66, 110, 140, 55, 66, 71}.
The step might be eliminated to improve the effectiveness of the algorithm, if not very significant.

Step 11   Locate the ASCII characters corresponding to apiece element of R` and formulate a string.

For instance, in this case, the string is   " ⊨) $) 6 / é - . [£ . ; ≠ L ⊢ j C l N [æ Z ⊣ G Q * 4 ` G Pts 3 # / L q 6 B a ä `) B n î 7 B G

Step 12   Fragment the intact message into blocks of 54 characters. If the number of characters is not a multiple of 54, then pad requisite quantity of null characters. Consider the six faces of a Rubik cube as shown in the subsequent figure. Insert 54 characters row-wise into apiece face and the rotation of face would be F, R, B, L, U, D.

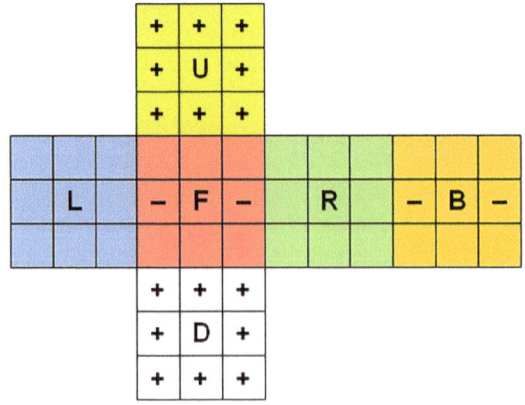

In the apparent case, the faces in order of Front, Right, Back, Left, Up and Down would appear like: " ⊧) $) 6 / é - . [£ . ; ⊯ L ⊦ j C 1 N [æ Z ⊣ G Q * 4 ` G Pts 3 # / L q 6 B a ä `) B n î 7 B G

"	⊧	)
$	)	6
/	é	-

.	[	£
.	;	⊯
L	⊦	j

C	1	N
[	æ	Z
⊣	G	Q

•	4	`
G	Pts	3
#	/	L

q	6	B
a	ä	`
)	B	n

î	7	B
G		

Step 13　Rubik cube rotation would be employed to alter the position of the set of characters to obtain the ciphertext. In real-time scenario, the number of rotations to be applied and the type of rotations to be applied should be decided prior and the key set could be employed to determine them. For the exhibited example, a left-to-right rotation would be employed once on row 2 and a down-to-up rotation would be employed to column 2.

"	$	)
3	é	)
/	7	-

.	[	£
6	.	;
L	⊢	j

C	6	N
⌐	1	æ
⊣	[	Q

*	4	`
Z	G	Pts
#	/	L

q	ä	B
a	B	`
)	⊨	n

î		B
G		
	G	

The encrypted data must then be " $) 3 é) / 7 - . [£ 6 . ; L ⊢ j C 6 N ⌐ 1 æ ⊣ [Q * 4 ` Z G Pts # / L q ä B a B !) ⊨ n î B G G by discarding the padding.

Note: In case of encryption and decryption, for apiece user, certain amount of stuff would be amassed in the Encryptor as well as the Decryptor. The amassed stuff is:

i) The set K.
ii) The integer's m and r.
iii) The function $y = f(x)$ in the Encryptor and $y = f^{-1}(x)$ in the Decryptor.

After receiving the coded message, the user would have to place it in the Decryptor for decoding.

Three values are to be sent to the Decryptor.

Improvement in the efficiency of the computed values could be ensured employing 'Pseudo Cryptography.'

Pseudo Cryptographic Algorithm

Let all the computed values be encrypted by the Encryptor ahead of being sent by employing Caesar cipher method.

Let the key values be embodied one after another, side by side, like AAAAAAAA, that is, four key values, apiece of 2 digits.

Let the 'm' value be represented as MMM, that is, one 'm' value of 3 digits.

Let the 'r' value be represented as NNN, that is, one 'r' value of 3 digits.

Let all of these be concatenated, making it look like

AAAAAAAAMMMNNN.

(The default format for the numbers to be sent from the Encryptor to the Decryptor)

Since d (L) is 3, that is, the length of the actual text to be encrypted before padding, the above-obtained numeric string should be shifted clockwise towards right by three and finally three would be added to the end.

In the Decryptor's end, the Decryptor should confiscate the last digit from the string and it should rotate the string left that many times as mentioned by the digit.

Note: This might only work if the number of digits for the respective key values 'm' value and 'r' value are fixed.

Decryption Algorithm

Step 1   Enter the cryptic message.

In the exhibited case, it is

In the exhibited case, it is " ⊨) $) 6 / é - . [£ . ; ⋆ L ⊦ j C l N [æ Z ⊣ G Q * 4 ` G Pts 3 # / L q 6 B a ä `) B n î 7 B

Employing it on the Rubik cube in order of faces—Front, Right, Back, Left, Up and Down, the subsequent flow is ensured.

"	$	)
3	é	)
/	7	-

.	[	£
6	.	;
L	⊢	j

C	6	N
⊠	1	æ
⊣	[	Q

*	4	`
Z	G	Pts
#	/	L

q	ä	B
a	B	`
)	⊨	n

î		B
G		
	G	

Step 2   Rotating the cube from right-to-left once for row 2 and up-to-down once for column 2.

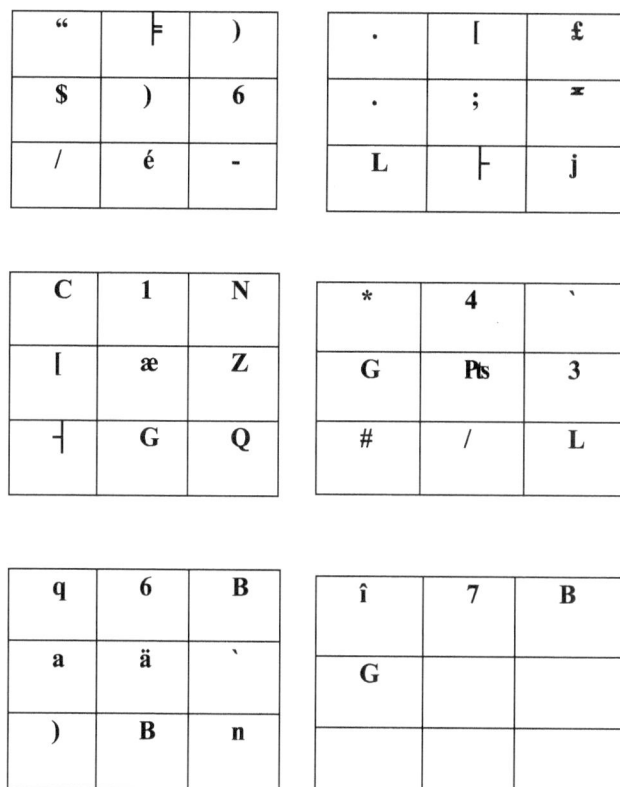

" ⊨) $) 6 / é - . [£ . ; ⍺ L ⊢ j C l N [æ Z ⊣ G Q * 4 ` G Pts 3 # / L q 6 B a ä `) B
n î 7 B

Step 3 Convert the cryptic message to its ASCII equivalent value. The code
becomes 34, 198, 41, 36, 41, 54, 47, 130, 45, 46, 91, 156, 46, 59, 39, 76,
195, 106, 67, 108, 78, 91, 145, 90, 180, 71, 81, 42, 52, 96, 71, 158, 51,
35, 47, 76, 113, 54, 66, 97, 132, 96, 41, 66, 110, 140, 55, 66.

Step 4 Subtract 32 from apiece values.

Obtained R_1 = {2, 166, 9, 4, 9, 22, 15, 98, 13, 14, 59, 124, 14, 27, 7, 44, 163,
74, 35, 76, 46, 59, 113, 58, 148, 39, 49, 10, 20, 64, 39, 126, 19, 3, 15, 44, 81, 22,
34, 65, 100, 64, 9, 34, 78, 108, 23, 34}.

Step 5 The set Q was already sent with the code, and the set K was amassed in
the Decryptor.

Using Q, K and R_1 obtain $Y` = \{y_i`\}$. $y`_I = k_i`*q_i + r_{Ii}$.

In this case, $Y` = \{$5534, 9094, 12749, 11614, 12289, 10318, 13217, 9494, 9293, 2318, 1694, 382, 3014, 1997, 11837, 12289, 9293, 11174, 11837, 1997, 9094, 10109, 1997, 10318, 2542, 10957, 2429, 11614, 13217, 9494, 12289, 2318, 12749, 10109, 8702, 11393, 9293, 10318, 7574, 11393, 12749, 11614, 2774, 10109, 9494, 2318, 1997, 9494$\}$.

Step 6   The inverse function which is unique to apiece user is amassed in the respective Decryptor.

In the stated example, the function is:

$X = y + 8000$.

Input $y = y_i`$ and find the corresponding x_i's to form the set X. Obtained $X = \{$13534, 17094, 20749, 19614, 20289, 18318, 21217, 17494, 17293, 10318, 9694, 8382, 11014, 9997, 19837, 20289, 17293, 19174, 19837, 9997, 17094, 18109, 9997, 18318, 10542, 18957, 10429, 19614, 21217, 17494, 20289, 10318, 20749, 18109, 16702, 19393, 17293, 18318, 15574, 19393, 20749, 19614, 10774, 18109, 17494, 10318, 9997, 17494$\}$.

Step 7   The integers 'm' and 'r' are already amassed in the Decryptor.

With the aid of X, m and r, the set $S` = \{s_i`\}$ is obtained.

$s_i` = sqrt(v_i-m*r)$. Obtained $S = \{$79, 99, 116, 111, 114, 105, 118, 101, 100, 55, 49, 33, 61, 52, 112, 114, 100, 109, 112, 52, 99, 104, 52, 105, 57, 108, 56, 111, 118, 101, 114, 55, 116, 104, 97, 110, 100, 105, 91, 110, 116, 111, 59, 104, 101, 55, 52, 101$\}$.

Step 8   Convert apiece element of $S`$ to its ASCII equivalent characters.

In this case, Octorived71! =4prdmp4ch4i9l8over7thandi [nto; he74e is obtained.

4   Result Analysis

Presume that a user has formulated a query in Oracle SQL format:

Select acc_no, password from dimension3 where c_id='C101', then the final operational modus operandi is established via Fig. 2.

From the above-stated diagram, the result is lucid, that whenever a query arrives to fetch the data from a particular dimension table, the data would always be returned in encrypted form.

In this example, as the query necessitates data from the dimension table 3; the data is sent to its associated two view tables after encryption (Fig. 3).

Since special characters have been employed, the encryption mechanism has been enhanced and tricky to pass through the most familiar cryptanalysis tests. A look at the frequency analysis of the letters from the above-mentioned ciphertext

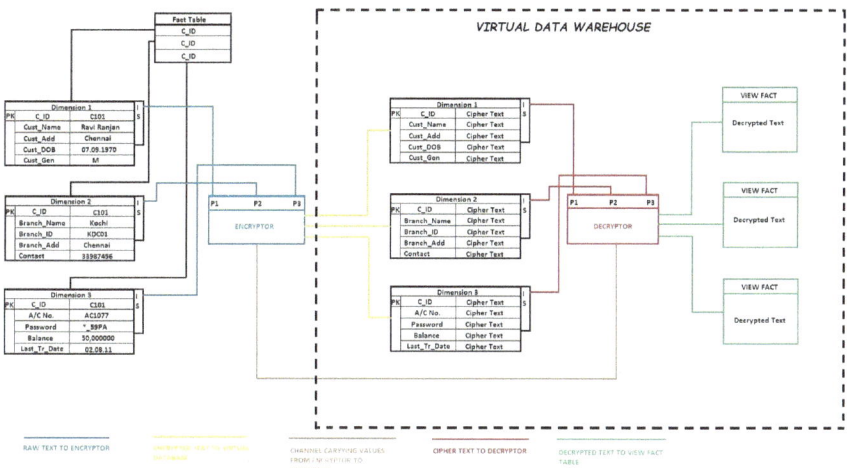

Fig. 2 Proposed virtual data warehouse model architecture

Fig. 3 Character frequency analysis

should endow with an insight into the strength of the cryptographic algorithm. The numbers in X-axis denote the letters of the English alphabet group.

5 Conclusion

Whenever the lexis safety approach in intelligence and deliberation, security is identical, but from instant to instant, employing and integrating security methodologies like cryptographic modus operandi, biometric and genetic algorithm formulation and appliance, quick response code appliance, etc. has not merely been strapping although rate inhibited. Distinctively, devise, execution and integration of any safekeeping method at interior formation of the data warehouse are amiable.

The devise and realization of projected Virtual Data Warehouse Model Employing Crypto–math Modus Operandi and Intelligent Sensor Algorithm for Cosseted Transference and Output Augmentation with incorporation of proposed Pseudo Cryptographic Algorithm for improved intrusion prevention is the nucleus of the bestowed safekeeping in the midst of quandary in offer.

The accessibility is unequivocally enhanced by employing Pseudo Cryptography, in that way ascertaining dwindling of admittance time and authenticating output augmentation for the data warehouse.

Throughout the compilation, deliberations as well as elucidation ascertain how safety could be realized and integrated at distinct layers employing the pioneering cryptographic modus operandi.

The amalgamation of projected architectural replica as well as distinct algorithms for secured transmission and performance enhancement ensure improved intrusion prevention and elevated level of security.

References

1. Pal, B., Chowdhury, R., Chatterjee, P., Dasgupta, S., & De, M. (2015). Performance enhancement of data warehouse using minimization of query processing proposal to improve ROI. *International Journal of Application or Innovation in Engineering and Management, 4* (12), 35–42. ISSN: 2319 4847. Indexed in Thomson Reuters Researcher ID–N–6095–2015.
2. Chowdhury, R., Datta, S., Dasgupta, S., & De, M. (2015). Implementation of central dogma based cryptographic algorithm in data warehouse for performance enhancement. *International Journal of Advanced Computer Science and Applications, 6*(11), 29–34, ISSN (Online): 2156 5570, ISSN (Print): 2158 107X.
3. Chowdhury, R., Dey, K. S., Datta, S., & Shaw, S. (2014). Design and implementation of proposed drawer model based data warehouse architecture incorporating DNA translation cryptographic algorithm for security enhancement. In *Proceedings of International Conference on Contemporary Computing and Informatics, IC3I 2014, Organized by Sri Jayachamarajendra College of Engineering, Mysore* (pp. 55–60). Proceedings in USB: CFP14AWQ-USB, ISBN–978-1-4799-6628-8, INSPEC Accession Number–14881472, Published and Archived in IEEE Digital Xplore, ISBN: 978-1-4799-6629-5.
4. Chowdhury, R., Chatterjee, P., Mitra, P., & Roy, O. (2014). Design and implementation of security mechanism for data warehouse performance enhancement using two tier user authentication techniques. *International Journal of Innovative Research in Science, Engineering and Technology, 3*(6), 165–172. ISSN (Online): 2319 8753, ISSN (Print): 2347 6710.
5. Chowdhury, R., Pal, B., Ghosh, A., & De, M. (2012). A data warehouse architectural design using proposed pseudo mesh schema. In *Proceedings of First International Conference on Intelligent Infrastructure, CSI ICII 2012, 47th Annual National Convention of Computer Society of India, Science City Auditorium, Kolkata* (pp. 138–141). Tata McGraw Hill Education Private Limited. ISBN (13)–978-1-25-906170-7, ISBN (10)–978-1-25-906170-1.
6. Chowdhury, R., & Pal, B. (2010). Proposed hybrid data warehouse architecture based on data model. *International Journal of Computer Science and Communication, 1*(2), 211–213. ISSN: 0973 7391.
7. Saurabh, A. K., & Nagpal, B. (2011). A survey on current security strategies in data warehouses. *International Journal of Engineering Science and Technology, 3*(4), 3484–3488. ISSN: 0975 5462.
8. Santos, R. J., Bernardino, J., & Vieira, M. (2011). A survey on data security in data warehousing: Issues, challenges and opportunities. In *International Conference on Computer as a Tool, EUROCON IEEE, Lisbon* (pp. 1–4). INSPEC Accession Number–12075581, Published and Archived in IEEE Digital Xplore, ISBN–978-1-4244-7486-8.
9. Vieira, M., Vieira, J., & Madeira, H. (2008). Towards data security in affordable data warehouse. In *7th European dependable computing conference.*

10. Patel, A., & Patel, J. M. (2012). Data modeling techniques for data warehouse. *International Journal of Multidisciplinary Research, 2*(2), 240–246.
11. Chaudhuri, S., & Dayal, U. (1997). An overview of data warehousing and OLAP technology. *Newsletter ACM SIGMOD Record, 26*(1), 65–74.
12. Farhan, M. S., Marie, M. E., El-Fangary, L. M., & Helmy, Y. K. (2011). An integrated conceptual model for temporal data warehouse security. *Computer and Information Science, 4*(4), 46–57.
13. Golfarelli, M., Maio, D., & Rizzi, S. (1998). The dimensional fact model: A conceptual model for data warehouses. *International Journal of Cooperative Information Systems, 7*(2–3), 215–247.
14. Golfarelli, M., & Rizzi, S. (1998). A methodological framework for data warehouse design. In *Proceedings of ACM First International Workshop on Data Warehousing and OLAP, DOLAP, Washington* (pp. 3–9).
15. Chowdhury, R., Bose, R., Sengupta, N., & De, M. (2012). Logarithmic formula generated seed based cryptographic technique using proposed alphanumeric number system and Rubik Rotation Algorithm. In *Proceedings of IEEE 2012 International Conference on Communications, Devices and Intelligent Systems, CODIS 2012, Organized by Jadavpur University, Kolkata* (pp. 564–567). Proceedings in CD: IEEE Catalog Number–CFP1207U-CDR, ISBN–978-1-4673-4698-6, Proceedings in Print: IEEE Catalog Number–CFP1207U-PRT, ISBN–978-1-4673-4697-9, INSPEC Accession Number–13285714, Published and Archived in IEEE Digital Xplore, ISBN–978-1-4673-4700-6.
16. Chowdhury, R., Ghosh, S., & De, M. (2012). String graphixification based asymmetric key cryptographic algorithm using proposed concepts of GDC and S-loop matrix. In *Proceedings of IEEE/OSA/IAPR International Conference on Informatics, Electronics & Vision 2012, ICIEV 2012, Organized by University of Dhaka, Dhaka, Bangladesh* (pp. 1152–1157). Proceedings in CD: IEEE Catalog Number–CFP1244S-CDR, ISBN–978-1-4673-1152-6, Proceedings in Print: IEEE Catalog Number–CFP1244S-PRT, ISBN–978-1-4673-1151-9, Conference Proceedings: ISSN: 2226 2105, INSPEC Accession Number–13058551, Published and Archived in IEEE Digital Xplore, ISBN–978-1-4673-1153-3.
17. Chowdhury, R., Dutta, S., & De, M. (2014). Towards design, analysis and performance enhancement of data warehouse by implementation and simulation of p2p technology on proposed pseudo mesh architecture. *International Journal of Innovative Research in Science, Engineering and Technology, An ISO 3297:2007 Certified Organization, 3*(6), 178–187. ISSN (Print): 2347 6710, ISSN (Online): 2319 8753.
18. Pal, B., Chattopadhyay, S., Mitra, S., Chowdhury, R., & De, M. (2012). Study and comparison of indexing models in data warehouse. *International Journal of Software Engineering Research & Practices, 2*(3), 1–8, ISSN (Print): 2231 2048, ISSN (Online): 2231 0320.

Framework for Geospatial Query Processing by Integrating Cassandra with Hadoop

S. Vasavi, M. Padma Priya and Anu A. Gokhale

Abstract Nowadays we are moving towards digitization and making all our devices such as sensors, cameras connected to Internet producing big data. This big data has variety of data and has paved the way for the emergence of NoSQL databases, like Cassandra for achieving scalability and availability. Hadoop framework has been developed for storing and processing distributed data. In this work, we mainly investigated on storage and retrieval of geospatial data by integrating Hadoop and Cassandra using prefix-based partitioning and Cassandra's default partitioning algorithm, i.e. Murmur3Partitioner techniques. Geohash value is generated that acts as a partition key and also helps in effective search. Hence, the time taken for retrieving data is optimized. When user requests for spatial queries like finding nearest locations, searching in Cassandra database starts using both partitioning techniques. A comparison on query response time is made so as to verify which method is more effective. Results showed that prefix-based partitioning technique is efficient than Murmur3 partitioning technique.

Keywords Big data · Spatial query · Geohash · Cassandra · NoSQL databases
Murmur3Partitioner · Prefix-based partitioning

S. Vasavi (✉) · M. Padma Priya
VR Siddhartha Engineering College, Kanuru, Andhra Pradesh, India
e-mail: vasavi.movva@gmail.com

M. Padma Priya
e-mail: padmapriya.mallela39@gmail.com

A. A. Gokhale
Illinois State University, Normal, IL, USA
e-mail: aagokha@ilstu.edu

© Springer Nature Singapore Pte Ltd. 2018
S. Margret Anouncia and U. K. Wiil (eds.), *Knowledge Computing and Its Applications*, https://doi.org/10.1007/978-981-10-6680-1_7

1   Introduction

Companies that use big data for business challenges can gain advantage by integrating Cassandra with Hadoop. Hadoop distributed file system framework can
process voluminous of data generated from various sources. Out of various NoSQL
databases, Cassandra supports linear scalability and high availability for ensuring
fault tolerance. As such, when integrated, Cassandra and Hadoop together increase
the processing capabilities to manage big data efficiently. Geospatial data helps
in identifying the geographic location of an object, its features and boundaries on
earth. Such data can be analysed to serve various purposes such as tourism, health
care, geomarketing and intelligent transportation system. There are two data types
for spatial data, vector and raster. Both data types store object reference as latitude
and longitude (vertices/paths or grid cells) as shown in Fig. 1. Raster data includes
remote sensing, photogrammetric, and vector data includes geographical positioning system (GPS), digitizing. Cassandra integration with Hadoop helps in querying
spatial data efficiently by reducing the query response time.

Raster data can be represented at its original resolution and form without generalization. But the location of each vertex needs to be stored explicitly. Advantage
of vector data is the geographic location of each cell is implied by its position in the
cell matrix. Disadvantage is it is difficult to adequately represent linear features
depending on the cell resolution. Figs. 2 and 3 present example for vector data and
raster data.

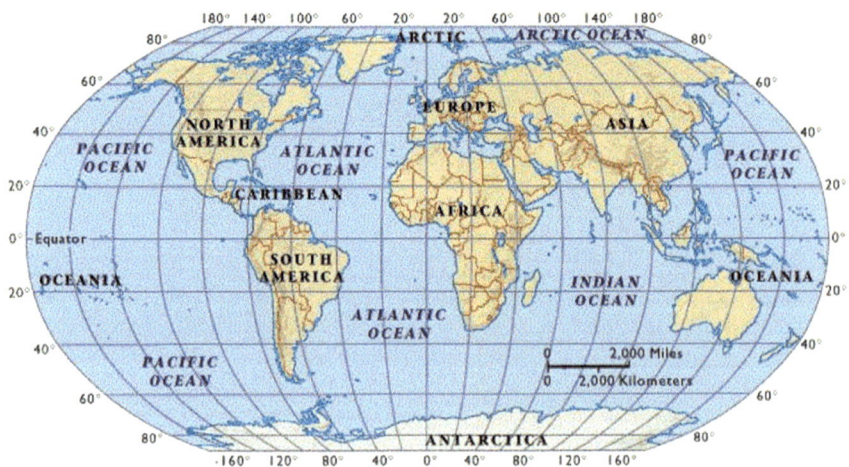

Fig. 1 Longitude and latitudes of earth [9]

Fig. 2 Geospatial vector data type [4]

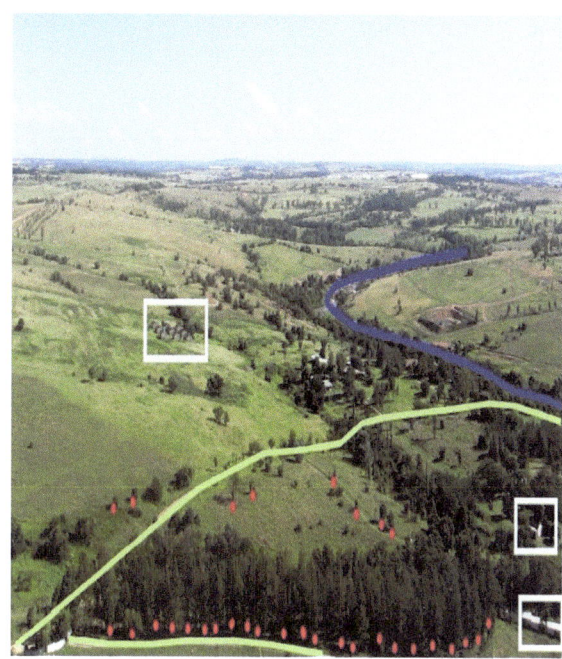

Fig. 3 Geospatial raster data type [4]

1.1 Motivation

Traditional databases (relational database) are suitable for storing and querying structured data that guarantees ACID properties. With the emergence of Internet, large amount of unstructured data is being produced. NoSQL databases that guarantee CAP properties are suitable for storing such unstructured data. Dynamo, MongoDB, BigTable, HBase, Cassandra are designed to handle the data storage and processing with less response time. Even though MongoDB suits for complex queries such as social networking applications where we have to be optimized for latency, HBase and Cassandra when integrated with Hadoop are equivalently good in such a scenario. Cassandra DB and its query language CQL support queries such as indexing, search libraries but not spatial queries. Existing works for labelling and retrieving Cassandra database are not efficient. This chapter aims at adding the functionality of spatial querying for Cassandra database by integrating Cassandra with Hadoop.

1.2 Problem Statement

Every location on this globe is an intersection of latitude and longitude co-ordinates as shown in Fig. 4. Hence, this point of representation is used to find a location on map. There are many different kinds of spatial operations such as reclassifying maps, spatial joins, raster analysis, topological overlay, measuring distance and connectivity, characterizing neighbourhood [3]. Proximity analysis queries such as "which parcels are within 100 m of the subway", contiguity analysis queries such as "which states share a border with Coorg", neighbourhood analysis queries such as "calculate an output value at a location from the values at nearby locations". Other GIS functions include zooming and panning, reordering layers and selecting features. Most commonly used operations are to find the nearest locations of a specified source location. But finding these locations based on latitude–longitude co-ordinate values is really a bit difficult task, especially when dealing with high precision values.

Geohashing technique can be used to overcome this problem. It takes latitude–longitude pair as input and produces a Geohash value, whose length is based on precision value specified. Searching the entire database sequentially for a required destination using this Geohash value may not deliver efficient results as expected. Thus, parallel processing must be done to get rid of this issue. To achieve this, Cassandra is integrated with Hadoop which is an efficient distributed parallel processing framework by using MapReduce programming model.

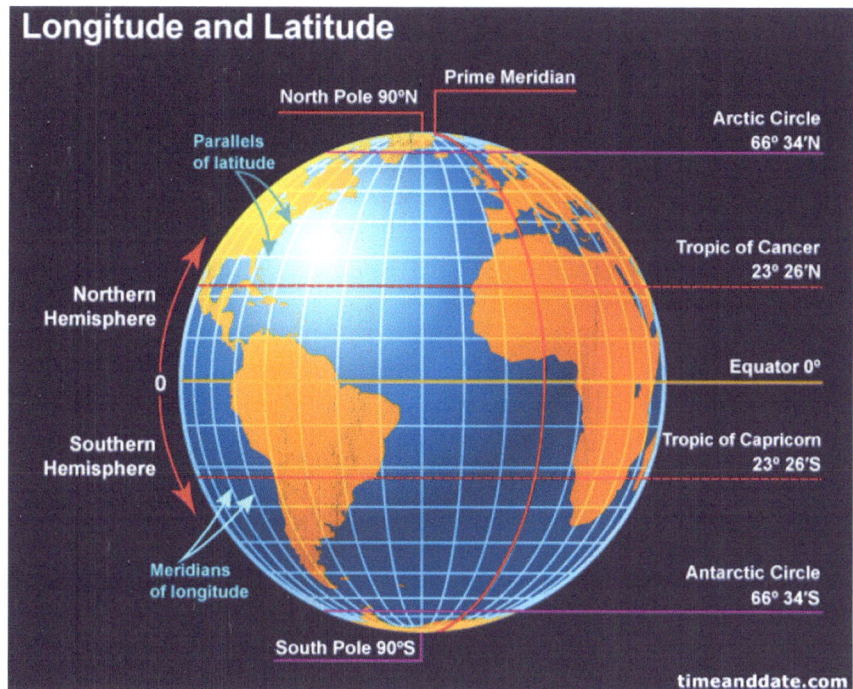

Fig. 4 Geographic co-ordinate system [21]

1.3 Spatial Query Processing Architectures

Figure 5 presents system design and architecture for geospatial data query processing.

Geospatial data is stored in a SQL Server 7.0 database. ArcGIS software is used to update spatial data and for further spatial analysis. It is also used to plot maps as needed. CADClient is used to perform data exchange between geodatabase and AutoCAD. ArcIMS software is used to deploy spatial information over the Web. ArcSDE will hold all the pieces together.

Figure 6 presents Hadoop-GIS (Hive) architecture that includes data partitioning, data storage, query language, query translation and query engine. The query engine consists of index building, query processing and boundary handling on top of Hadoop.

Hive is used for feature queries on top of MapReduce. HiveQL is used for spatial data processing. But this architecture will be difficult as we were unable to do traditional MapReduce job. The architecture shown in Fig. 7, SpatialHadoop, is directly integrated into the Hadoop making. It is more efficient than Hadoop-GIS.

Problem with using Hive for geospatial data processing is that Hive will do cross-join that will not suit to big data. In this chapter, we propose Cassandra-based

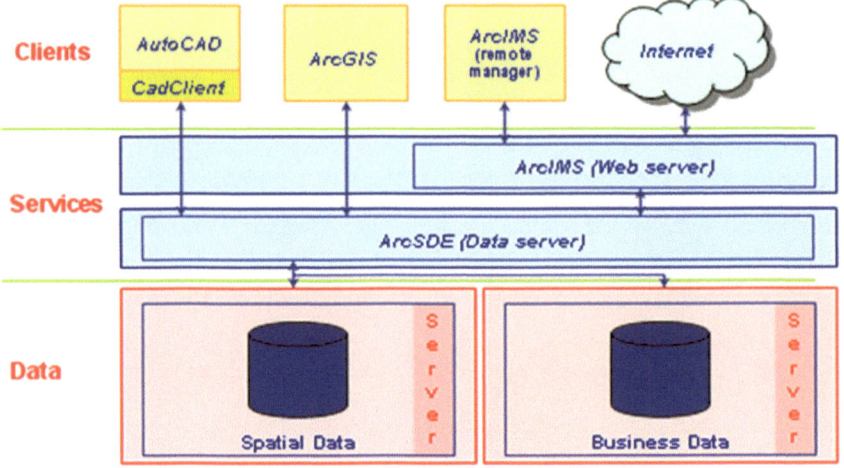

Fig. 5 System design and architecture for Geospatial query processing [2]

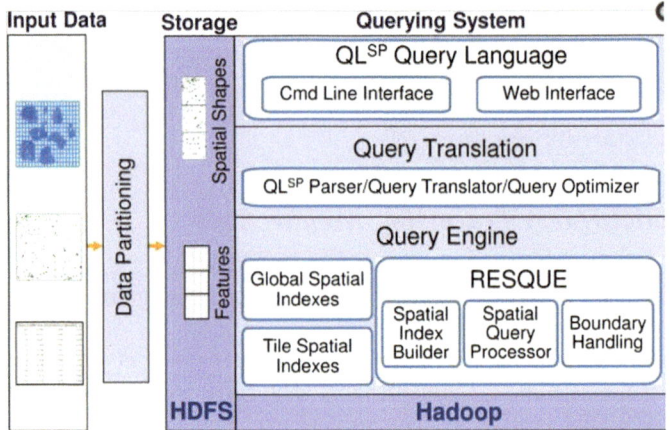

Fig. 6 Architecture overview of Hadoop-GIS (Hive) [1]

framework that uses indexing for query processing. Geospatial data processing architectures SpatialSpark, GeoSpark, Magellan, Simba and LocationSpark based on Spark have been proposed in the literature. Advantage of using Spark framework is that it achieves fault tolerance because it works on distributed in-memory data. In experiments conducted by Tang et al. [20], LOCATIONSPARK performed better than Magellan in query execution, because the spatial index used by it avoids visiting unnecessary data partitions, better than SpatialSpark in speed, faster than GeoSpark in spatial range. Problem with LOCATIONSPARK is that it generates false positives because of embedding Spatial Bloom Filter into global spatial index.

Fig. 7 Architectural view of
GeoSpark [15]

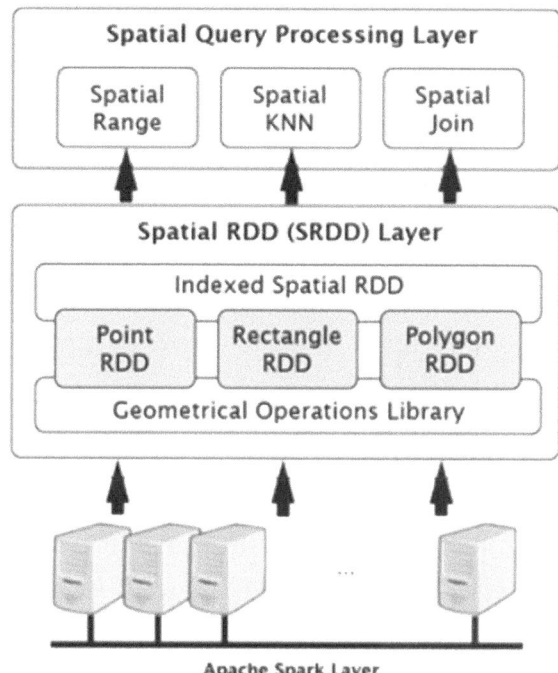

1.3.1 Apache Cassandra

Cassandra is one of the open source frameworks for distributed database management of NoSQL systems, developed at Facebook and built on Amazon's Dynamo and Google's BigTable. It is consistent, scalable and fault-tolerant column-oriented database. It can accommodate variety of data and supports ACID properties. It can work on commodity hardware with quick response time. Figure 8 presents physical deployment of Cassandra cluster. Nodes of the cluster contain some chunks of data and are shown as a ring. Data is partitioned based on primary key with two parts: partition key based on hash function and second part will store information on clustering and sorting the data for a given partition.

1.3.2 Integrating Cassandra with Hadoop

Hadoop is a distributed system for processing variety of data, and Cassandra aims at high availability. By combining both Hadoop and Cassandra, we can process spatial operations such as joins, indexing, Geohashing on massive amounts of geospatial data effectively. Figure 9 presents MapReduce process in a Cassandra/Hadoop cluster.

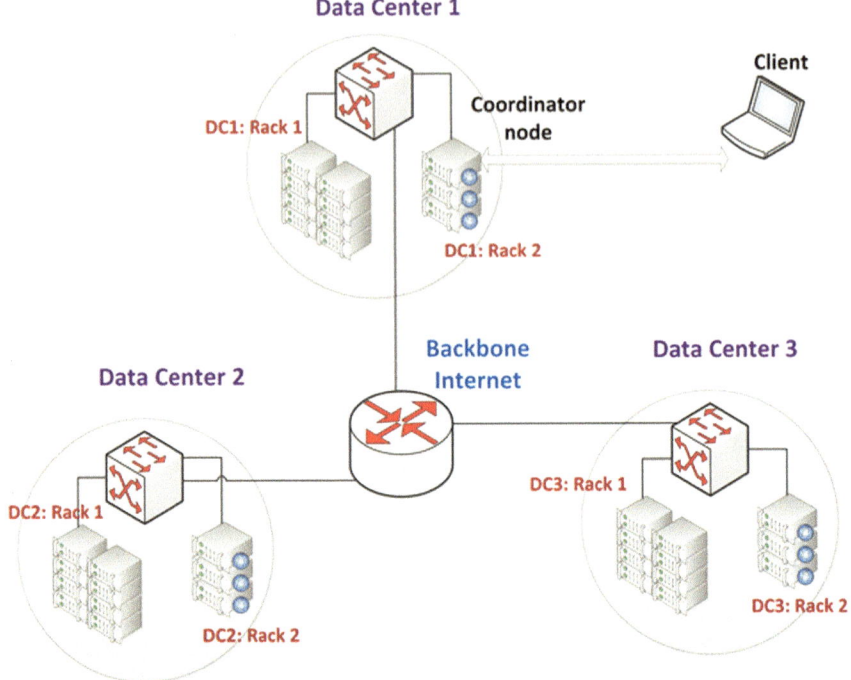

Fig. 8 Physical deployment of Cassandra cluster [12]

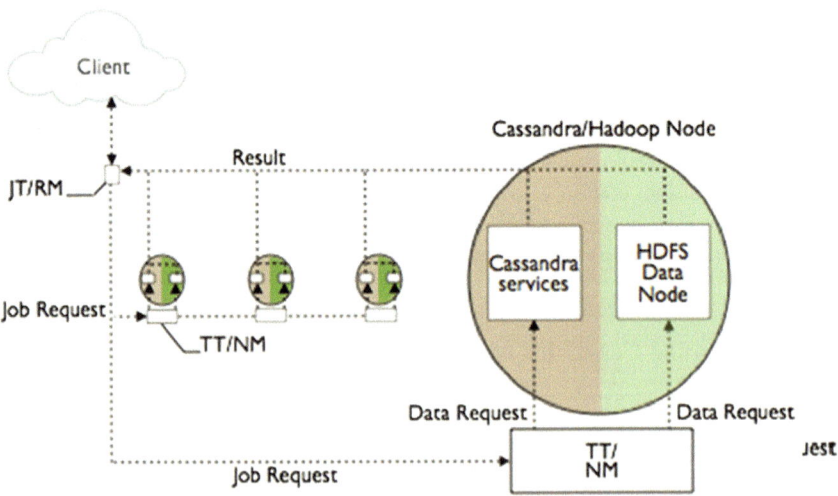

Fig. 9 MapReduce process in Cassandra/Hadoop cluster [10]

This architecture will have Hadoop cluster over the Cassandra nodes. Each Cassandra node will have Hadoop TaskTracker (TT) and Data Node. When client submits the job, JobTracker/ResourceManager (JT/RM) initiates the task execution by first interacting with NameNode (NM). JT/RM queries Cassandra for DataNodes information, further sends a MapReduce job request to the Task Trackers/Node Managers (TT/NM). Cassandra nodes run a TT process on each node for processing Map and Reduce task.

1.4 GeoHashing

Geohashing is a way to express a location using latitude and longitude or alphanumeric string. Geographic coordinate system (GCS) uses a 3D spherical surface to define locations on the earth. It forms a grid-like structure called graticule. Equator will have graticule (0,0) to define 0 latitude and 0 longitude. The Geohash identifies a particular grid cell defined by latitude and longitude. Figure 10 presents how the world is divided into two halves "0" and "1". For example if the location in question is in the half "0", divide portion "0" into two halves again as "0" and "1", making two-bit precision "00" and "01" as shown in Fig. 11.

A greater precision will give more exact location. That is portion "00" can further be divided as "000" and "001", and the process continues until we come across with exact location as shown in Fig. 12.

As the process of subdividing is repeated, Geohash of 32 bit looks like: 0011110001011001011010100011110110110101.

Each 5 bits are converted to one character as shown in Table 1.

00111 10001 01100 10110 10100 01111 01101 10101.

Finally, Geohash value is "**7jdqngep**".

It is clearly illustrated that greater precision will give longer strings as shown in Fig. 13.

Fig. 10 Subdividing the earth surface up to one bit

Fig. 11 Subdividing the
earth surface up to two bits [9]

Fig. 12 Subdividing the
earth surface up to three bits
[9]

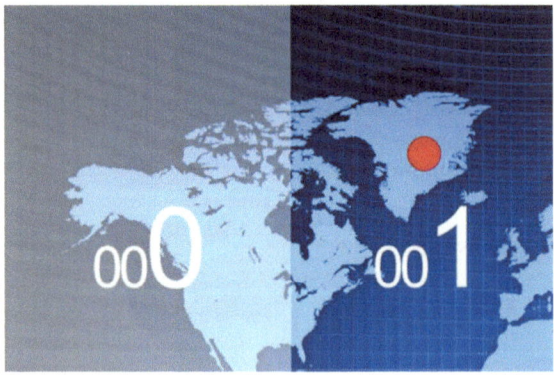

Table 1 Base-32 character map [9]

Decimal	0	1	2	3	4	5	6	7	8	9	10	11	12	13	14	15
Base 32	0	1	2	3	4	5	6	7	8	9	B	c	d	E	f	g
Decimal	16	17	18	19	20	21	22	23	24	25	26	27	28	29	30	31
Base 32	h	j	k	m	N	p	q	r	s	t	u	v	w	x	y	z

Fig. 13 Moveable typescripts: enter latitude, longitude and precision to obtain Geohash [19]

1.5 Data Partitioning in Cassandra

Cassandra distributed database runs on multiple nodes. Partitioning schemes such as Random Partitioner (MD5-based distribution of data), Byte Ordered Partitioner (key bytes-based data distribution) and Murmur3Partitioner (MurmurHash values-based data distribution) will determine in which node data should be stored. Cassandra data storage is shown in Fig. 14 such as a circular ring. The ring will have ranges that are equal to the number of nodes, where each node is responsible for other ranges of data. A node is first assigned with a token (position) before joining the ring.

Cassandra uses random partitioning technique specifically ByteOrderedPartitioner (that in turn uses consistent partitioning technique) for assigning data to cluster nodes. MD5 hash value (range is from 0 to 2^127) is

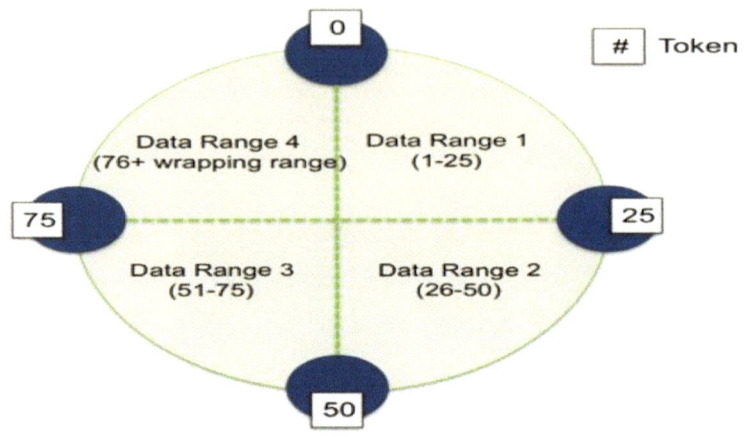

Fig. 14 Four-node cluster partitioned with token ranges [6]

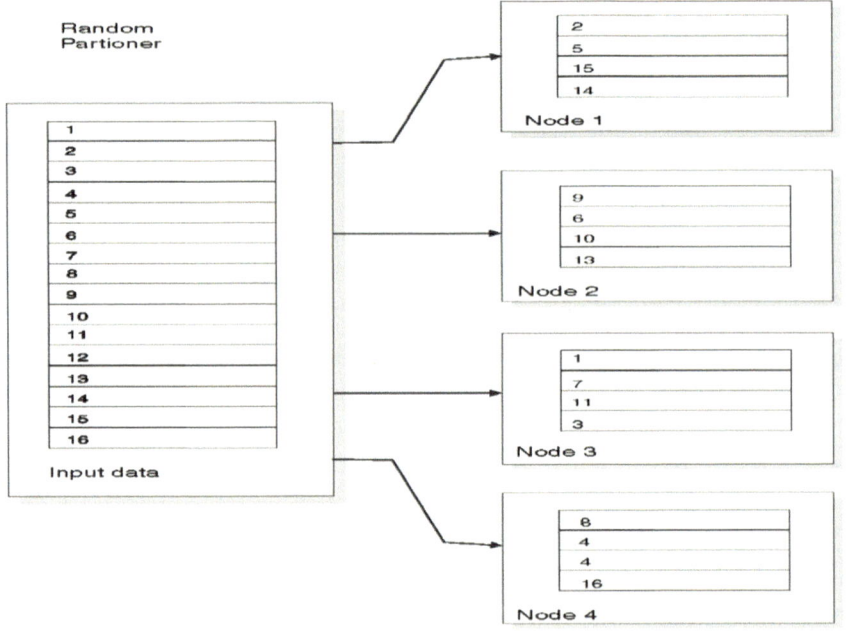

Fig. 15 Random partitioner [6]

generated for each of the record. A simple example for random partitioner is shown in Fig. 15 [6].

There are three types of Byte Ordered Partitioner (BOP). In the first type of partitioner (ByteOrderedPartitioner), instead of using encoded strings, keys (hexadecimal representation) are stored in the order of their raw bytes. Second type of partitioner (OrderPreservingPartitioner) UTF-8 encoded value of the keys is used to store the data. Last type of partitioner (CollatingOrderPreservingPartitioner) storage of row keys is based on US English locale (EN_US). Latest versions of Cassandra use Murmur3Partitioner that in turn uses Murmurhash function and virtual nodes for storing the data. Hash function generates a 64-bit hash value (in the range -2^{63} to $+2^{63} - 1$) as partition key. Murmur3Partitioner also uses consistent hashing that manages distribution of data based on partition key across the cluster.

The main objective of this chapter is to propose a novel geospatial computing system "GeoHadoop Architecture" for efficient processing of spatial queries in a distributed in-memory environment that takes advantage of integrating Cassandra with Hadoop for geospatial querying by using prefixing, spatial indexing and partitioning techniques. This objective can be achieved with the following methodology which is summarized as follows:

1. Understanding architecture and data model of Cassandra.
2. Collection of geographical datasets according to the requirements.
3. Partitioning of Cassandra data using Murmur3Partitioning technique.
4. Apply spatial query using Cassandra Query Language (CQL).
5. Process the same data using Hadoop MapReduce framework.
6. Apply "prefix-based partitioning algorithm".
7. Compare the performance evaluation values between Murmur3Partitioning technique and prefix-based partitioning algorithm.

2 Literature Survey

Centralized systems have more response time and slower data access. Since 1980s there is a huge demand for decentralized databases. Spatial query processing has been used in many GIS-based applications where significant conclusions can be made from end-user location information. But spatial data is voluminous and requires scalable techniques, distributed computing platforms and technologies for efficient processing of spatial queries. Even though techniques given by Hadoop MapReduce, Hadoop Distributed File System (HDFS) and NoSQL databases for spatial queries have been proposed, but requires modification of systems or frameworks to implement indexing techniques [14]. Spatial query processing techniques as mentioned in the previous section can be classified into two categories basing on the amount of data selected: huge selection such as k-NN queries and low data selection such as k-NN join.

Work reported in Zhang et al. [22] describes MapReduce and showed how spatial queries can be expressed without explicitly addressing any of the details of parallelization. Their experimentation on sample spatial queries proved that

MapReduce is also appropriate for small-scale clusters and computing intensive applications.

R-tree based approach is proposed in Liao et al. [16]. It constructs hierarchical index structures to organize datasets in dimensional space so as to improve the query performance on HDFS. Authors concluded that when data is stored in HDFS where index structures, buffer management are supported, query response time can be reduced.

In another work by Liu et al. [17], summarized that even though HDFS can handle big data, could not perform well for large amount of small files. Because of this, Web-based applications could not be benefited from Hadoop. An approach is proposed to optimize the I/O performance of small files in HDFS environment. This is done by combining small files into a single large file and to create index for each file. Experiment results showed that their approach achieved good performance. Their work also focused on methods for parallelizing task computation and distributing the data over clusters.

Finding nearest location is one of the important spatial queries that find the nearest location(s) from the user location. Existing methods for finding nearest location such as zip code method (zip code of the places are matched and those locations whose zip code belong to the same region or which are nearer are extracted), k-d tree method (a binary tree that finds nearest locations in $\log(n)$ time), polygon method (boundaries are defined on the map, and queries are performed on points belonging to this polygon), Voronoi diagram (Voronoi cells are defined as shown in Fig. 16a and based on seed point as shown in Fig. 16b nearest locations are identified) could not process large data and cannot find nearest location with high precision [7]. Work reported by Lee [13] is on finding nearest locations using Voronoi diagrams in O(nlogn) time.

There are other spatial partitioning techniques such as grid, quad-tree(spatial indexing for RDBMS), STR/STR+, Z-curve, Hilbert curve. In the work reported by Moniruzzaman and Hossain [18], most of the NoSQL databases such as Cassandra databases are not supporting geospatial indexes and hence spatial queries. Work described by Brahim et al. [5] handles spatial data retrieval within Cassandra

(a)   **(b)**

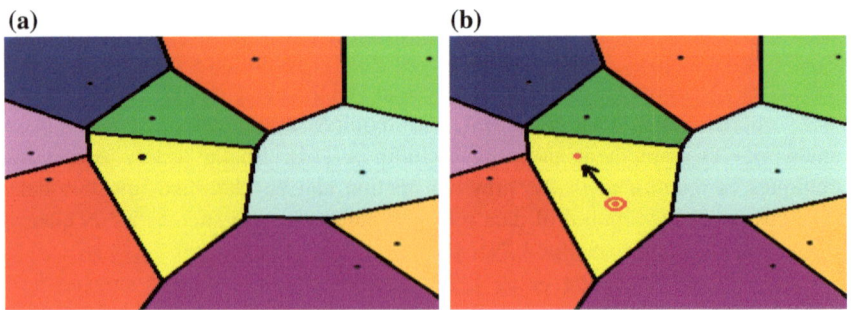

Fig. 16 **a** Voronoi cells [13]. **b** Nearest location using seed point [13]

NoSQL database. A framework is proposed and implemented for spatial queries. Work described by Fox et al. [8] proposed a spatio-temporal index structure for achieving horizontal scalability of NoSQL databases.

2.1 Drawbacks of Existing System

The main drawback of existing system is time complexity for processing large datasets. Our work mainly focuses on reducing the query response time by using integration framework and implementing various effective techniques. Also, random partitioning technique poses several drawbacks such as uneven load balancing, sequential writes that does not distribute data across the cluster. As such Murmur3 partitioning and prefix-based partitioning are used in our framework. We also made an attempt at using multiple partitioners to determine which partitioner is performing well after using Geohashing technique to improve geospatial query processing.

3 Proposed System

This section presents the detailed functionality of the proposed system to perform geospatial querying in the GeoHadoop Architecture. In this architecture, Cassandra is integrated with Hadoop MapReduce framework as shown in Fig. 17 so that query response time for spatial data analysis could be optimized.

3.1 Methodology for Processing Spatial Queries

(i) SPATIAL DATA STORAGE

(a) Geospatial data is collected from various external devices or sensing systems. Collection of data can also be through event messages or current geolocation of end users. This data is in the form of < Latitude, Longitude > pair for different locations, where spatial queries are processed on this pair. A dataset is constructed with this pair of data.

(b) Geohashing technique (that converts the longitude, latitude pair into a single value which is represented in binary format) is applied for each tuple in the dataset which takes < Latitude, Longitude > pair as input and produces "Geohash" value as output. Thus, a new dataset is formed by adding an additional column containing Geohash values. Data is stored according to Cassandra schema, i.e. columnar database schema.

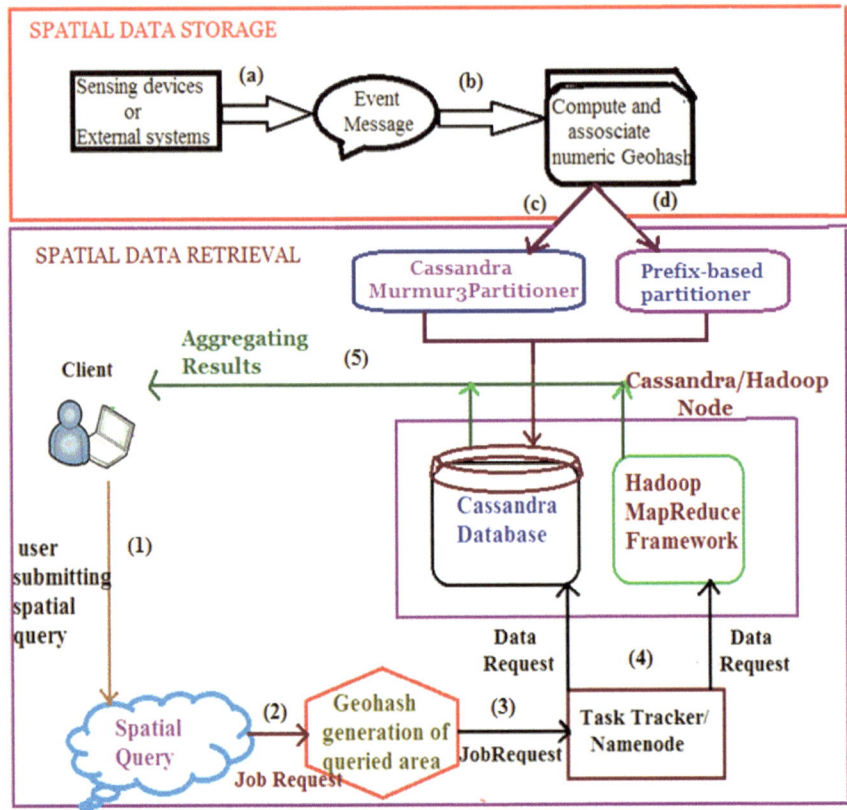

Fig. 17 Architecture for processing spatial queries by integrating Cassandra with Hadoop MapReduce framework

(ii) SPATIAL QUERYING

(a) User submits spatial query.
(b) Compute Geohash value for the queried area using Geohashing technique. MapReduce framework is used for parallel implementation of the input query. The input query logic is executed in parallel over the independent chunks and over different mappers.
(c) Hadoop MapReduce framework performs Map and Reduce task as described in the algorithms of Sect. 4.4. Initially data is retrieved from Cassandra database for input to map function, after it performs reduce function it writes back to the Cassandra database.
(d) Finally the output result is delivered to the end user.

3.2 Hadoop MapReduce Framework

Data processing using Hadoop framework is divided into two phases:

- Mapper phase: Performs mapping of required attributes using Map( ) function.
- Reducer phase: Reduces the number of entries using Reduce( ) function and returns required tuples.

Here, both mapper and reducer use "prefix-based matching algorithm" as shown in Fig. 18.

3.2.1 Prefix-Based Partitioning Algorithm

Geohash code which is a string represents a region on the earth. This Geohash is used in spatial indexing, for latitude and longitude [18]. It provides a spatial hierarchical data structure that will remove characters from the end of the string so as to reduce the precision. This implies that if the Geohash code is longer then it represents the smallest region on the earth. Further, nearby places will have similar Geohash code.

Mapper job mainly takes Geohash, Place_name as <key, value> pair and produces Place_name and its prefix matched count as shown in Fig. 19. The output of mapper phase is taken as input to reducer phase as shown in Fig. 20. Basing on the length of prefix matched, each entry in data node is compared whether it is greater than or equal to the precision value. If this condition is satisfied then entry is stored otherwise discarded. Finally, the result set is sorted in decreasing order of their length of prefix matched and displays all the places that are nearer to the location specified by the client. Figure 21 presents the flowchart for processing spatial queries using data presented in Cassandra database. Firstly it receives the spatial queries from the user through any interface. These queries are submitted to the query interpreter which interprets the syntax of CQL queries. After approving that

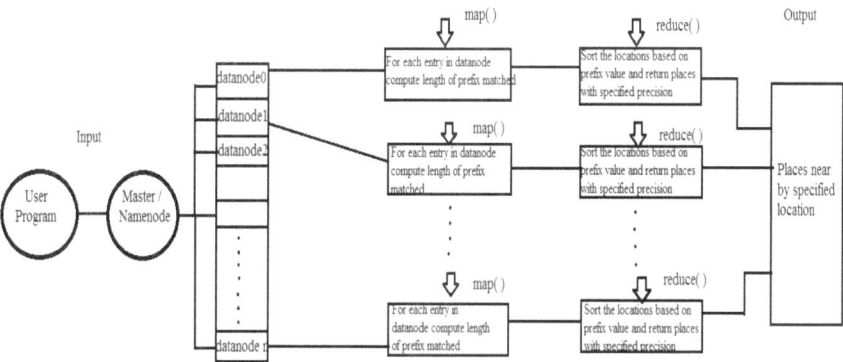

Fig. 18 Phases of Hadoop MapReduce framework for spatial search

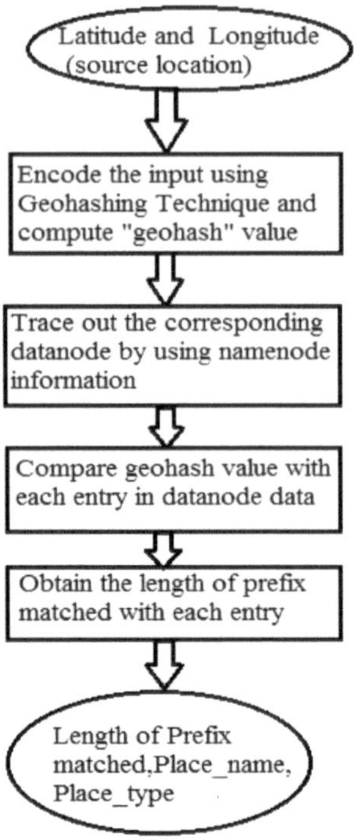

Fig. 19 Mapper task

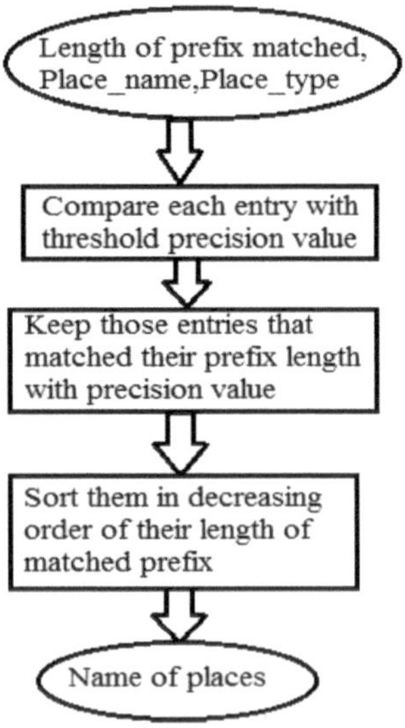

```
      ┌─────────────────────────┐
     (  Latitude and Longitude   )
      (  (source location)        )
       └──────────┬──────────────┘
                  ⇓
      ┌─────────────────────────┐
      │ Encode the input using  │
      │ Geohashing Technique and│
      │ compute "geohash" value │
      └──────────┬──────────────┘
                 ⇓
      ┌─────────────────────────┐
      │ Trace out the corresponding│
      │ datanode by using namenode│
      │ information             │
      └──────────┬──────────────┘
                 ⇓
      ┌─────────────────────────┐
      │ Compare geohash value with│
      │ each entry in datanode data│
      └──────────┬──────────────┘
                 ⇓
      ┌─────────────────────────┐
      │ Obtain the length of prefix│
      │ matched with each entry │
      └──────────┬──────────────┘
                 ⇓
      ┌─────────────────────────┐
     (  Length of Prefix         )
      ( matched,Place_name,       )
      ( Place_type                )
       └─────────────────────────┘
```

Fig. 20 Reducer task

```
      ┌─────────────────────────┐
     (  Length of prefix matched, )
      ( Place_name,Place_type     )
       └──────────┬──────────────┘
                  ⇓
      ┌─────────────────────────┐
      │ Compare each entry with │
      │ threshold precision value│
      └──────────┬──────────────┘
                 ⇓
      ┌─────────────────────────┐
      │ Keep those entries that │
      │ matched their prefix length│
      │ with precision value    │
      └──────────┬──────────────┘
                 ⇓
      ┌─────────────────────────┐
      │ Sort them in decreasing │
      │ order of their length of│
      │ matched prefix          │
      └──────────┬──────────────┘
                 ⇓
      ┌─────────────────────────┐
     (  Name of places           )
       └─────────────────────────┘
```

Fig. 21 Spatial query
processing

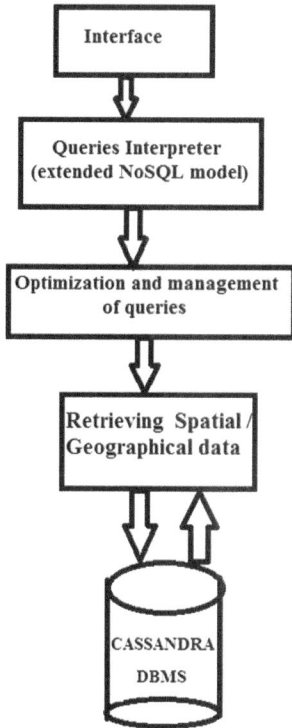

queries are syntactically correct, query optimization is performed for reducing
query response time. Finally, the matched data is retrieved from Cassandra database
containing spatial/geographical data.

3.2.2 Murmur3 Partitioning

Figure 22a, b shows mapper and reducer phases that can be performed using
Murmur3Partitioner technique. During mapper phase, Geohash value is calculated
using Geohashing technique for the required place based on its latitude and longitude. At the back end, Cassandra is used as a storage database and it uses
Murmur3Partitioning technique to partition the data. Hence, data is distributed
across multiple datacenters (or) nodes equally. As a result, data is fetched based on
its Murmur3 hash value corresponding to Geohash value. During reducer phase,
data such as length of prefix matched, Place name and Place type is collected and
sorted in decreasing order to display nearby places.

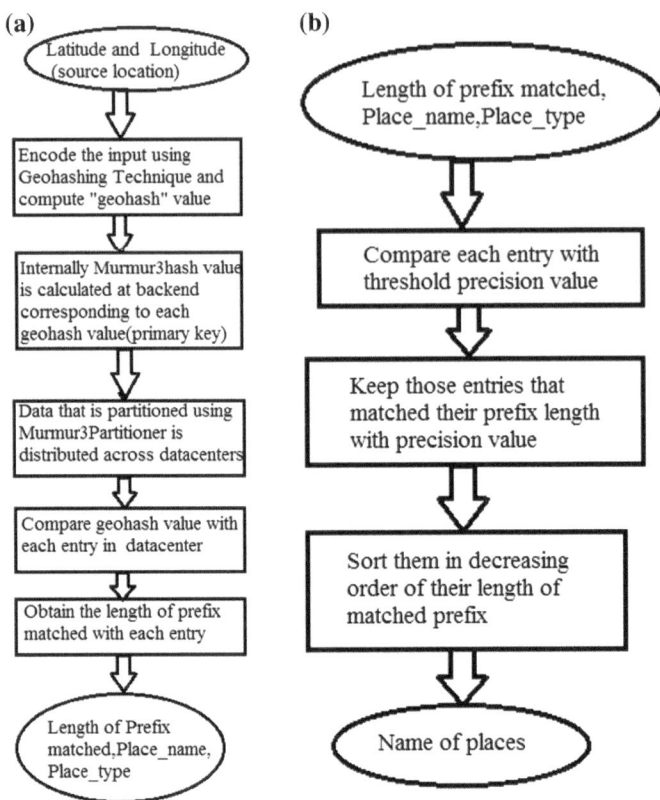

Fig. 22 **a** Mapper phase; **b** reducer phase

4 Results and Analysis

4.1 Dataset Used

Data requirements which involve collection of required data are the first step for the database analytics projects. Partitioning techniques on geographical dataset of India are performed which contains about 50,000 tuples of data available at http://www.latlong.net/country/india-102.html. The dataset is taken in the form of .csv format, which contains Place_name along with its latitude and longitude values. The Place_type is also considered for processing the dataset. Figure 23 presents a snapshot of database. These tables are in denormalized form, since denormalization is best practice for modelling NoSQL columnar databases. Table 2 presents list of attributes along with data types. Place_name represents the name of the location. Place_type refers to the type of the place like village, town, garden, hospital. Latitude and longitude columns represent the position of location on the globe.

Place_Name	Country	Place_Type	Latitude	Longitude
Indramani-Park--Churu--Rajasthan	India	Garden	28.292093	74.971504
Bharathiyar-University--Coimbatore--Tamil-Nadu	India	Garden	11.039022	76.876419
Rathore-Karshi-Farm--Kuchera--Rajasthan	India	Garden	26.962351	73.961838
Green-Park--Dhule--Maharashtra	India	Garden	20.888765	74.771164
Kheda--Gujarat	India	Garden	22.75065	72.684669
Chandabali--Bhadrak--Odisha	India	Town	20.773993	86.743698
Jalesar--Uttar-Pradesh	India	Town	27.472601	78.306557
Ottapidaram--Tamil-Nadu	India	Town	8.907723	78.020966
Hanumangarh-Town--Rajasthan	India	Town	29.5748	74.332977
Sahajbahal--Odisha	India	Town	20.634178	83.328171
Jhusi--Uttar-Pradesh	India	Town	25.425602	81.917152
Mahidharpur--Odisha	India	Town	20.694019	85.179703
Dhasa--Gujarat	India	Town	21.786919	71.517563
Yermala--Maharashtra	India	Town	18.39283	75.870705
Niali--Odisha	India	Town	20.141563	86.059158
Redhakhol--Odisha	India	Town-	21.074162	84.341057
Magarload--Chhattisgarh	India	Town-	20.748428	81.850487
Padmapur-Town--Odisha	India	Town-	21.000067	83.064713
Barkote--Odisha	India	Town	21.535925	85.015167
Eranellur--Kerala	India	Town	10.618336	76.126198
Boudh--Odisha	India	Town	20.828445	84.327133
Birmitrapur--Odisha	India	Town	22.407398	84.7174
Dankaur--Uttar-Pradesh	India	Town-	28.347696	77.553345
Kantamal--Odisha	India	Town	20.660383	83.744194

◄ ► ►│ DATASET

Fig. 23 Geographical dataset of India [4]

Table 2 List of attributes with data types	Attribute	Data type (CQL)
	Place_name	Text
	Country	Text
	Place_type	Text
	Latitude	Decimal
	Longitude	Decimal

Figure 24 shows the conceptual data model representation of World Map entities, whereas Fig. 25 presents physical data model representation of the database.

4.2 Software Requirements

- Operating systems: Windows 8.1 and above, Ubuntu 10.0.14+.
- Java versions 1.7 and above.
- Hadoop versions 2.6 and above.
- Cassandra versions 3.0 and above.
- Maven version 3.0 and above.
- Eclipse IDE.

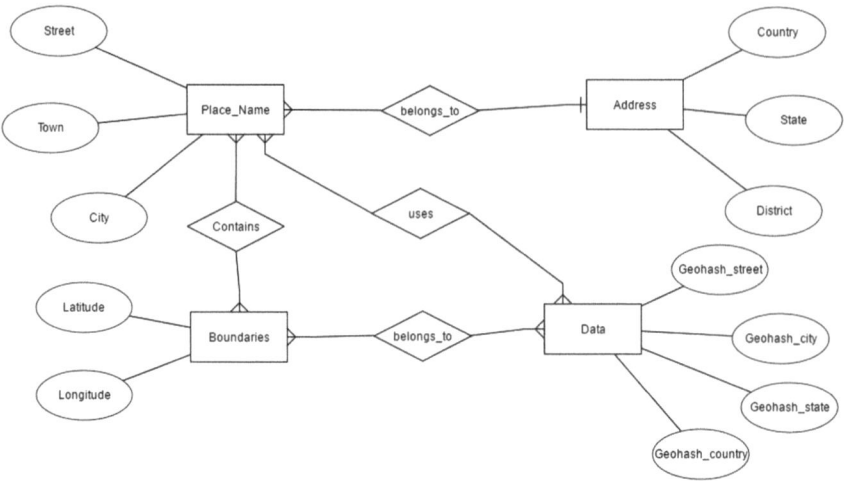

Fig. 24 Entity–relationship (ER) diagram for World Map dataset

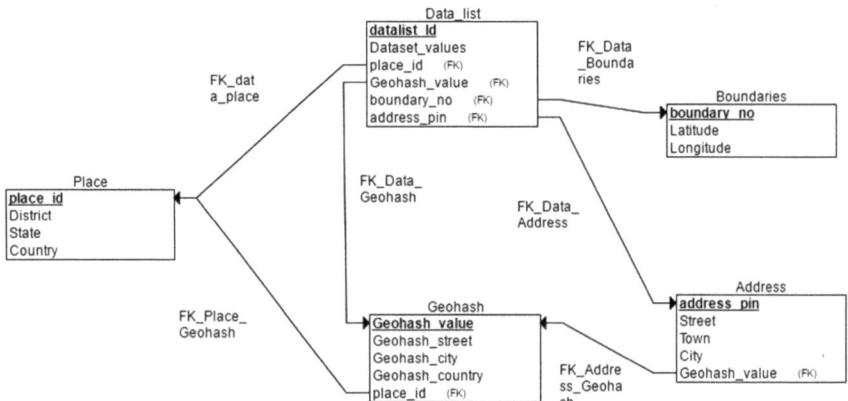

Fig. 25 Physical data model representation on World Map dataset

As all Hadoop and Cassandra projects support Linux operating system, we can install any Linux flavoured OS. As a groundwork we implemented Java and Cassandra Installation, and Hadoop 2.7.3 is installed and integrated with Cassandra environment.

4.3 Implementation Overview

1. Initially a dataset containing places with their <latitude, longitude> is collected and stored in .csv format.
2. As shown in Fig. 25, a new dataset is formed by adding an additional column named "Geohash" which contains Geohash values computed using Geohashing technique.
3. Connect to the Hadoop and Cassandra server.
4. .csv geospatial data will execute in parallel because of underlying MapReduce framework. Input to the mapper and output from the mapper will be of the form <key, value> pair, i.e. <Place_Name, geohash> where Place_name stands for name of the location, Geohash stands for hash code of the place. This output is given to reducer for further processing.

As shown in Fig. 26, each location has unique (longitude, latitude) pair and as such used in our implementation. Attribute type tells the category of the place such

	Place_Nar	Country	Place_Typ	Latitude	Longitude	GeoHash
1						
2	Indramani	India	Garden	28.29209	74.9715	ttj2fn2t6smh
3	Bharathiyi	India	Garden	11.03902	76.87642	t9ywv3edyque
4	Rathore-K	India	Garden	26.96235	73.96184	tsu97h375dx8
5	Green-Par	India	Garden	20.88877	74.77116	tetnv6wtykhb
6	Kheda--Gt	India	Garden	22.75065	72.68467	ts59mvxk5s93
7	Chandaba	India	Town	20.77399	86.7437	tgtwjvuc68vp
8	Jalesar--U	India	Town	27.4726	78.30656	tszsmcjpe00b
9	Ottapidari	India	Town	8.907723	78.02097	t9x6xmt5hrbv
10	Hanumani	India	Town	29.5748	74.33298	ttkb5rzteykc
11	Sahajbaha	India	Town	20.63418	83.32817	tgem2hmdz1ed
12	Jhusi--Utt	India	Town	25.4256	81.91715	tud28hd4zgxx
13	Mahidhar|	India	Town	20.69402	85.1797	tgstdr1zquwp
14	Dhasa--Gt	India	Town	21.78692	71.51756	tefggqjw0004
15	Yermala--	India	Town	18.39283	75.87071	tembwkr65840
16	Niali--Odi	India	Town	20.14156	86.05916	tgt4w6dm5erp
17	Redhakho	India	Town-	21.07416	84.34106	tgezzhrmchd3
18	Magarloac	India	Town-	20.74843	81.85049	tgdnn91sqe6d
19	Padmapur	India	Town-	21.00007	83.06471	tgep6nvyk05p
20	Barkote--(	India	Town	21.53593	85.01517	tgu6w86pugrh
21	Eranellur-	India	Town	10.61834	76.1262	t9yhkm12kcr3
22	Boudh--O	India	Town	20.82845	84.32713	tgeyqztnkxef
23	Birmitrapi	India	Town	22.4074	84.7174	tguprz1kd3xz
24	Dankaur--	India	Town-	28.3477	77.55335	ttp1kb897ct3

Fig. 26 Modified dataset with Geohash values

	M1		▼		f_x		
	A	B	C	D	E	F	G
1	Place_Nar	Country	Place_Typ	Latitude	Longitude	GeoHash	Murmur3Hash(x64 128-bit)
2	Muttom--	India	Town	8.126215	77.31964	t9qyprq6whr0	4041dac4 89c06ed4 2a7c5878 3b16dcfa
3	Kattuvilai	India	Town	8.158933	77.33675	t9qyrv6chz37	06b919e1 d604d7e9 d0685af4 1f8816bb
4	Colachel--	India	Town	8.17862	77.2561	t9qyw0b5vt22	c467a5fe 73561f0e ddff0021 0433e959
5	Poovar--K	India	Town	8.317691	77.07084	t9qz3f1q09yz	867c1b89 2f14b9f8 bff9413c 8934c6ea
6	Kaliakkavi	India	Town	8.33444	77.17101	t9qzkj4pwujz	5c32d095 ecfef590 b925b436 c768c4c7
7	Vendalico	India	Village	8.351273	77.3073	t9qzx0m3ww4e	3a0efa17 55528776 326d8919 43d90a00
8	Kadayal--	India	Town	8.404483	77.26662	t9qzy1zzdtjm	b706c5d2 0d1804ed 54707548 3758028f
9	Puthenthi	India	Village	8.103346	77.41927	t9rn1envrv4j	a8fb9fd3 441ef77b dde8e16c d4195b82
10	Pallam--Ti	India	Town	8.099403	77.43418	t9rn443y7nnh	c0e4120a e3f03af6 9e151bac 02abb845
11	Parakkai--	India	Town	8.136897	77.45216	t9rn63qby593	eddb322b 3b53baea 2c72f3f9 f4697a54
12	Suchindra	India	Town	8.155	77.465	t9rn6u8634pt	bdb3e4e0 eba8cafd 1426543c 0e22c68e
13	Mylaudy--	India	Town	8.154739	77.50174	t9rn7se0c3tb	78a1890e 6c087ae6 ef69f1b9 f7549166
14	Anjugram	India	Town	8.157288	77.57738	t9rnmkfpeusy	8d1c4573 644a3266 c1b92321 65d1733b
15	Vallioor--	India	Town	8.385437	77.60722	t9rptyxbequ2	47d32f7f 8ca27d3d 958e7d4a a8dad0eb
16	Thisayanv	India	Town	8.334946	77.86531	t9rr7v7f9681	8095efa9 d7e46148 c8e5854c 1faa8acf
17	Mudalur--	India	Town	8.419547	77.94959	t9rrvu9pmjm6	e407c488 407394a9 cf5e42e8 50797d9d
18	Vazhayur-	India	Town	11.21753	75.89909	t9vzz4h4c5d2	87e820cb 360311ed 86b937bb 06e80c18
19	Anchuthe	India	Town	8.675045	76.75864	t9w96ejzy197	d4a718d6 efd5894b 17ec58bf 58ac8453
20	Puthusser	India	Village	8.767477	76.80924	t9w9gu4f301v	c4c7d8af 9fedabeb f98f8f46 a47ee5df
21	Mangalap	India	Town	8.626767	76.84615	t9w9hdmwt1uv	8796befd c1b3813f c7ee905d 2940a6d7
22	Attingal--|	India	Town	8.695034	76.81788	t9w9knch4rf4	5c794e6c d3e85a6e 67e1f7d3 de3e115b
23	Maranallo	India	Town	8.471725	77.07249	t9wb1y4rrh58	681d44ef 7e73630a 94b67a5c 56ebc876
24	Ottasekha	India	Village	8.482012	77.1327	t9wb70n761sh	02719a43 eb422753 bdd613de cbf6da7c
25	Anad--Ker	India	Town	8.633505	77.01109	t9wc07tvb9bz	94877614 484d2adc 4f71148a df15e0b3

Fig. 27 Murmur3Hash values for corresponding Geohash value

as restaurant, university, hospital, shopping mall, garden, temple, church, museums. Figure 27 depicts calculation of Murmur3 hash values for corresponding Geohash value for the given dataset.

4.4   Algorithms

As explained in the earlier sections, integrating Cassandra with Hadoop increases the performance efficiency, and the Map reducer tasks can be performed by using various partitioning techniques. The default partitioning technique for Cassandra is Murmur3Partitioner. But when query response time criteria is considered, prefix-based partitioning technique gives faster response when compared to the Murmur3Partitioning algorithm as proved in the next section and is shown in Fig. 28. The reason behind this is prefix-based partitioning will partition the data based on prefix of Geohash value, which makes spatial search easier.

4.4.1 Using Murmur3 Partitioning Technique

(a) Algorithm for Geohashing Technique
   Procedure Geohashing (longitude, latitude)

1. Each time world is divided with a vertical line into two halves, the first half is given value 0 and the other half is given the value 1.
2. Each such divided region is further divided with a horizontal line into two parts. This time first half is assigned "0" and "1" is assigned to the other half.
3. This division process is repeated until a very long series of bits is obtained.
4. Finally this bitwise series is converted into the base-32 format.
5. The Geohash code of each region is assigned basing on whichever region the location is falling.

(b) Algorithm For Mapper

```
Procedure Map(  place_type,              /*input key*/
                Murmurhashvalue,              /*input value*/
                Place_type            /*output key*/
                Prefix_match        /*output value*/
                Geohash             /*geohash value for latitude longitude co-ordinates*/
)


{
  For each column in dataset
  Murmuhash = values[6];     /* Murmurhash value of each entry in datacenter */
  hash=values[5]; /* Geohash value of corresponding MurmurHash value*/
  if(Murmurhashvalue !=Murmurhash )
     For ( i=0; i< geohash.length; i++)
             If(geohash.charAt[i]==hash.charAt[i])
                   Count++;
          Context.write(place_name, count)
  }
```

Output of Mapper: Key-value pair of   place_name with its prefix Matched.

(c) Algorithm For Reducer

```
Reducer (    Place_Name              /* key */
             Count                    /*value*/
             Output Place_Name    /*output*/
          )
{
    For every column in dataset
              If(count>=5)          /*  assume precision for matching prefix be 5 */
                   Context.write(Place_name)
}
```
Output: Places nearby the given Place_Type.

The drawback of the above specified method is that Cassandra database will store data randomly according to its 64-bit token generated internally. As a result, the required queried data must be searched on all nodes for finding nearest neighbours. Hence, the efficiency of query response time may be reduced.

4.4.2 Using Prefix-Based Partitioning Method

(a) Partitioning method

1. Consider 32 data nodes, where each node contains values starting with each character in table.
   Node-0 : geohash values starting with '0'
   Node-1 : geohash values starting with '1'
   Node-2 : geohash values starting with '2'
   Node-3 : geohash values starting with '3'

             .   .
             .   .
             .   .

   Node-z : geohash values starting with 'z'
2. Consider a node array and assign node starting values to that particular array
   For example:
   Node[0]='0';
   Node[1]='1';
   Node[2]='2';
   ....
   Node[31]='z';

(b) Algorithm For Mapper

```
Procedure Map(   place_type,           /*input key*/
                          Geohash,             /*input value*/
                          dataNode             /* node [corresponding geohash
value letter ]*/
                          Place_type           /*output key*/
                          Prefix_match         /*output value*/
                     )
{
   For each row in node
Hashvalue= node[ i];
if( geohash !=hashvalue )
      For ( i=0; i< geohash.length; i++)
            If(geohash.charAt[i]==hashvalue.charAt[i])
                                    Count++;
                  Context.write(place_name, count)
   }
```
Output: Key-Value pair of   place_name with its prefix matched.

(c) Algorithm For Reducer

```
Procedure Reducer (   Place_name       /* key */
                          Count                /*value*/
                          Place_name       /*output*/
                     )
{
For every row in dataNode
            If(count>=5)          /* assume precision for matching prefix be 5 */
            Context.write(place_name)
}
```
Output: Prints corresponding place_name

4.5 *Performance Evaluation*

This section presents results of query processing both using Murmur and prefix-based partitioning methods. Sample queries for finding nearby places of specified locations are shown as follows:

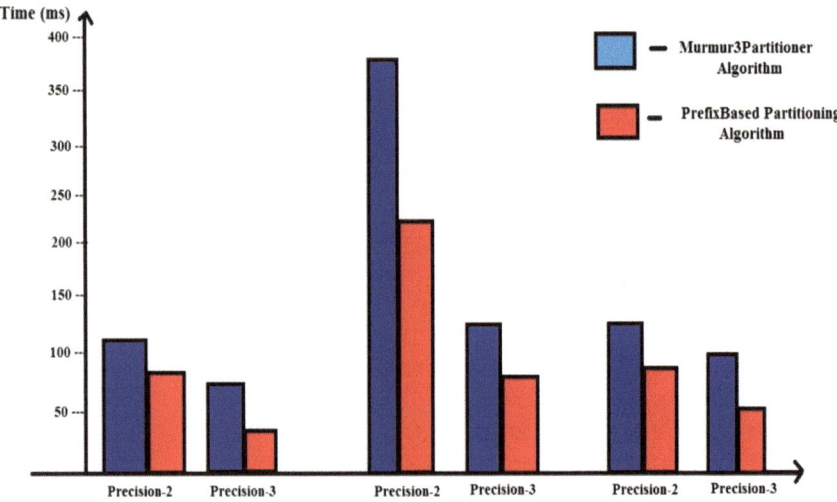

Fig. 28 Performance evaluation of Murmur3Partitioner and prefix-based partitioner

Query 1: Find places nearby Muttom Town in TamilNadu
Query 2: Find nearest places of Kandla
Query 3: Find all the places nearer to Jairampur in Arunachal Pradesh

For each query, we are going to give nearest locations based on the precision values. If the user wants only the nearest places then we will consider higher precisions as threshold values. But if user specifies to find nearer places then the threshold value can be reduced. Hence considering all the possibilities, we have calculated the query response time for both precisions 2 and 3 (according to our dataset values). As prefixes matching with precision-2 will be more, the time taken to retrieve data from database will be more compared to higher precision values. Here, we have plotted a graph as shown in Fig. 28 for more clear explanation considering precision values on *x*-axis and time (in milliseconds) taken for querying on *y*-axis.

From the Fig. 28, it is very clear that the performance of prefix-based partitioning technique is more comparative. The reason behind this is splitting of data into different data nodes, and performing MapReduce functionality reduces the number of searches as well as increases the efficiency which in turn reduces the query response time. On an average, Murmur3 partitioning took 238 milliseconds whereas prefix-based partition took 103.5 ms.

5 Conclusions and Future Work

Finally, Cassandra works as an effective storage system providing scalability, high performance and wide applicability. Cassandra can support a very high update throughput while delivering low latency. Murmur3Partitioner plays a major role in providing better performance than other databases because of its 32- or 128-bit hash values. But considering geospatial search use case another user-defined partitioning technique called prefix-based partitioning is giving better results comparatively. And from our work we can conclude that Cassandra can also support for spatial queries with user-defined functions. Hence when voluminous amount of data is to be processed and analysed, Cassandra may play a best role in deriving expected results as fast as possible within seconds. Even though Hadoop itself provides HDFS for storing data, we preferred another database Cassandra externally, reason being Cassandra is suitable for online applications whereas Hadoop is suitable for batch-oriented analytical tasks [11]. Integrating Hadoop and Cassandra made us to process spatial query data stored in Cassandra without having to move data into Hadoop environment. It took some time for us to create the dataset, because Hadoop can store any formatted data whereas Cassandra needs structured data. Geohash helped to distribute data and also enabled to improve query performance.

In future, we are planning to set up configurations for Hadoop as well as Cassandra at multi-node clusters which can produce best results for processing large datasets especially geographical data. We can also propose some other effective searching technique for finding nearest places as Geohash search fails in some edge cases like when same point belongs to different domains. Finding nearest locations can also be used as an application for various organizations like telecom industry to implant cell towers.

References

1. Aji, A., Wang, F., Vo, H., Lee, R., Liu, Q., Zhang, X., et al. (2013). Hadoop-GIS: A high performance spatial data warehousing system over MapReduce. *Proceedings of VLDB Endowment, 6*(11), 1009.
2. Benkirane, M., & Kettani, D. (2017). www.aui.ma/personal/ ~ D.Kettani/courses/gis/GDB-benkirane.ppt Last accessed April 12, 2017.
3. Berry, J. K. (1987). Fundamental operations in computer-assisted map analysis. *International Journal of GIS, 1,* 119–136.
4. Bobov, R. (2017). Spatial data visualization spatial data. https://portal.opengeospatial.org/files/?artifact_id=73214. Last accessed April 12, 2017.
5. Brahim, M. B., Drira, W., Filali, F., & Hamdi, N. (2016). Spatial data extension for Cassandra NoSQL database. *Journal of Big Data, 3,* 11.
6. DataStax Apache Cassandra Documentation. (2016). http://www.odbms.org/wp-content/uploads/2013/11/cassandra10.pdf. Last accessed October 20, 2016.
7. Dubey, N. K., & Agrawalan, S. (2015). Efficient approach to find nearest location using geohashing on Hadoop and Pig. *International Journal of Engineering Research-Online, 3*(3), 771–777.

8. Fox, A., Eichelberger, C., Hughes, J., & Lyon, S. (2013). *Spatio-temporal indexing in non-relational distributed databases*. Commonwealth Computer Research, Inc. IEEE.
9. Geohash and Its Format. http://geohash.org/site/tips.htmlLast. Accessed January 3, 2016.
10. Hadoop Support. (2017). https://docs.datastax.com/en/cassandra/2.1/cassandra/configuration/configHadoop.html. Last accessed March 11, 2017.
11. Hadoop vs. Cassandra. (2017). https://www.datastax.com/nosql-databases/nosql-cassandra-and-hadoop. Last accessed April 12, 2017.
12. Lakhshman, A., & Malik, P. (2010). Cassandra: A decentralized structured storage system. *ACM SIGOPS Operating System Review, 44*(2), 35–40.
13. Lee, D. T. (1982). On k-nearest neighbor Voronoi diagrams in the Plane. *IEEE Transactions Computers*.
14. Lee, K., Ganti, R. K., Srivatsa, M., & Liu, L. (2014). Efficient spatial query processing for big data. In *ACM SIGSPATIAL '14*, November 04–07, 2014.
15. Lenka, R. K., Barik, R. K., Gupta, N., Ali, S. M., Rath, A., & Dubey, H. (2016). Comparative analysis of SpatialHadoop and GeoSpark for geospatial big data analytics. Cornell University Library.
16. Liao, H., Han, J., & Fang, J. (2010). Multi-dimensional index on Hadoop distributed file system. In *Proceedings of IEEE Fifth International Conference on Networking, Architecture, and Storage* (pp. 240–249).
17. Liu, X., Han, J., Zhong, Y., Han, C., & He, X. (2009). Implementing WebGIS on Hadoop: A case study of improving small file I/O performance on HDFS. In *Proceedings of IEEE International Conference on Cluster Computing and Workshops* (pp. 1–8).
18. Moniruzzaman, A. B., & Hossain, S. A. (2013). Nosql database: New era of databases for big data analytics—Classification, characteristics and comparison. *International Journal of Database Theory and Application, 6*(4), 1–13.
19. Movable Type Scripts: Geohashes. http://www.movable-type.co.uk/scripts/geohash.html. Last accessed April 12, 2017.
20. Tang, M., Yu, Y., Aref, W. G., Mahmood, A. R., Malluhi, Q. M., & Ouzzani, M. (2016). *In-memory distributed spatial query processing and optimization*. Purdue Technical Report 2016.
21. What are Longitudes and Latitudes. https://www.timeanddate.com/geography/longitude-latitude.html. Last accessed April 11, 2017.
22. Zhang, S., Han, J., Liu, Z., Wang, K., & Feng, S. (2009). Spatial queries evaluation with MapReduce. In *Proceedings of GCC '09*.

Knowledge Computational Intelligence in Network Intrusion Detection Systems

Neeraj Kumar and Upendra Kumar

Abstract An Intrusion Detection System (IDS) is acting as first line of intrusion detection for providing network security in various areas like defence, e-commerce, autonomous systems etc. For network administrator, it plays an important role in understanding details about packet arriving and numerous activities involved within network. It helps the administrator in taking decisions at every stage of network life cycle. It can never access the IDS without knowing the performance of measure. Two most popular issues involved to exploit security are virus and attacker, generally termed as hacker. Generally, hacker is stimulated by adventure of importance. Hacker societies are strong in abstraction and their status is firm by their ability level. Gentle intruder gets through assets and slows the outcomes for genuine user like IDS and intrusion prevention systems (IPSs) are intended to help in frustrating hacker terrorization which can restrict remote logons to specific IP addresses and can utilize within a virtual private network technology (VPN). Many intrusion detection methodologies are proposed so far to resolve such issues but the main problem was performance of network and accuracy in detection of intrusion. To achieve this, there should be a strong mechanism for having true knowledge about the data which is flowing over the network. Proposed work done with study and analysis of various existing intrusion detection techniques and found lack in true prediction of intrusion. Considering these challenges, proposed work focused on handling both types of intrusion either anomaly based or signature based as complete hybrid model. Through knowledge extraction using soft computing and minimizing false alarm problem, proposed hybrid model found as true result-oriented intrusion detector.

Keywords NIDS · FLC · MLP · KDD99 · Alarm · Attack · ANFIS
Genetic · Expert system · Soft computing

N. Kumar (✉) · U. Kumar
Department of Computer Science and Engineering,
Birla Institute of Technology, Mesra, India
e-mail: javaneeraj@gmail.com

U. Kumar
e-mail: upendrakumarphdp@gmail.com

© Springer Nature Singapore Pte Ltd. 2018
S. Margret Anouncia and U. K. Wiil (eds.), *Knowledge Computing
and Its Applications*, https://doi.org/10.1007/978-981-10-6680-1_8

1   Introduction

It is important to know intelligent knowledge computing to increase detection rates and reduction in false alarm rates as an application of robust network intrusion detection. Aim of learning by this chapter is to have survey in speedy growth into Network Technologies system in I.T. age enormous flow of data from network every instant, so observably there should be a strong Network Intrusion Detection system (NIDS) [1] is an vital intrusion finding system that is used as a counter quantity to avoid data reliability and system accessibility from attack or a vigorous mechanism need to differentiate between relevant and non-relevant data particularly acting as an attack. Also an IDS which is fault tolerance, runs continuously, imposes a minimal overheads on system, and adapts to change in system and users. Thus to deliver total network security from intrusion, this chapter will contribute to recommend an Inventive Intrusion Detection model using computational intelligence to comprise strong Networks System. Therefore in this chapter, objective is to find out the optimize intrusion detection mechanism techniques [2], challenges as research gap and proposed a model to overcome performance issues. Intrusion Detection System should be treated as an expert system when any abnormal behavior is encountered and then auto update this activity as a new intrusion into training data-set for future reference [3–5].

Intrusion detection has remained one of the main zones of computer security whose impartial is to classify these mischievous actions in network circulation and highly defend the resources from the threat. Most of the IDS strain to achieve their job in actual period, but due to environments like degree of examination and computation it needs to suffer, the real-time performance is not always probable. IDS may be either host based or network based [6].

Intrusion detection method can be characterized into two types based on its examination methods that are as follows:

- **Anomaly based intrusion detection method:**

Anomaly detection in which deviance from usual performance show the attendance of purposefully or in voluntarily motivated attacks or faults. Anomaly detection methodologies are based on structure prototypes of normal data and detect deviances from the normal model in practical data. Anomaly detection algorithms have the advantage that they can detect new kinds of intrusions as deviances from standard practice. In this work, in order to model network transportation, each linking record is inspected and basic transportation features are extracted. After preprocessing, the objective of the intrusion detection algorithm becomes to sequence the system with normal data and model standard network transportation from the given set of standard data. Then, the task would be to define whether the test data belongs to "normal" or to an "abnormal" performance from a given new exam data. The proposed anomaly detection module is composed of three sub-modules [4, 6–8].

1. Preprocessing
2. Anomaly Analyzer and
3. Communication.

- **Misuse or signature-based intrusion detection method**:

In order to notice intrusions, the anomaly detection examines the deviance from the usual actions at user or system level but the abuse detection matches sample data to known insensitive designs. Machine education generally forms the base for anomaly method, and it can detect innovative attacks by comparing undecided ones with standard transportations but has high false alarm rate due to trouble of modeling practical standard behavior profiles for protected system. With misuse method, pattern matching on known signs front-runners to high accuracy for detecting threats, but it cannot detect innovative attacks as innovative attack signs are not available for pattern matching. Most of the current IDSs follow a sign based or mismanagement method which is similar to virus scanners, where events are detected after matching with exact predefined patterns known as signs. The restriction recognized with signature-based IDS is their fall to detect new attacks and also abandonments light differences of known patterns. Besides that, it is found to have important organisational overhead cost attached with it in order to maintain signature [6, 8, 9].

Although the research on fraud and its detection has been carried out for many years and telecommunication companies have spent relatively more money on this than the research community, still their efforts do not reach beyond the limit of the companies and have not been made accessible by the public research community. The various faces of digital transmission of wireless communication have been discussed in. Tumbling is the vulnerability of wireless communication to a wireless fraud, keeping in mind that the fraud could allow the fraudsters to access and steal telephone services and digital technology. This work has also mentioned about clone frauds and attempts to stop cloners. Increase in the incidents of phone frauds with corporation and telecommunication companies are reported in. It has talked about the alliance formed to stop phone frauds [6].

On prosperous system produces alarm.

- Intrusions are the actions that violate the security policy of a system.
- Intrusion Detection is the process used to identify intrusions.
- Intrusion detection is a network security mechanism to defend the computer network system from attacks.
- Protect a network or systems from malicious activity or strategy violations. A vital component for an effective defense-in-depth security strategy.

This chapter is framed as follows: In Sect. 2 Discussed various algorithms to deal with anomaly and signature based intrusion detection. Section 3 is dedicated to survey on different literature on IDs and their performance. At the end of this section discuss about the various challenges as research gap, and their expected solutions is in Sects. 5 and 6 as an conclusion and summary.

2 Network Intrusion Detection System

Intrusion detection is an imperative method in network security. Lots of intrusion detection approaches are present to resolution such issues, but the main difficult is presentation. It is imperative to increase the discovery rates and reduce false alarm rates in the area of intrusion detection. Therefore, in this research objective is to optimize intrusion detection mechanism using soft computing techniques is planned to incredulous performance issues. Intrusion Detection System (IDS) should be treated as an expert system when any abnormal behavior is encounters, and then auto update this activity as a new intrusion into training data-set for future reference.

Intrusion detection techniques can be categorized into three main types [6] such as statistical-based, knowledge based, and machine learning based. In the statistical-based case, the behavior of the system is signified from a accidental standpoint. Another work is knowledge-based NIDS techniques which try to capture the demanded behavior from existing system data like protocol stipulation, network congestion. Two main roles calculation, and then the comparison, of the performance of another intrusion detection approaches: these are the competence of the discovery procedure, and the charge concerned in the process. Without underestimating the reputation of the charge, at this point the competence portion must be emphasizing. Four states occur in this equivalent to the relative among the effect of the detection for an analyzed traffic ("normal" vs. "intrusion") and its real scenery ("abnormal" vs. "fictitious") [6]. These states are false positive (FP), if the analyze event is not guilty (or "ok") from the perception of security, but it is categorized as malicious; true positive (TP), if the analyzed event incorrectly classified as intrusion/malicious; false negative (FN), if the examined event is malicious but it is categorized [6].

2.1 Classification of Modeling Intrusion Detection Techniques

Arrangement of the intrusion detection techniques permitting to the handing out nature is involved in the behavioral model contemplation (Fig. 1).

Normal and true negative (TN), if the analyze occasion is suitably distinguish as usual/inoffensive. It is clear that low FP and FN rates, together with high TP and TN rates, will result in high-quality efficiency values. The fundamentals for trend analysis, knowledge and machine learning-based NIDS, along with the major sub-types of each, are described below:

i. **Statistical-Based Techniques**:
   In statistical-based techniques [10], the network transportation movement is taken and an outline representative its stochastic activities is formed. This outline is built on metrics such as the traffic rate, the number of packets for

Fig. 1 Classification of intrusion detection techniques

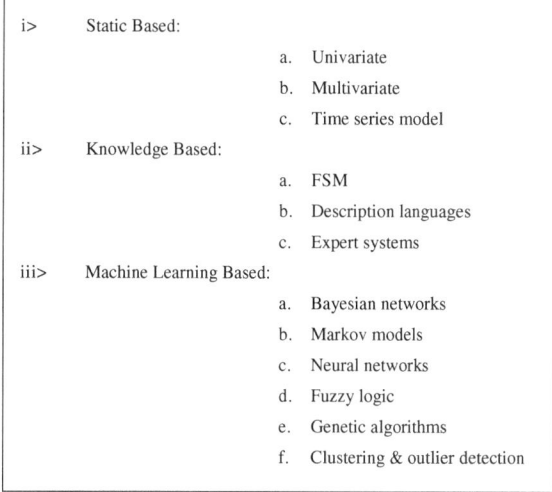

i>	Static Based:	
		a. Univariate
		b. Multivariate
		c. Time series model
ii>	Knowledge Based:	
		a. FSM
		b. Description languages
		c. Expert systems
iii>	Machine Learning Based:	
		a. Bayesian networks
		b. Markov models
		c. Neural networks
		d. Fuzzy logic
		e. Genetic algorithms
		f. Clustering & outlier detection

each protocol, the rate of links, the number of different IP addresses, etc. Two data-sets of network traffic are considered during the anomaly detection process: one parallel to the presently detected profile over time, and the other is for the before-trained arithmetic outline. As the network actions occur, the present profile is determined and an anomaly score is expected by judgment of the two behaviors [2, 11]. The gain generally happening of an anomaly when the score a guaranteed inception. However, some shortcoming also be keen out. First, this kind of NIDS is vulnerable to be trained by an attacker in such a way that the network transportation generated during the attack is reflected as normal. Second, setting the values of the different parameters/metrics are a hard task, particularly because the balance among false positives and false negatives is affected. Moreover, a statistical distribution per variable is assumed, but not completely performances can be molded by using stochastic methods. Likewise, maximum of these structures trust on the guess of a quasi-stationary process, which is not always realistic [12].

ii. **Knowledge-Based Technique**:
The so-called practiced system approach is one of the greatest widely used knowledge-based IDS schemes. However, like other A-NIDS methods, practiced systems can also be classified into other, dissimilar classifications [13]. Professional systems are proposed to catalog the review data affording to a set of rules, containing three steps. First, dissimilar elements and modules are recognized from the exercise records. Second, a set of cataloguing rules, factors, or processes are realized. Third, the check data is categorized consequently. More preventive/actual in some senses are requirement-based anomaly methods, for which the desired model is physically created by a human expert, in terms of a set of rules (the provisions) that seek to decide

reasonable system behavior. If the stipulations are complete enough, the model will be able to detect illegitimate behavioral patterns [13].

Additionally, the amount of false positives is bargain, mostly because this kind of system escapes the problem of anodyne activities, not formerly detected, being reported as intrusions. Stipulations could also be developed by using some kind of proper implement. For example, the finite state machine (FSM) methodology—a sequence of states and transitions among them—seems suitable for molding network protocols [14]. For this persistence, standard description languages such as N-grammars, UML, and LOTOs can be considered. The most substantial benefits of recent methodologies to anomaly detection are those of strength and elasticity.

Their main disadvantage is that the growth of high-quality knowledge is often difficult and time-consuming [14]. This problem, however, is common to other NIDS methods for which the notion of normalcy is obtained absolutely by examining training data.

iii. **Machine Learning-Based Technique**:
Machine learning usually discusses to the deviations in systems that implement tasks attendant with expert system and AI. Such tasks implicate appreciation, control, extrapolation, etc. The "change" may be whichever improvement to previously performance arts systems fusion of new systems. Machine learning techniques are established on forming a categorical or contained model that enables the patterns analyzed to be catalog. An unusual specific of these structures is the essential for characterized data to train the developmental model, a process that places severe strains on resources. Several machine learning-based schemes have been applied to NIDS [6]. Some of the most essential is quoted below:

- **Bayesian networks**
- **Clustering and outlier detection**
- **Markov models**
- **Neural networks**
- **Fuzzy logic techniques**
- **Genetic techniques**
- **Soft computing technique**

3 Literature Survey

Year	Ref. no.	Study description approach	Signature detection	Signature prevention	Anomaly detection	Anomaly prevention	Technique	Advantages	Remark
2002	[15]	Network based IDS	Yes	No	Yes	No	Signature based and anomaly based	Modeling or analysis of different attacks and their techniques	Not fully secured, still have huge risk of attack
2003	[16]	Host based IDS	Yes	Yes	No	No	Signature based	Automated response to malicious	Unable to detect and respond to anomaly behavior
2003	[17]	Network based IDS	Yes	Yes	No	No	Signature based and anomaly based	Centralized architecture	No mechanism of protection
2003	[18]	Host based	Yes	No	Yes	No	Signature based and anomaly based	Less false positive, efficient detection	No mechanism of protection
2005	[19]	Host based IDS	Yes	No	No	No	Signature based	Efficient and faster	Memory and implementation issues
2005	[20]	Host based IDS	Yes	No	No	No	Signature based	Flexibility of self-configuration	Large amount of memory and training staff is required
2005	[21]	NIDS self-organizing map and J.48	Yes	No	No	No	Misuse based	Good for signature based for limited attack	For anomaly Low false positive rate Time complexity is very high Limited attack
2006	[22]	Host based multi layered firewall	Yes	Yes	No	No	Signature based	Flexibility of self-configuration	Cannot detect anomaly behavior of intrusion
2007	[23]	Host based	Yes	Yes	Yes	No	Signature based and anomaly based	Reliable trusted and efficient	Memory and implementation issue
2007	[24]	Host based	Yes	No	No	No	Signature based	Cost effective, efficient	Unable to detect anomaly behavior
2007	[25]	Work on data mining approach, by determining the confidence ratio factor, able to predict about the intrusion	Yes	No	Yes	No	Average performance	Autotuning	Less accuracy in case Dynamic and real time
2008	[26]	Host based IDS	Yes	No	Yes	No	Signature based and anomaly based	Flexibility of customize, Cost effective	High rate of false positive, well trained analysts are required

(continued)

(continued)

Year	Ref. no.	Study description approach	Signature detection	Signature prevention	Anomaly detection	Anomaly prevention	Technique	Advantages	Remark
2008	[27]	Host based IDS	Yes	Yes	Yes	Yes	Signature based and anomaly based	Secured infrastructure	Well trained analysts are required
2008	[28]	Network based IDS	Yes	Yes	No	No	Signature based	Flexibility of self-configuration	Cannot detect anomaly behavior of intrusion
2008	[29]	Host based IDS	Yes	Yes	Yes	Yes	Signature based and anomaly based	Real-time response, reduce human effort	Security of mobile agent, needs to adopt some other techniques
2008	[30]	Network based IDS using fuzzy	Yes	Yes	Yes	No	Signature based and anomaly based	Autotuning adaptive	Less accuracy in case of anomaly No dynamic and real time
2009	[31]	Host based IDS	Yes	Yes	Yes	Yes	Signature based and anomaly based	Strong detection and protection mechanism	A large amount of memory are required High rate of false alarm
2010	[32]	Host based IDS	Yes	Yes	Yes	Yes	Signature based and anomaly based	Automatic response, reduce human effort	Cost ineffective, implementation, updating, monitoring issues
2011	[33]	Network based IDS using neural, SVM, PCA, GA	Yes	Yes	No	No	Signature based	Use of refinement of data using PCA, SVM and for optimization using genetic	But lacking in the detection of misuse, anomaly based detection
2013	[34]	NIDS cloud based	Yes	Yes	No	No	Fuzzy Theory policy based	Real-time response, in anomaly based	Lack of robustness, scalability, alarm rate in signature based
2013	[35]	NIDS FLC KDD99 Co operative	Yes	No	Yes	Yes	Anomaly based	False alarm rate: 1% Anomaly detection: 99%	It should be hydride
2014	[12]	Intelligent Intrusion Detection System in wireless sensor network	Yes	No	Yes	Yes	Signature based	Good detector	Problem of packet drop
2015	[36]		Yes	No	Yes	Yes	Anomaly based		Should be intelligent

(continued)

(continued)

Year	Ref. no.	Study description approach	Signature detection	Signature prevention	Anomaly detection	Anomaly prevention	Technique	Advantages	Remark
		Intrusion Detection System (IDS): anomaly detection using outlier detection approach						Good approach using data mining, efficient	
2015	[37]	IPS prevention system	Yes	No	Yes	Yes	Anomaly based	Real-time intrusion detection and prevention system	Hybrid
2016	[38]	Semi-supervised learning approach for Intrusion Detection System, apply semi-supervised-learning feed forward neural approach. Also develop a fuzzy MF maintain accuracy	Yes	Yes	Yes	Yes	Signature based and anomaly based	Reliable	Using J48 in this paper is again a time consuming and complex task, it increase the time complexity

Conclusion as Research Gap:

- High false alarm and limited by training data by anomaly detection method.
- Assurance of anomaly free dataset for train the network.
- Detection of digh rate of false positive attack alarm.
- Real-time anomalies detection.
- ID model consists of Anomaly and signature attacks both.
- Proactive intrusion detection technology.
- Proactive and hybrid intrusion detection model.

Problem and Challenges:

- **Problem**: Many IDS is proposed so far to resolve such issues, but the main problem is performance and accuracy.
- **Challenges**: To achieve this, there should be a strong mechanism for having true knowledge about the data which is flowing over the network. It is important to intelligent knowledge computing in increase detection rates and reduction in false alarm rates as an application of robust network intrusion detection.

4 Brainstorming on Methodology

In network intrusion detection research, one standard approach for discovery attacks is watching a network's action for anomalies: deviancies from rest of the details of regularity before well-read from benevolent traffic typically known using tools hired from the machine learning community. However, despite widespread academic research one finds a striking gap in terms of actual distributions of such systems: compared with other intrusion detection approaches, machine learning a computational intelligence is rarely engaged in operative "real world" settings. Evaluation in which learn measurable characteristics of different IDSs and Kdd Cup99 as a test bed. Generally, IDS principle is based on assumption that assuming intruder behavior differs from legitimate users, but overlap in behavior s causes problem like false positive and false negative [6].

In network intrusion detection research, single standard approach for discovery attacks is checking a network's movement for anomalies: deviances from rest of the details of normalcy beforehand learned from benign traffic classically identified using tools borrowed from the machine learning community. However, despite extensive academic research one finds a striking gap in terms of actual deployments of such systems: compared with other intrusion detection approaches, machine learning a computational intelligence is rarely employed in operative "real world" settings.

Methodology for achieving above gap

1. By preprocessing the data
2. Strong anomaly analyzer
3. By assurance of anomaly free dataset for train the network
4. Design a robust NIDS model
5. Design fuzzy logic controller for to detect False—Positive—Alarm
6. For accuracy in network-train—using BPN
   A complete solution of intrusion detection—using soft computing.

- **Proposed Intelligent and Hybrid NIDS Model**:
  This chapter found that an Intrusion Detection System thru the integration of multi computing systems and intelligence techniques such as fuzzy logic controller (FLC), multi-layer perception (MLP) and adaptive Neuro-Fuzzy inference system (ANFIS) Neural-Fuzzy-Genetic a complete soft computing approach. The Chapter introduces Network Intrusion Detection Systems (NIDS), which monitor the network traffic and detect any likely attacks.
- **Intrusion Detection System with the integration of multicomputing systems and intelligence techniques such as**: (a complete soft computing approach:)

  - Fuzzy logic controller (FLC),
  - Multi-layer perception (MLP) and adaptive neuro-fuzzy inference system (ANFIS)
  - Neural-fuzzy-genetic.

- **The system consists of three modules**:

  - **Collector**: The collector module works to gather and filter network traffics.
  - **Analyzer**: The analyzer agent uses fuzzy and expert system to classify the data.
  - **Predictor modules**: Predictor maker agent uses fuzzy logic controller (FLC) to make the final decision in inferring knowledge about improvement of true attack detection and false alarm detection accuracy up to 99% rate of 1%.

- **False Positive Rate (FPR)**:
  False positive rate refers to the percentage of normal data which is wrongly recognized as an attack and is defined as follows:

$$\text{Rate of False Positive} = \text{FP}/(\text{FP} + \text{TN})$$

- **False Negative Rate (FNR)**:
  False negative rate refers to the percentage of attack data which is wrongly recognized as normal and is defined as follows:

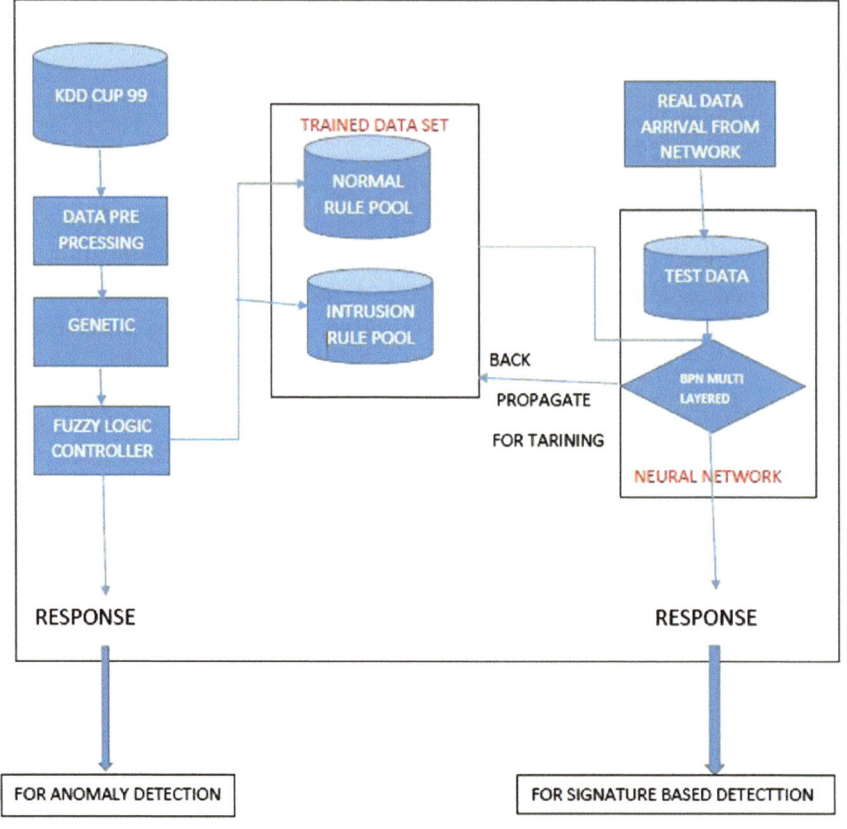

Fig. 2 Hybrid solution of both type of intrusion either misuse/signature based or anomaly based

$$\textbf{Rate of False Negative} = \textbf{FN}/(\textbf{FN} + \textbf{TP})$$

- **Proposed NIDS Model using Soft Computing**

The system consists of three modules: collector, analyzer, and predictor modules. The collector module works to gather and filter network traffics. The analyzer agent uses fuzzy and expert system to classify the data. Finally, the predictor maker agent uses fuzzy logic controller (FLC) and genetic optimization to make the final decision in inferring knowledge about improvement of true attack detection and false alarm detection accuracy up to 99% rate of 1%. The proposed system can be simulated using MATLAB and KDD Cup 1999 dataset as benchmark for authenticity of result (Fig. 2).

Algorithm: Classify KDD Dataset, Feature Extraction
Input: KDD dataset
Output: Dataset into two classes, i.e., rule pool (Normal and attack)

1. Pick KDD dataset
2. categorize dataset into "normal" and "attack" class
3. Change attributes to numeric value
4. Uncover maximum value for each attribute/feature
5. Choose important attribute/features
6. Store rules in rule pool

Genetic Approach:

Algorithm: Rule Pool Generation Using Genetic Algorithm
Input: set of preprocessed data, generations (G), and size of population.
Output: many more rules in the new rule pool.

1. Initialize population
2. N is size of population T, i.e., (minimum fitness value) = 0
3. Choose two chromosome (or rules) from domain of population
4. Apply crossover on chromosome
5. Apply mutation operator to the chromosome
6. If rule present in pool of rule then step7
   Else
       Determine fitness value of new chromosome
7. If fitness i ≥ T then
   Add newly generated chromosome to rule pool
8. Exhibit number of rules generated for the input generations
   Go to fuzzy rule extraction algorithm

5 Conclusion

Proposed hybrid model is capable to deal with not only the anomaly and signature based intrusion but also for new arrivals/activities of intrusion. To having true knowledge about data proposed feature extraction algorithm and to obtained pool of relevant and appropriate fuzzy rule, used Genetic algorithm, which builds more up proposed model even stronger. After rigorous analysis come of conclusion to enhanced performance with less false alarm, in intrusion detection approaches, achieve through soft computing approaches (like: Neural-based Intrusion–Detection System, Bayesian-Based Intrusion Detection System, Fuzzy Logic Based Intrusion Detection System, Rule-Based Intrusion Detection System, Immune Based Intrusion Detection System), as a test-bed for intrusion—like Kdd cup99, DARPA etc.

6 Summary

This chapter covers to reasonable extent the above-mentioned issues and draws some conclusion and the discussion is focused to develop Intrusion Detection System which is capable to deal both kinds of issues associated with hackers i.e., pattern-based behavior and violation of rules using intelligent expert system like soft computing, compare and contrast with various methodologies using real intrusion dataset and compute true knowledge about intrusion with enhanced accuracy in such a way to assist network administrator to better understand the IDS they use. Beauty of using soft computing is that it can not only able to trap anomaly and misuse kind of intrusion but also if new kind of intrusion is encountered then it can train or customize the network accordingly into the intrusion_data_set with new nomenclature for future use.

Therefore in this research, objective is to optimize intrusion detection method using soft computing techniques is projected to defeat performance issues. Intrusion Detection System (IDS) should be treated as an expert system when any abnormal behavior is encountered then auto update this activity as a new intrusion into training dataset for future reference.

Have true knowledge about the data is important factor and found using soft computing we can achieve true knowledge computing, which had major impact on intrusion detection.

References

1. Su, M. Y., Yu, G. J., & Lin, C. Y. (2009). A real-time network intrusion detection system for large-scale attacks based on an incremental mining approach. *Computers & Security, 28*(5), 301–309.
2. Azodi, A., Cheng, F., & Meinel, C. (2016, March). Towards better attack path visualizations based on deep normalization of host/network IDS alerts. In *IEEE 30th International Conference on Advanced Information Networking and Applications (AINA), 2016* (pp. 1064–1071). IEEE.
3. Hubballi, N., & Suryanarayanan, V. (2014). False alarm minimization techniques in signature-based intrusion detection systems: A survey. *Computer Communications, 49*, 1–17.
4. Folino, G., Pisani, F. S., & Sabatino, P. (2016, March). A distributed intrusion detection framework based on evolved specialized ensembles of classifiers. In *European Conference on the Applications of Evolutionary Computation* (pp. 315–331). Cham: Springer.

5. Reddy, R. R., Ramadevi, Y., & Sunitha, K. V. N. (2016). Data fusion approach for enhanced anomaly detection. In *Innovations in computer science and engineering* (pp. 275–285). Singapore: Springer.
6. Jabez, J., & Muthukumar, B. (2015). Intrusion detection system (IDS): Anomaly detection using outlier detection approach. *Procedia Computer Science, 48*, 338–346.
7. Ashfaq, R. A. R., Wang, X. Z., Huang, J. Z., Abbas, H., & He, Y. L. (2017). Fuzziness based semi-supervised learning approach for intrusion detection system. *Information Sciences, 378*, 484–497.
8. Jogdand, P., & Padiya, P. (2016, March). Survey of different IDS using honey token based techniques to mitigate cyber threats. In *International Conference on Electrical, Electronics, and Optimization Techniques (ICEEOT)* (pp. 802–807). IEEE.
9. Kenkre, P. S., Pai, A., & Colaco, L. (2015). Real time intrusion detection and prevention system. In *Proceedings of the 3rd International Conference on Frontiers of Intelligent Computing: Theory and Applications (FICTA) 2014* (pp. 405–411). Cham: Springer.
10. Kumar, V., Srivastava, J., & Lazarevic, A. (Eds.). (2006). *Managing cyber threats: Issues, approaches, and challenges* (Vol. 5). Berlin: Springer Science & Business Media.
11. Shah, B., & Trivedi, B. H. (2015, February). Improving performance of mobile agent based intrusion detection system. In *Fifth International Conference on Advanced Computing & Communication Technologies (ACCT), 2015* (pp. 425–430). IEEE.
12. Ahmad, I., Abdullah, A., Alghamdi, A., & Hussain, M. (2013). Optimized intrusion detection mechanism using soft computing techniques. *Telecommunication Systems*, 1–9.
13. Mehra, L., Gupta, M. K., & Gill, H. S. (2015, September). An effectual & secure approach for the detection and efficient searching of network intrusion detection system (NIDS). In *International Conference on Computer, Communication and Control (IC4), 2015* (pp. 1–5). IEEE.
14. Toumi, H., Talea, A., Marzak, B., Eddaoui, A., & Talea, M. (2015). Cooperative trust framework for cloud computing based on mobile agents. *International Journal of Communication Networks and Information Security, 7*(2), 106.
15. Wagner, D., & Soto, P. (2002, November). Mimicry attacks on host-based intrusion detection systems. In *Proceedings of the 9th ACM Conference on Computer and Communications Security* (pp. 255–264). ACM.
16. Kozushko, H. (2003). Intrusion detection. Host-based and network-based intrusion detection systems. *Independent Study*.
17. Bai, Y., & Kobayashi, H. (2003, March). Intrusion detection systems: technology and development. In *17th International Conference on Advanced Information Networking and Applications, 2003, AINA 2003* (pp. 710–715). IEEE.
18. Bai, Y., & Kobayashi, H. (2003, March). New string matching technology for network security. In *17th International Conference on Advanced Information Networking and Applications, 2003, AINA 2003* (pp. 198–201). IEEE.
19. Han, S. J., & Cho, S. B. (2003). Combining multiple host-based detectors using decision tree. *AI 2003: Advances in Artificial Intelligence*, 208–220.
20. Tan, L., & Sherwood, T. (2005, June). A high throughput string matching architecture for intrusion detection and prevention. In *ACM SIGARCH Computer Architecture News* (Vol. 33, No. 2, pp. 112–122). IEEE Computer Society.
21. Mrdović, S., & Zajko, E. (2005). *Secured intrusion detection system infrastructure*.
22. Depren, O., Topallar, M., Anarim, E., & Ciliz, M. K. (2005). An intelligent intrusion detection system (IDS) for anomaly and misuse detection in computer networks. *Expert Systems with Applications, 29*(4), 713–722.
23. Carlson, M., & Scharlott, A. (2006) Vol 2 http://citeseerx.ist.psu.edu/showciting?cid= 4669055.
24. Carlson, M., & Scharlott, A. (2006). Intrusion detection and prevention systems. In *CS536 Data Communication and Computer Networks Final Paper*.
25. Janakiraman, R., Waldvogel, M., & Zhang, Q. (2003, June). Indra: A peer-to-peer approach to network intrusion detection and prevention. In *Proceedings of Twelfth IEEE International*

Workshops on Enabling Technologies: Infrastructure for Collaborative Enterprises, 2003, WET ICE 2003 (pp. 226–231). IEEE.

26. Laureano, M., Maziero, C., & Jamhour, E. (2007). Protecting host-based intrusion detectors through virtual machines. *Computer Networks, 51*(5), 1275–1283.

27. Yu, Z., Tsai, J. J., & Weigert, T. (2007). An automatically tuning intrusion detection system. *IEEE Transactions on Systems, Man, and Cybernetics, Part B (Cybernetics), 37*(2), 373–384.

28. SANS (2008).

29. Sams.

30. Guimaraes, M., & Murray, M. (2008, September). Overview of intrusion detection and intrusion prevention. In *Proceedings of the 5th Annual Conference on Information Security Curriculum Development* (pp. 44–46). ACM.

31. Shibli, M. A., & Muftic, S. (2008). In *International Conference on Security & Cryptography*, IEEE.

32. Yu, Z., Tsai, J. J., & Weigert, T. (2008). An adaptive automatically tuning intrusion detection system. *ACM Transactions on Autonomous and Adaptive Systems (TAAS), 3*(3), 10.

33. Awodele, O., Idowu, S., Anjorin, O., & Joshua, V. J. (2009). A multi-layered approach to the design of intelligent intrusion detection and prevention system (IIDPS). *Issues in Informing Science & Information Technology, 6.*

34. Patel, A., Qassim, Q., & Wills, C. (2010). A survey of intrusion detection and prevention systems. *Information Management & Computer Security, 18*(4), 277–290.

35. Ahmad, I., Abdullah, A., Alghamdi, A., & Hussain, M. (2011). *Journal of Business Media, LLC* (27 July 2011, Springer).

36. Patel, A., Taghavi, M., Bakhtiyari, K., & JúNior, J. C. (2013). An intrusion detection and prevention system in cloud computing: A systematic review. *Journal of network and computer applications, 36*(1), 25–41.

37. Feizollah, A., Shamshirband, S., Anuar, N. B., Salleh, R., & Kiah, M. L. M. (2013, August). Anomaly detection using cooperative fuzzy logic controller. In *FIRA RoboWorld Congress* (pp. 220–231).

38. Sarda, A. R. (2014) In *Springer Proceedings of the 3rd International Conference on Frontiers of Intelligent Computing: Theory and Applications* (FICTA) 2014.

39. Garcia-Teodoro, P., Diaz-Verdejo, J., Maciá-Fernández, G., & Vázquez, E. (2009). Anomaly-based network intrusion detection: Techniques, systems and challenges. *Computers & Security, 28*(1), 18–28.

40. Malhotra, A., & Bajaj, K. (2016). A survey on various malware detection techniques on mobile platform. *International Journal of Computers and Applications, 139*(5), 15–20.

41. Das, N., & Sarkar, T. (2014). Survey on host and network based intrusion detection system. *International Journal of Advanced Networking and Applications, 6*(2), 2266.

42. Gautam, S. K., & Om, H. (2017). Comparative analysis of classification techniques in network based intrusion detection systems. In *Proceedings of the First International Conference on Intelligent Computing and Communication* (pp. 591–601). Singapore: Springer.

43. Ahmad, A., & Senga, B. P. S. (2017). Instruction detection system based on support vector machine using BAT algorithm. *International Journal of Computer Applications, 158*(8).

44. Modi, C., Patel, D., Borisaniya, B., Patel, H., Patel, A., & Rajarajan, M. (2013). A survey of intrusion detection techniques in cloud. *Journal of Network and Computer Applications, 36*(1), 42–57.

45. Denning, D., & Neumann, P. G. (1985). *Requirements and model for IDES-a real-time intrusion-detection expert system.* SRI International.

46. Estevez-Tapiador, J. M., Garcia-Teodoro, P., & Diaz-Verdejo, J. E. (2003, March). Stochastic protocol modeling for anomaly based network intrusion detection. In *Proceedings of First IEEE International Workshop on Information Assurance, 2003, IWIAS 2003.* (pp. 3–12).

Efficacy of Knowledge Mining and Machine Learning Techniques in Healthcare Industry

H. Shaila Koppad and S. Anupama Kumar

Abstract Knowledge mining is the process of discovering the knowledge from the larger database. As the size of the data is increasing enormously in the healthcare industry knowledge mining techniques are used to extract and mine the dataset to acquire new knowledge. Machine learning is a technique of training the system. In connection with artificial intelligence, statistics, and computer science, it is also known as statistical learning or predictive analytics. In recent years, application of machine learning and knowledge mining methods is been used everywhere in daily life. Healthcare system can cater prime diagnosis data of human healthcare details and reference to the doctors. Historical medical records afford other healthcare providers to access quickly and recognize the patients past and current health status. Chronic obstructive pulmonary disease is becoming one of the causes for leading deaths. An experiment is conducted to predict the presence and severity of the chronic obstructive pulmonary diseases (COPDs) using knowledge mining and machine learning techniques. Logistic and multinominal regression has been implemented to predict the prevalence of the disease using attributes from various sources and structures.

Keywords Chronic obstructive pulmonary disease (COPD) · Machine learning Logistic regression · Multinomial logistic regression · Regression analysis

H. Shaila Koppad · S. Anupama Kumar (✉)
Department of MCA, R V College of Engineering, Bengaluru, India
e-mail: anupamakumar@rvce.edu.in

H. Shaila Koppad
e-mail: shailahk@rvce.edu.in

© Springer Nature Singapore Pte Ltd. 2018
S. Margret Anouncia and U. K. Wiil (eds.), *Knowledge Computing and Its Applications*, https://doi.org/10.1007/978-981-10-6680-1_9

1 Introduction

The recent trends in advancement of technology have brought in a lot of knowledge to humankind. Huge amount of medical data is available in various forms across the country which can be efficiently mined and analyzed to bring out new knowledge which will be used for the betterment of mankind [1].

The healthcare sector or industry in history has generated huge amounts of data driven by record maintaining, patient care, compliance and regulatory necessities [2, 3]. Although most data are stored as document, the latest trend is toward swift digitization of these huge amounts of dataset [4]. Determined by required necessities and potential to progress the excellence of healthcare service by dropping the expenses, these immense quantities of data hold the assurance of supporting a wide range of medical and healthcare functions to originate previously untouched intelligence and intuition from data to report several new and essential questions. In the health sector, it affords stakeholders with new intuitions that have the potential to advance tailored care, improve patient conclusions, and avoid pointless expenses [5].

In recent years, technological improvements have brought in a lot of software systems which can be used to store and analyze the huge amount of data available in the medical repositories. Knowledge mining deals with extraction of interesting models from data in large databases. Knowledge mining with machine learning techniques can be used in healthcare sector for the benefit of patients and consumers, providers, and payers.

Knowledge mining and machine learning algorithms will be applied to the massive amount of data available with

- Regulatory bodies to understand and analyze the data pertaining to laws, coding, regulations, best practices, guidelines, performance, reporting, and costs.
- Pharmaceuticals, genetics, medical devices, diagnostic test decision support, drug–drug interaction, biomedical systems to analyze numerous related issues.
- Stakeholder to understand and analyze the relationship between software development, security and access, interoperability, information/data storage, "big data" to small data, usability, data transfer, educating/training/ communicating, reporting the various stakeholders.

The data available in the health repositories can be extracted and preprocessed using various data mining techniques. Data mining has initiated extensive applicability in the healthcare sectors, wherein it can be used in organizing optimum treatment methods, finding efficient cost structures of patient care, and predicting disease risk factors.

Research via data mining models has been applied to categorize the numerous parameters that cause the disease and predict the prevalence/severity of the disease. The different data mining algorithm/techniques like classification, clustering, and association mining can be applied to develop models in healthcare research to predict the occurrence of the disease, classify the symptoms, and find interesting relations that lead to the morality due to a disease and detect fraud in the health insurance sectors, etc.

The machine learning algorithms can be used to recognize complex patterns within rich and massive data to make intelligent data-driven decisions. Learning algorithms can be used to predict the prevalence/severity of the disease. The various machine learning algorithms which can be used to analyze the data are decision tree-based algorithms, rule-based algorithms, Naïve Bayes classification techniques, SVM classifiers, etc. Machine learning techniques are used less by the researchers which would yield a more accurate and efficient result than the statistical techniques. It helps the doctors to predict the prevalence of the disease with better accuracy, to visualize the output in a better way, and help the decision makers to prevent the disease.

The clinic-epidemiological data can be logically structured to help the healthcare centers [6]. It will help them to classify the data to diagnose the disease in patient, prone to get the disease, prevent the disease at an initial stage, and to find severity of the disease.

In order to classify and predict the occurrence of the disease, it becomes important to collect and structure the attributes in a logical manner. The various attributes that would be included in the disease study should be made available in the clinical datasets. All the different factors associated with disease should be brought in together using the technological advancement and analyzed in an efficient manner. The severity of the diseases can be predicted based on the interaction with the data directly using different data mining tools and analyzed using different machine learning algorithms. The relationship and patterns between the attributes can be visualized such that the presence of the disease can be predicted accurately.

This chapter discusses the application of different machine learning algorithms in diagnosing patient disease and predicts the severity of the disease. The accuracy and efficiency of the machine learning techniques in predicting the disease are also discussed.

2 Problem Definition

According to the report of the World Health Organization, by the year 2030, COPD will kill more people compared to other diseases in the world [7]. Chronic obstructive pulmonary disease, or COPD, aligns to a set of diseases that cause breathing-related problems and airflow blockage which is present among the people of all age groups. Being an important disease to be addressed by the nation, it has been an interest to the clinico-epidemiological and epidemiological agents have been addressed using wide range of concerns from etiology to health services utilization [8]. Chronic obstructive pulmonary disease (COPD) (which was 11% during 2012) prevalent in India is rated the second reason causing morality and is estimated to the major cause of death during 2030. Epidemiology of COPD in India is the scientific study of the frequency/pattern distribution and causes or risk factors of health-associated states and trials using the recent technological advancements, and the output of the study would be to predict the prevalence of the disease in India and prevent it for the betterment of the people [9]. Conventionally, it was assumed tobacco smoking like cigarettes is the main risk factor to get COPD. In India, tobacco smoking is in diverse forms like hookahs, bidis, and chillums which becomes a risk factor for COPD [7, 9, 10]. COPD leftovers wrongly or poorly diagnosed by healthcare experts to recognize patients at great risk to be victim of COPD fortunately awareness of the patient's healthcare clinical history assure to be relaxed for patients. Knowledge mining is the method of investigating huge datasets to pull or extract concealed and unfamiliar patterns, knowledge and relationships that are problematic to classify with outdated statistical methods.

In the existing system, it is very difficult to diagnose COPD disease in the patients due to lack of clinical history and different risk factors [11]. The patient's illness data and visit to different hospitals are not centralized. If patient visits new doctor, the previous information is not shared with the doctor. Due to these reasons, there are higher possibilities to misinterpret the patient data and give a wrong diagnosis. This experiment uses the clinical data to analyze the symptoms of the patient and predict the prevalence of the disease [12].

3 Data Collection

Healthcare data can be accumulated by various methods such as by observing, interviewing, using existing information, and managing written questionnaire. For this experiment, the data available in various forms like flat files collected from the symptoms of the patients are used. The data in the chest x-rays are preprocessed according to the need and stored in the file. The data are preprocessed and converted into variables and constants as the predictors of the disease. These variables and constants are used in defining various values to recognize the study of COPD disease in healthcare domain. A variable is something that can take on dissimilar values, and it can also be described as characteristic of an object, phenomenon, or person which takes on various values. For example, weight, height, and age are variables because there are different weights, heights, and ages. A constant is anything which cannot take on various values or vary its existing value.

3.1 Types of Variables

The variable types are been categorized by depending upon the value it is defining. Below-mentioned instances are the various types of variables:

1. Categorical variables are variables that can take on precise values only within a defined array of values. For instance, "gender is considered as categorical variable". Your age should be between (in the middle of) births to death: you can't be mortal; there is no middle ground when it comes to gender; you can either be female or male; you must be one, and you cannot be both.

   a. Ordinal variables: Ordinal variables are associated with particular sequence in ordering or ranking in them by ascending or descending sequence as present in Tables 1 and 2.
   b. Nominal variables: Nominal variable is mutually exclusive means they do not contain ordering and ranking in them.

2. Numerical variables: These variables are articulated in numbers, e.g., quantity of cigarette smoked by a person per day. The variable "cigarette" can take dissimilar values for smoker to chain smoker. Other instance of variables is as follows: clinic-home distance, height of person, body mass index, weight of person, monthly income, and so on. Numerical variables can either be discrete or continuous.

   a. Continuous: Continuous types of variables have fractions. One can develop more precise values depending on the instrument considered. For instance: Number of Packs Smoked by a person in a year (6.5 packs or 35.7 packs), height are in centimeters (2.5 cm or 2.546 cm or 2.543216 cm), FEV1% (58.3 or 68.34).

Table 1 Ordinal variables and its values

Variable	Value
Stages of COPD	Stage I Stage II Stage III Stage IV
Severity of COPD	Mild Moderate Severe Very severe
Night walking disability	No disability Partial disability Total disability

Table 2 Nominal variables and its values

Variable	Value
Sex	Male Female
Smoking	Smoker Non-smoker
Patient type	In-patient Out-patient

b. Discrete: Discrete variables have only integer's values without decimal or fraction values, e.g., many times the patient has visited to a clinic (0, 1, 2, 3, etc.), pulse rate of a person (70, 86, 94, etc.).

3. Background Variables: In most of the studies, background variables, such as educational level, age, socioeconomic status, sex, religion, and marital status, are considered as variables. Background variables are repeatedly associated to a many of independent variables; ultimately, it will impact the problem. In background variables, measures are only considered which are vital to the study. These background variables are notorious "confounders".

4 Experimental Setup

Knowledge mining is carried out using various tasks. After preprocessing, the machine learning algorithm should be applied over the data. An experimental setup is made using R. To work with R, RStudio should be installed. The software can be downloaded from the https://www.rstudio.com. Once RStudio is installed, open the application and create a New RScript file as shown in Fig. 1.

In this experiment, the dataset is stored in comma separated values (csv) file. Load the COPD dataset. To read the file, use the function read.csv ("path") by providing the specific path and result should be stored in a variable.

When working with real-time dataset the data might contain missing values, NA values, or corrupted values. This kind of dataset has to be cleaned and preprocessed for the analysis. Now let us consider in the COPD dataset, where "age" is considered as one of the attributes. This "age" attribute can hold different numeric

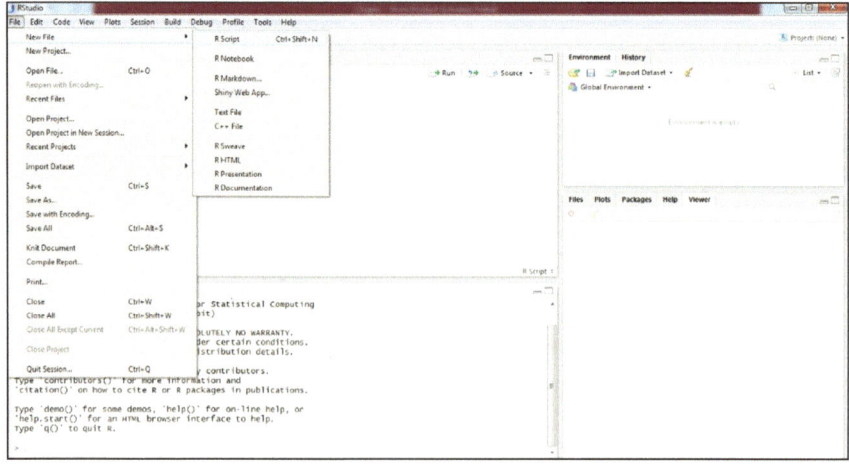

Fig. 1 Screen of R

values. For example, the age of the person is entered ":seven" in words as shown in Fig. 2 which is considered as a corrupted value.

The corrupted value has to be corrected for the further processing. These kinds of values will be replaced by the mean value of that attribute. But the mean value cannot be calculated since it contains string "seven". Now replace "seven" with some negative value, so that the data is not affected as shown in Fig. 3, and calculate the mean value as shown in Fig. 4.

Replace the negative value with the mean as shown in Fig. 5, and if any missing or NA values are present, they are also been processed in the same manner. But age cannot be in decimal value, so we have to convert it again to integer values only. Similarly, take care of remaining corrupted and missing values.

Now cleaned and preprocessed COPD dataset is ready for the analysis using machine learning algorithms.

[1] 11	52	65	59	13	63	70	40	15	33	65	38	62	10	25	32
[17] 62	36	11	33	13	seven	64	3	65	59	15	622	37	17	35	15
[33] 50	40	35	35	36	57	16	68	54	77	72	58	20	17	62	20
[49] 52	9	50	8	35	48	25	11	74	12	44	53	37	24	69	7

Fig. 2 Age attribute values of the COPD dataset

```
> Mydata$age
 [1] 11  52  65  59  13  63  70  40  15  33  65  38  62  10  25  32  62  36  11  33  13  -11 64  3   65
[26] 59  15  622 37  17  35  15  50  40  35  35  36  57  16  68  54  77  72  58  20  17  62  20  52  9
[51] 50  8   35  48  25  11  74  12  44  53  37  24  69  7   17  8   11  17  36  21  60  18  14  52  45
47 Levels: 10 11 12 13 14 15 16 17 18 20 21 24 25 3 32 33 35 36 37 38 40 44 45 48 50 52 53 54 57 ... -11
```

Fig. 3 Age attribute corrupted value replaced with the negative value of the COPD dataset

```
> agemean=mean(as.numeric(as.character(Mydata$age)),rm.na=T)
> agemean
[1] 44.04
```

Fig. 4 Calculation of mean value of age attribute of the COPD dataset

```
> Mydata$age<-as.numeric(as.character(Mydata$age))
> Mydata$age[which(Mydata$age==-11)]=agemean
> Mydata$age
 [1]  11.00  52.00  65.00  65.00  59.00  13.00  63.00  70.00  40.00  15.00  33.00  65.00  38.00  62.00  10.00
[15]  25.00  32.00  62.00  36.00  11.00  33.00  13.00  44.04  64.00   3.00  65.00  59.00  15.00 622.00
[29]  37.00  17.00  35.00  15.00  50.00  40.00  35.00  35.00  36.00  57.00  16.00  68.00  54.00  77.00
[43]  72.00  58.00  20.00  17.00  62.00  20.00  52.00   9.00  50.00   8.00  35.00  48.00  25.00  11.00
[57]  74.00  12.00  44.00  53.00  37.00  24.00  69.00   7.00  17.00   8.00  11.00  17.00  36.00  21.00
[71]  60.00  18.00  14.00  52.00  45.00
```

Fig. 5 Replacing mean value with the corrupted value of age attribute

5 Regression Analysis

Regression analysis is study of predictive modeling technique which examines the relationship between independent variable and dependent variable. Regression analysis method is used for forecasting, time series modeling, and finding the fundamental influence of the relationship between the variables. There are different types of regression models, but in this research work, we are using logistic regression to predict the diagnosis of the disease and multinomial logistic regression for predicting the severity of the disease in the patient.

5.1 Logistic Regression Model

Regression curve can be fit for the model y = f(x) when y is a categorical variable. Typical reason to use this model is to predict *y*, given a set of predictors *x*. Predictors *x* can also be categorical, continuous, and may also be a mix of categorical and continuous.

The categorical variable y, has dissimilar value. In the diagnosis of COPD, y is considered as patients where, f(x) can be either the value 0 or 1 stating the patient suffering from the disease or not. A diagnosis of COPD disease among the patient can be studied using machine learning classification: assumed a set of attributes for respectively COPD patient such as number of FEV1%, BMI, different symptoms values, chest X-ray report values, the algorithm should decide whether the patient is diagnosed with COPD (1) or not (0). We can also predict a dependent variable on logistic regression which can assume one or more values.

Function glm() in R language helps to be relaxed to fit logistic regression method or model. We will be using logistic regression machine learning technique for prediction of COPD disease in the patients on the clean and preprocessed COPD dataset.

Since the dataset contains lots of columns, choose only the relevant columns and creating subset from the original dataset using the subset () function. For example, COPD dataset may contain height of a patient which is not necessary to be considered as predictor for the disease since it cannot be the cause or symptoms for the disease. Now the data has to be analyzed by splitting the data into two sets called as training and testing set. The training group or set will be used to fit our model which we will be testing over the testing group or set. To fit the model using glm() function we must be certain to specify the parameter family = binomial in the function.

From the logistic regression model, we are able to analyze and interpret the fitting value. These values not statistically significant variables. In this research, the null deviance shows how the response variable or dependent variable is predicted by the model only by the intercept given values. If we don't consider any of the explanatory variables, the null deviance is 405.63 units on 318 degrees of freedom.

The residual deviance is the result got after including the explanatory variables. After adding the explanatory variable, the deviance has reduced to 58.69 on 311 degree of freedom. The deviance is reduced to 349.94 units with the loose of 7 degrees of freedom. From the result, we analyzed that BMI is not statistically significant and FEV is statistically significant since it has the lowest p value proposing a robust connotation of FEV related to patient with the probability of having COPD. In the logit model, the response variable is log odds: $\ln(odds) = \ln(p/(1-p)) = a*x1 + b*x2 + \cdots + z*xn$. When the FEV of patient increases by 1 unit the log of odds of COPD = 1 rises by 0.05 units. When Dyspnea grade rises by 1 unit the log of odd COPD = 1 will fall by 1.44 units. We predict y on a group of subjective data by setting the proper parameter type = "response". The likelihoods are measured in the form of $P(y = 1|X)$, and the decision boundary is considered as 0.5, i.e., $P(y = 1|X) > 0.5$, if $y = 1$ which can be examined that the chance of patient likely to COPD disease is more and when $y = 0$ which can be examined the chance of patient likely to disease is less.

The accuracy obtained on COPD dataset from this research work is 0.9481796 as shown in Fig. 6, which is obtained by considering the classification error of 0.04388715. The classification error can be calculated as shown in Fig. 7.

From Fig. 8, the following observation is made. The receiver operating characteristic (ROC) and area under curve (AUC) are the characteristic performance measurement for a binary classifier [6]. AUC value obtained from this research work is 0.948 since it is nearer to 1 we can consider the model as a best predictive ability model. The ROC is a curve created by plotting the true positive rate (TPR) against the false positive rate (FPR) at numerous thresholds [6].

From the above experimental results, the presence of the disease could be easily predicted [13]. But the severity of the disease could not be predicted using logistic regression since it is a two-value predictor. To overcome this problem, the experiment is further continued using multinominal regression. The following section explains the working of the multinominal regression over the dataset.

```
> auc <- performance(pr, measure = "auc")
> auc <- auc@y.values[[1]]
> auc
[1] 0.9481796
```

Fig. 6 Calculation of the accuracy

```
> misClasiticError <- mean(results != Mydata$COPD)
> misClasificError
[1] 0.04388715
```

Fig. 7 Calculation of the classification error

Fig. 8 ROC plot

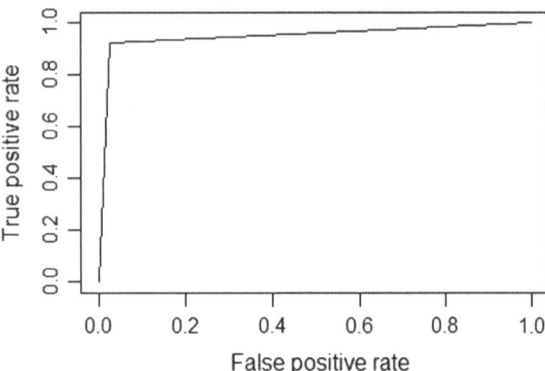

5.2 Multinomial Logistic Regression

Multinomial logistic regression (MLR) is a method of linear regression analysis led whenever dependent variable is nominal with more than two levels. Most of the times, it will be used to define the data and describe the relationship among more than one continuous-level (interval or ratio scale) independent variables and one dependent nominal variable. You are able to comprehend nominal variable as a variable where there will be no intrinsic ordering.

In this experiment, we will be using MLR on COPD dataset to predict the severity of COPD disease since diagnosis of the disease among the patient is not sufficient. It is also necessary to find the severity of the disease where the outcome variable is the severity type, and the predictor variables are chest X-ray status which has three-level categorical variable and Body Mass Index (BMI) a continuous variable.

To work with MLR in R, install foreign, nnet, ggplot2, reshape2 packages and implement them using library() function or require() function. The multinom() function from nnet package is used to estimate the model. Model execution output Fig. 9 shows the iteration history with the final negative log-likelihood 805.563840. To get the residual deviance value, multiply the log likelihood by two and displayed in the model summary the value is 1611.128.

The summary output of the model is calculated by the summary() function, which contains coefficient block and standard errors block. According to COPD dataset, we get three rows in each block corresponding to the model equation. In the set of coefficients, we understand that the first row of above table is being associated to COPD = "Mild" to our reference point COPD = "Very Severe" and the

Fig. 9 Result of model execution

```
initial  value 1089.627368
iter  10 value 821.519565
iter  20 value 805.651390
final  value 805.563840
```

```
> z <- summary(copd_test)$coefficients/summary(copd_test)$standard.errors
> z
         (Intercept) CHEST.X.RAYHYPERINFLATION CHEST.X.RAYNormal        BMI
Mild        -5.428354                10.9560041         49.94741 -1.1975941
Moderate    -7.631004                13.1773114         61.56454 -0.4157493
Sever        1.610495                 0.1978997         19.15044 -1.0128875
```

Fig. 10 Result of Z-score

second row to COPD = "Moderate" to our reference point COPD = "Very Severe", and the third row to COPD = "Severe" to our reference point COPD = "Very Severe". A one-unit increase in body mass index declines the log odds of being in Mild COPD versus Very Severe COPD by 0.1169, Moderate COPD versus Very Severe COPD by 0.0355, Severe COPD versus Very Severe COPD by 0.0797. The log odds of being in Mild COPD versus Very Sever increased by 10.7297, if moving from Chest X-ray = "CARDIOMEGALY" Chest X-ray = "HYPERINFLATION". Calculate Z-score as shown in Fig. 10 and p value for the variables in the model.

To see the risk ratios exponent the coefficients of the model as shown in Fig. 11.

A one-unit increase in BMI, the relative risk ratio is 0.8895 for being in Mild COPD versus Very Severe COPD, relative risk ratio is 0.9650 for being in Moderate COPD versus Very Severe COPD and risk ratio is 0.9233 for being in Severe COPD versus Very Severe COPD.

The fitted() function can be used to calculate the predicted probabilities as shown in Fig. 12 for each outcome level.

The prediction values can be obtained using the predict() function where the parameter type = "probs" as shown in Fig. 13.

```
> exp(coef(copd_test))
         (Intercept) CHEST.X.RAYHYPERINFLATION CHEST.X.RAYNormal       BMI
Mild    5.564545e-04              4.569532e+04      5451729034.7 0.8895893
Moderate 1.106868e-04             1.123366e+05      8193424853.3 0.9650806
Sever   2.150264e+01              1.057454e+00         111562.1 0.9233120
```

Fig. 11 Calculation of exponent coefficient

```
    Very Sever          Mild       Moderate        Sever
1 1.008768e-01  1.200559e-01  4.948085e-01  0.2842588
2 9.702535e-07  2.138884e-01  3.967493e-01  0.3893614
3 8.556550e-02  1.436394e-01  4.659376e-01  0.3048576
4 1.035265e-06  2.046915e-01  4.095686e-01  0.3857388
5 2.591559e-01  7.481604e-06  1.167534e-05  0.7408250
```

Fig. 12 Result of fitted()

```
> predict(copd_test, newdata = dses, "probs")
  Very Sever          Mild      Moderate        Sever
1 2.396417e-01 8.061325e-06 1.130959e-05 0.7603389
2 1.082208e-06 1.984674e-01 4.184665e-01 0.3830650
3 8.933483e-02 1.373209e-01 4.736163e-01 0.2997280
```

Fig. 13 Using predict() to calculate prediction values

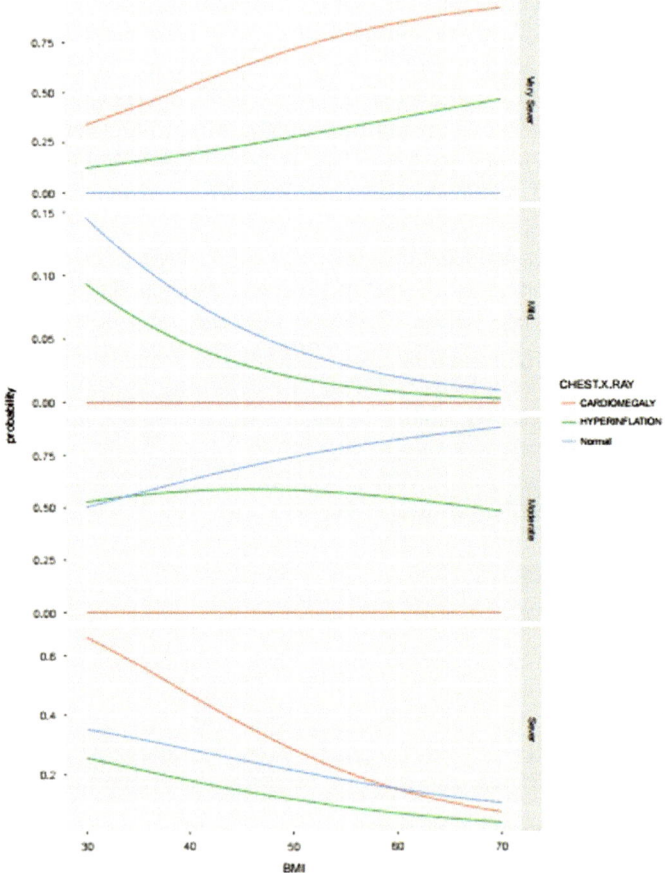

Fig. 14 Predicted probabilities plot

The predicted probabilities are plotted using ggplot() function to explore the distribution of dependent variable vs independent variables, and the chest X-ray types are differentiated by plotting them in different colors as shown in Fig. 14.

The multinominal regression is found very efficient in finding the probability of the severity of the disease. The probability of the severity of the disease from one band to another is also discussed.

6 Conclusion

Knowledge mining methods were used to extract and create COPD dataset from the larger databases. The created COPD dataset was cleaned and preprocessed. Machine learning algorithms were used to train the system over COPD dataset. These machine learning algorithms helped to predict the presence and severity of COPD in the patient. The experiment helped to diagnose COPD disease among the patient using the logistic regression and severity of the disease in the patient using multinomial logistic regression machine learning algorithms. The efficiency of the logistic regression is analyzed using ROC and AUC and found the curves to be closer to 1. The multinomial algorithm was found very effective in depicting the probabilities of the severity of the disease.

Acknowledgements We would like to express our special thanks of gratitude and deep regards to Dr. Yunus Sheriff pursuing DNB Pulmonology for his classic guidance, appreciated feedback and constant encouragement in understanding COPD and analyze the complications. His valuable suggestions were of immense help to us in getting this work done.

References

1. Unstructured Data: A Big Deal in Big Data. (n.d.). Retrieved from http://www.digitalreasoning.com.
2. Sun, J., & Reddy, C. K. (2013). Big Data analytics for healthcare. In *SIAM International Conference on Data Mining*.
3. Liu, W., & Park, E. K. (2014). Big Data as an e-health service. *International Conference on Computing, Networking and Communications, 3*(6), 982–988.
4. Feldman, B., Martin, E. M., & Skotnes, T. (2012). Big Data in healthcare hype and hope. Dr. Bonnie.
5. Mukherji, A. (n.d.). Indian healthcare system: Challenges and opportunities. Retrieved from http://tejas.iimb.ac.in/interviews/41.php.
6. Kumar, R., & Indrayan, A. (2011). Receiver operating characteristic (ROC) curve for medical researchers. *Indian Pediatrics, 48*, 277–287.
7. The Global Burden of Disease. (n.d.). Retrieved December 22, 2011, from www.who.int/healthinfo/global_burden_disease/projections/en/index.html.
8. Salvi, S. (2011). In S. K. Jindal (Ed.), *COPD: The neglected epidemic: Pulmonary and Critical Care Med* (pp. 971–974). Jaypee.
9. Adeloye, D., Basquill, C., Papana, A., Chan, K., Rudan, I., & Campbell, H. (2015). An estimate of the prevalence of COPD in Africa. A systematic analysis. *COPD: Journal of Chronic Obstructive Pulmonary Disease, 12*(1), 71–81.

10. Liang, Y., & Kelemen, A. (2016). Big Data science and its applications in health and medical research: Challenges and opportunities. *Journal of Biometrics & Biostatistics, 7,* 307. https://doi.org/10.4172/2155-6180.1000307.

11. Peek, N., Holmes, J. H., & Sun, J. (2014). Technical challenges for Big Data in biomedicine and health: Data sources, infrastructure, and analytics. *Yearbook of Medical Informatics, 9*(1), 42–47. http://doi.org/10.15265/IY-2014-0018.

12. Global Strategy for the Diagnosis. (2014). *Management and prevention of chronic obstructive pulmonary disease*, Global Initiative for Chronic Obstructive Lung Disease.

13. Koppad, S., & Kumar, A. S. (2016). Application of Big Data analytics in healthcare system to predict COPD. In *International Conference on Circuit, Power and Computing Technologies [ICCPCT] IEEE*, ISBN-16 978-1-5090-1276-3.

14. Futrel, K. (2013, October). Structured data: Essential for healthcare analytics & interoperability. MT(ASCP).

15. Raghupathi. (2014). *Health Information Science and Systems, 2*(3). Retrieved from http://www.hissjournal.com/content/2/1/3.

Cryptanalysis of Protocol for Enhanced Threshold Proxy Signature Scheme Based on Elliptic Curve Cryptography for Known Signers

Raman Kumar

Abstract The proxy signature is the elucidation to the entrustment of signing capabilities in any secure electronic milieu. Numerous schemes are prophesied, but they are chattels of information security. In this, I anticipate an enhanced secure threshold proxy signature scheme based on elliptic curve cryptography. I compare the performance of scheme(s) with the performance of a scheme has been anticipated by the writer of this article formerly. I investigate enhanced threshold proxy signature scheme for diverse parameters like entropy, floating frequencies/intuitive synthesis, ASCII histogram, autocorrelation, histogram analysis and vitany. Consequently, the enhanced threshold proxy signature scheme based on elliptic curve cryptography is safe and effective against infamous conspiracy attack(s).

Keywords Proxy signature · Unforgeability · Secret sharing · Time constraint Elliptic curve cryptography · Non-repudiation and threshold scheme for known signers

1 Introduction

Nowadays Internet is close part of our life. The data going transversely via Internet may be unsafe. There are a lot of examples for its illustrations. When I discuss ATM PIN, SSN, or some other secluded information,it becomes a complete diverse story. So, security engineering plays vital role for this.

Nowadays commercial milieu, creating a framework for the validation between notions and arenas, is quite difficult. Elliptic curve cryptography is one of the most powerful but slightest understood types of cryptography. An increasing number of Web sites evolve extensive usage of elliptic curve cryptography to protect all as of customer's HTTPS acquaintances to know how they pass data among data centers.

R. Kumar (✉)
Department of Computer Science and Engineering, I. K. Gujral
Punjab Technical University, Kapurthala, Punjab, India
e-mail: er.ramankumar@aol.in; dr.ramankumar@ptu.ac.in

© Springer Nature Singapore Pte Ltd. 2018
S. Margret Anouncia and U. K. Wiil (eds.), *Knowledge Computing and Its Applications*, https://doi.org/10.1007/978-981-10-6680-1_10

Fig. 1 Elliptic curve
cryptography for enhanced
threshold proxy signature
scheme

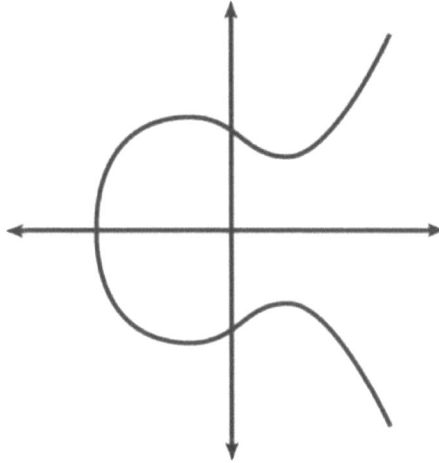

Principally, it is vital for end users to comprehend the technology behind any security system to trust it. Figure 1 displays elliptic curve cryptography for enhanced threshold proxy signature scheme.

RSA is the most prominent algorithm used in public key cryptography technique for encryption and decryption and digital signatures. The key difference among RSA and ECC is that, ECC offer the equal level of security for minor key sizes. It is highly scientific in nature. The authors Koblitz and Miller who practical implemented discrete logarithm on elliptic curves (ECDLP) over finite fields by which elliptic curve cryptography plays a vital role in public key cryptography [1].

1.1 Elliptic Curve

The elliptic curve is similar to normal curve drawn on X- and Y-axes. The aforementioned curve may have points. A piece can be nominated by an (x, y) coordinate just like other graphs.

For example points $(4, -9)$ which means 4 units on right hand side of x-axis from center and 9 below y-axis from center.

Consider elliptic curve (E) with points (P) now produce random no (d) as:

$$Q = d \times P$$

Now E, P, Q are public values and challenge is to find d.

Elliptic Curve Cryptography

Message M is sent to x, y at P_m.

Table 1 NIST suggested key sizes

Bits of security	Symmetric key algorithm	Equivalent hash function	Equivalent RSA key size	Equivalent ECC key size
80	Triple DES (2 keys)	SHA-1	1024	160
112	Triple DES (3 keys)	SHA-224	2048	224
128	AES-128	SHA-256	3072	256
192	AES-192	SHA-384	7680	384
256	AES-256	SHA-512	15,360	512

I cannot encode message as (x, y) not all coordinates $E(a, b)$ curve on point G (Table 1).

A's private key generates public key

$$P_A = n_A \times G$$

$$P_B = n_B \times G$$

Encryption

$$C_m = \{KG, P_M + KP_B\}$$

Decryption

$$P_m = \{KP_B - n_B(KG)\}$$

The design and analysis of security is an attractive concern for the end user. Furthermost cryptographic connoisseurs acclaim that present systems bid at least 128 bits of security. Usually RSA executions with 1024 or 2048 bit keys provides less security than AES-128. Figure 2 displays Radar chart shows key sizes suggested by NIST.

This chart offerings alternative way to RSA and ECC (Fig. 3).

Key lengths commonly rise over time as the calculation accessible to assailants [2].

Mambo et al. (1996) presented notion of a proxy signature.

In this, I investigate threshold proxy signature schemes. The performance evaluation for the same has been discussed below.

NIST Recommended Key Sizes

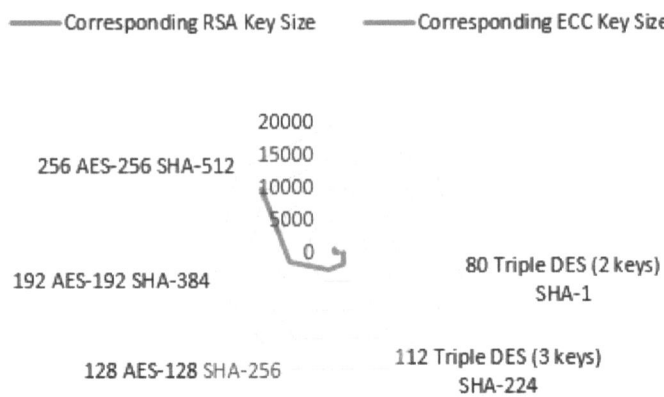

Fig. 2 Radar chart shows key sizes suggested by NIST

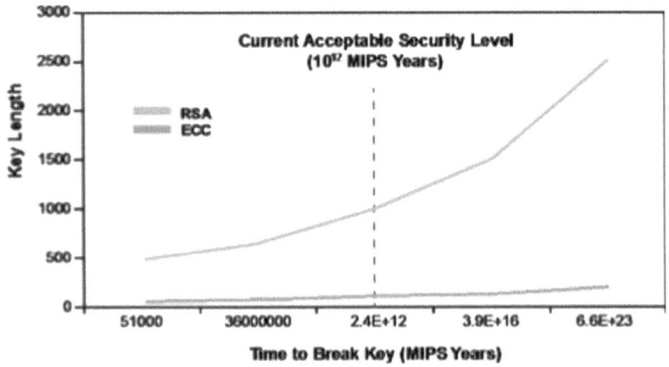

Fig. 3 Key sizes shadowed by Elliptic curve cryptography for enhanced threshold proxy signature scheme

2 Review of Threshold Proxy Signature Schemes

2.1 History of Threshold Proxy Signature Schemes

In the antiquity of proxy signature scientific expansion, the authors calculated threshold proxy signature which works with $(1, n)$, (t, n), and (n, n) threshold delegations.

The history of threshold proxy signature schemes is thru up in Table 2.

Denmedt and Frankel (1991) modified the ElGamal public key cryptosystem and RSA and Lagrange Coefficient to yield security.

Table 2 History of threshold proxy signature schemes

Sr. no.	Scheme	Method
1	Shamir and Blakley (1979)	Lagrange interpolating polynomial and linear projective geometry
2	Elgamal [3]	Discrete logarithms
3	Miller [4]	Uses of elliptic curves in cryptography
4	Koblitz [1]	Elliptic curve cryptosystems
5	Desmedt and Frankel [5]	RSA and Lagrange coefficient
6	Zhang [6]	Discrete logarithms
7	Kim et al. [7]	Discrete logarithms
8	Sun [8]	Discrete logarithms
9	Lee (2001)	Discrete logarithms
10	Hwang et al. [9]	RSA and Lagrange coefficient
11	Wang et al. [10]	RSA and Lagrange coefficient
12	Kuo and Chen [11]	RSA and Lagrange coefficient
13	Jiang (2007)	RSA and Lagrange coefficient
14	Fanyu (2007)	RSA and Lagrange coefficient
15	Li et al. [12]	RSA and Lagrange coefficient
16	Geng et al. [13]	RSA and Lagrange coefficient
17	Kumar et al. (2007–2016)	RSA and Lagrange coefficient
18	Kumar et al. (2017)	ECC and Lagrange coefficient

To make proxy signature be pertinent to group oriented states, Zhang [6] and Kim et al. [7] planned (t, n) threshold proxy signature in 1997.

The perception of the proxy signature scheme has planned in 1996 that obtains countless courtesy.

Regrettably, the Zhang's scheme revealed apprehensive by Lee et al. [14].

Kim-Park-Won scheme determined that scheme is not having the property of non-repudiation.

The antiquity of each proxy signature scheme is thru up in Table 3. Founded on these types, I try to assimilate all these slants in a meta-proxy signature scheme.

- In Zhang's schemes, the parameter α, β are nominated by proxy signer.
- $h()$ means that cryptographic one-way hash function.
- In Kumar et al.'s scheme, $T = c_0 + \sigma x e' = f''(0) + f(0)xe'$ can be calculated by smearing the Lagrange formula.

Table 3 The history of each proxy signature scheme

Proxy signing key	Verification key	References
$x_o + kK$	$y_o K^K$	Mambo et al. [15, 16]
$x_o + kK + x_p y_p$	$y_o K^K y_p{}^{yp}$	Mambo et al. [15, 16]
$x_o h(W, K) + k$	$y_o{}^{h(W,K)} + K$	Kim et al. [7]
$x_o h(W, K) + k + x_p h(W, K)$	$(y_o y_p)^{h(W, K)} K$	Kim et al. [7]
$x_o r + (k + \alpha)$	$y_o{}^r r$	Zhang [6]
$x_o r + (k\beta)$	$y_o{}^r r$	Zhang [6]
$x_o h(r, \text{ProxyID}) + k + x_p k$	$y_o{}^{h(r,\text{ProxyID})} r$	Lee et al. [14]
$x_o + kh(W, K) + x_p y_p$	$y_o K^{h(W,K)} y_p{}^{yp}$	Sun and Hsieh (1999)
$x_o h(W, K, y_p) + k + x_p$	$y_o{}^{h(W,K,yp)} K y_p$	Sun and Hsieh (1999)
$x_o h(W, K, y_o, y_p) + k + x_p K$	$y_o{}^{h(W,K,yo,yp)} K y_p{}^K$	Yen et al. (2000)
$x_o h(W, K, y_p) + k + x_p h(W, K, y_o)$	$y_o{}^{h(W,K,yp)} K y_p{}^{h(W,K,yp)}$	Sun (2000)
$y = g^{c0}, C_1 = g^{c1}, C_2 = g^{c2}, \ldots, C_t$ $_{-1} = g^{ct-1} \pmod{p}$	$y' = g^T x((y_0)^{h(mw,K)} K)^{-ei)} \bmod$ p and $e' = h(y', m)$	Kumar et al. (2007–2016)
$d_p = d_p + s_o \pmod{t}$	$\sigma = \text{Sign}_{d_p}(m)$	Kumar et al. (2017)

2.2 Elliptic Curve Discrete Log Problem (ECDLP)

The strong point of the Elliptic Curve Cryptography deceits in the Elliptic Curve Discrete Log Problem (ECDLP). The announcement of ECDLP is as follows.

Say E be an elliptic curve and $P \in E$ be a point of order n. Specified a point $Q \in E$ by

$$Q = d \times P, \quad \text{for a certain } m \in \{2, 3, \ldots, d - 2\}$$

Treasure the d for which the above equation holds.

2.3 Security Analysis of Kim et al. and Related Schemes

The Kim et al.'s [7] revealed insecure by Sun et al. [17] update attack. Kim et al. They mainly emphasized on threshold delegation.

Sun et al. [17] reviewed Kim et al.'s [7] and also identified the concept of random numbers.

Inopportunely, the Zhang's scheme [6] has also exposed to be self-doubting by Lee, Hwang and Wang.

Hwang et al. [9] protracted Desmedt and Frankel's scheme.

Sun et al. [17], in 1999, also worked on both the Kim-Park-Won and Mambo-Usuda-Okamoto schemes.

Hwang et al. [9] have publicized that Sun's scheme has a safety measures flaw.

Lee et al. [18] have projected ($t_1/n_1-t_2/n_2$) proxy signature scheme based on factorization of the square root modulo of a composite number. They have little work on message warrant.

Tzeng et al. [19] have discussed the concept of batch verification scheme.

Hwang et al. [20] have anticipated a multi-proxy multi-signature scheme permits investigation on each proxy reuirements.

Lu et al. [21] have planned the concept of authentication server with the time stamp.

Yang et al. [22] likened with Hsu et al.'s scheme, where the secret shares premeditated is not required.

Tzeng et al. [23] have wished-for a scheme with ($t3$, $n3$) shared verification. They planned the security based on one-way hash function. They also discussed discrete logarithm problem.

Tzeng et al. [24] have accessible security analysis of Hwang-Lin-Lu scheme. Hwang-Lin-Lu scheme is susceptible to forge attack.

Hwang et al. [25] have generalized version of the ($t_1/n_1-t_2/n_2$) proxy signature scheme based on elliptic curve discrete logarithm problem only.

In 2001, Hsu et al. incorporated a non-repudiable threshold proxy signature scheme. Tsai et al. [26] looked into the weakness of the Hsu-Wu-Wu scheme. They were unable to found the malicious proxy signer.

Li et al. [27] deliberated three kinds of proxy signature schemes: the ($t/n-1$), the ($1-t/n$), and ($1-1$). The real signer cannot entrust the warrant.

Hwang et al. [28] have deliberated the Hwang and Shi's scheme without using one-way hash function. Here they created proxy signer with secure channels.

Hwang et al. [29] have offered a cryptanalysis of Sun's threshold proxy signature scheme. It has been observed that the secret key can be compromised by collusion attack.

All analysis signposted that the scheme fails to placate all the necessities except a few. So, it must placate altogether the following rudimentary requirements [9, 11, 13, 12]:

1. Secrecy
2. Proxy Protected
3. Unforgeability
4. Non-repudiation
5. Time Constraint
6. Known Signers
7. (t_n, n_n) threshold verifying.

3 Enhanced Elliptic Curve Authentication Encryption/ Decryption Scheme

The concept of elliptic curve authentication encryption/decryption scheme is as follows:

3.1 Elliptic Curve Authentication Encryption Scheme (ECAES) Encryption

A's province constraints $D = (q, FR, a, b, G, n, h)$ and public key Q. B has the province constraints D. B's public key is Q_B, and private key is d_B. The Elliptic Curve Authentication Encryption Scheme (ECAES) deviced is as follows.

A accomplishes the following steps in subsequent phases as listed below:

Step 1: Chooses a random integer r in $[1, n - 1]$
Step 2: Calculates $R = rG$
Step 3: Calculates $K = hrQ_B = (K_x, K_y)$, checks that $K \neq O$
Step 4: Calculates keys $k_1 \| k_2 = KDF(K_x)$ where KDF is a key derivation function
Step 5: Calculates $c = ENC_{k1}(m)$ where m is the message to be directed and ENC stands for a symmetric encryption algorithm
Step 6: Calculates $t = MAC_{k2}(c)$ where MAC is message authentication code
Step 7: Demonstrations (R, c, t) to B.

3.2 Elliptic Curve Authentication Encryption Scheme (ECAES) Decryption

To decrypt a cipher text, B accomplishes the subsequent phases as listed below:

Step 1: Accomplish a partial key rationale on R
Step 2: Calculates $K_B = h.d_B.R = (K_x, K_y)$, check $K \neq O$
Step 3: Calculates $k_1, k_2 = KDF(K_x)$
Step 4: Confirms that $t = MAC_{k2}(c)$
Step 5: Calculates $m = ENC_{K_1^{-1}}(c)$.

I can realize that $K = K_B$, since $K = h.r.Q_B = h.r.d_B.G = h.d_B.r.G = h.d_B. R = K_B$.

4 Security Analysis and Performance Evaluation of the Proposed Scheme

The analysis of the wished-for supposition for enhanced threshold proxy signature scheme is given below.

4.1 Entropy

In this case, the worth of entropy is the measure of the inclination of a process, to be entropically *chosen*, or to evolution in a direction. Additionally, entropy delivers a suggestion for a precise encryption technique. I have observed supposition based on entropy produced.

Figure 4 displays entropy for enhanced threshold proxy signature scheme. Figure 5 displays compression ratio required for each scheme. Table 4 lists the name and compression ratio required in each scheme.

4.2 Floating Frequencies/Intuitive Synthesis

It is accomplished by three-part entirety which revenues full benefit of the time complexity, space complexity, and communication overhead provided through the digital standard. Figure 6 displays floating frequencies/intuitive synthesis for enhanced threshold proxy signature scheme.

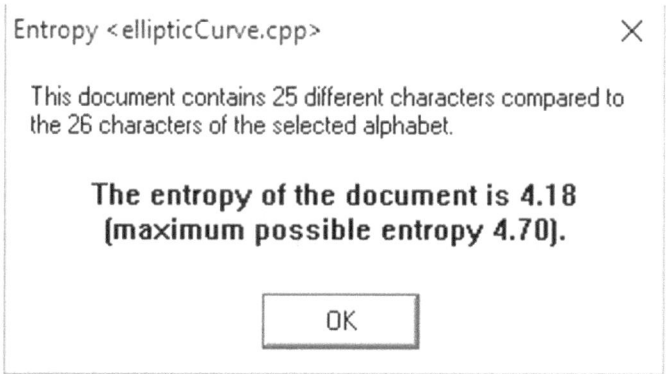

Fig. 4 Entropy for enhanced threshold proxy signature scheme

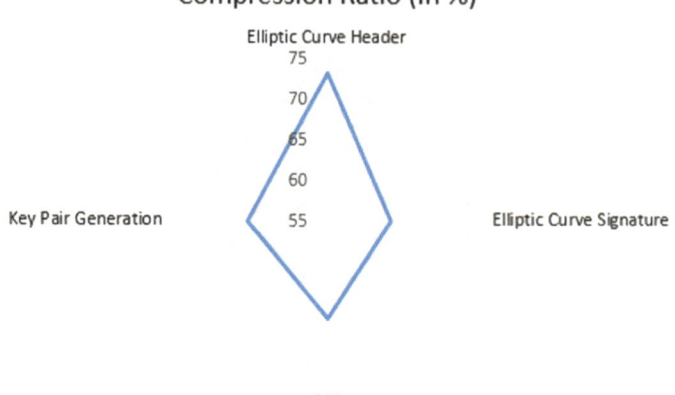

Fig. 5 Radar chart showing compression ratio required in each schemes

Table 4 Compression ratio (in %) for threshold proxy signature schemes

Threshold proxy signature scheme	Compression ratio (in %)
Elliptic curve header	73
Elliptic curve signature	63
ECC	67
Key pair generation	65

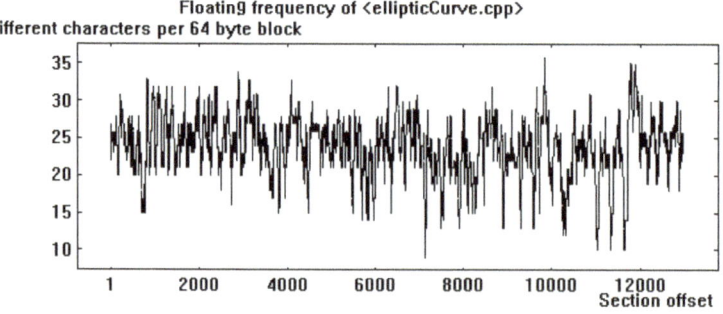

Fig. 6 Floating frequencies/intuitive synthesis for enhanced threshold proxy signature scheme

4.3 ASCII Histogram

It is demonstrated only for enormously in debugging code concerning probability calculations. Figure 7 displays ASCII histogram for enhanced threshold proxy signature.

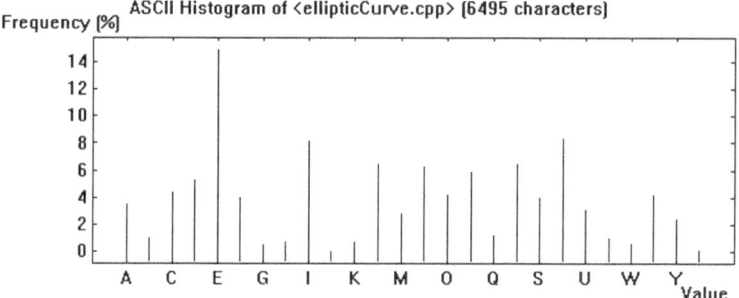

Fig. 7 ASCII histogram for enhanced threshold proxy signature scheme

4.4 Autocorrelation

A scientific depiction of resemblance between a given time series and a lag version of succeeding time intervals. It is another example of correlation. It is also known as serial correlation. Figure 8 displays autocorrelation for enhanced threshold proxy signature scheme

4.5 Histogram Analysis

A histogram is a graphical illustration that displays a diagrammatical impression of the distribution of data. I have observed histogram of for threshold proxy signature scheme.

Detailed View

The comprehensive view of the histogram analysis of all schemes can be signified as follows:

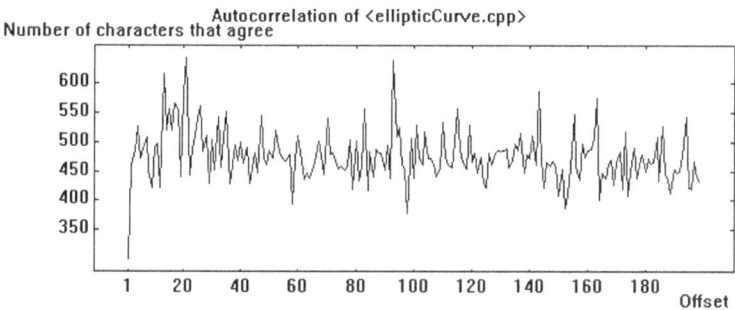

Fig. 8 Autocorrelation for enhanced threshold proxy signature scheme

Experiment:

Histogram Analysis of ⟨ECC.cpp⟩. File size 4890 bytes. Descending sorted on frequency.

Figure 9 displays Radar chart showing Histogram Analysis for ECC threshold proxy signature scheme. Table 5 lists the Histogram Analysis for ECC threshold proxy signature scheme.

Experiment:

Histogram Analysis of ⟨attacksECC⟩. File size 5321 bytes.

Figure 10 displays Radar chart showing attacks for ECC-enhanced proxy signature scheme. Table 6 lists attacks for ECC-enhanced proxy signature scheme.

Experiment:

Histogram Analysis of ⟨manual⟩. File size 95,154 bytes.

Figure 11 displays Radar chart showing ECC analysis for ECC-enhanced proxy signature scheme. Table 7 lists the ECC analysis threshold proxy signature scheme.

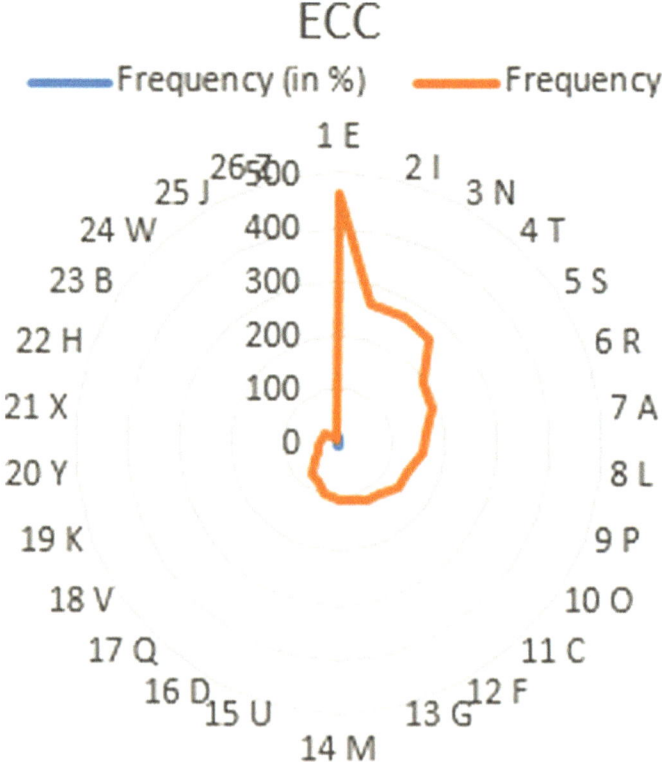

Fig. 9 Radar chart showing histogram analysis for ECC threshold proxy signature scheme

Table 5 Histogram analysis for ECC threshold proxy signature scheme

No.	Substring	Frequency (in %)	Frequency
1	E	14.5136	464
2	I	8.3203	266
3	N	8.2577	264
4	T	8.0075	256
5	S	6.1933	198
6	R	5.9431	190
7	A	5.2236	167
8	L	5.036	161
9	P	4.3791	140
10	O	4.3478	139
11	C	3.7848	121
12	F	3.6597	117
13	G	3.3469	107
14	M	3.253	104
15	U	3.0654	98
16	D	2.5336	81
17	Q	2.4711	79
18	V	1.564	50
19	K	1.2512	40
20	Y	1.1261	36
21	X	1.0322	33
22	H	0.9384	30
23	B	0.9071	29
24	W	0.5005	16
25	J	0.1877	6
26	Z	0.1564	5

Figure 12 displays Radar chart for Annexure.

Figure 13 displays Vitany diagram for ECC-enhanced proxy signature scheme.

Figure 14 displays Vitany diagram for attacks on ECC-enhanced proxy signature scheme.

Figure 15 displays Vitany diagram for analysis on ECC-enhanced proxy signature scheme.

Figure 16 displays scenario for encryption/decryption of ECC-enhanced proxy signature scheme.

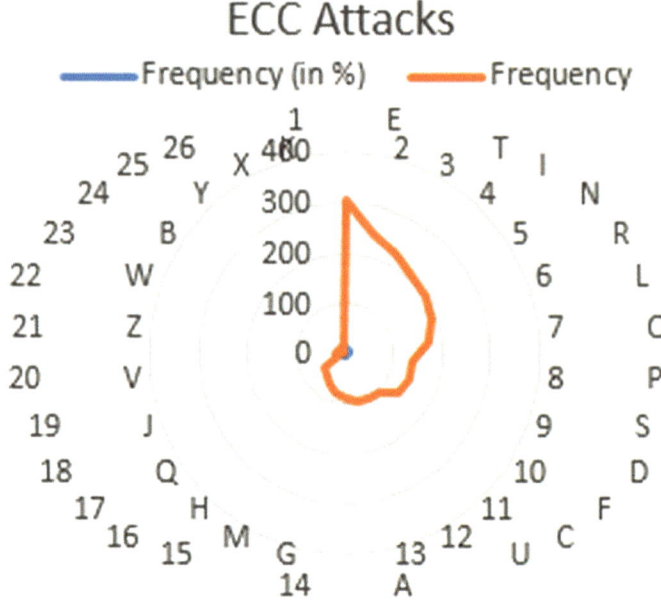

Fig. 10 Radar chart showing attacks for ECC-enhanced proxy signature scheme

Table 6 Histogram analysis showing attack for ECC threshold proxy signature scheme

No.	Substring	Frequency (in %)	Frequency
1	E	11.1765	304
2	T	8.8971	242
3	I	8.3088	226
4	N	7.6103	207
5	R	7.5	204
6	L	6.9118	188
7	O	6.4338	175
8	P	5.1838	141
9	S	5.1103	139
10	D	5	136
11	F	3.8235	104
12	C	3.6029	98
13	U	3.6029	98
14	A	3.3824	92
15	G	2.9779	81
16	M	2.6103	71

(continued)

Table 6 (continued)

No.	Substring	Frequency (in %)	Frequency
17	H	2.1691	59
18	Q	1.875	51
19	J	0.6618	18
20	V	0.6618	18
21	Z	0.6618	18
22	W	0.5515	15
23	B	0.4779	13
24	Y	0.4779	13
25	X	0.2941	8
26	K	0.0368	1

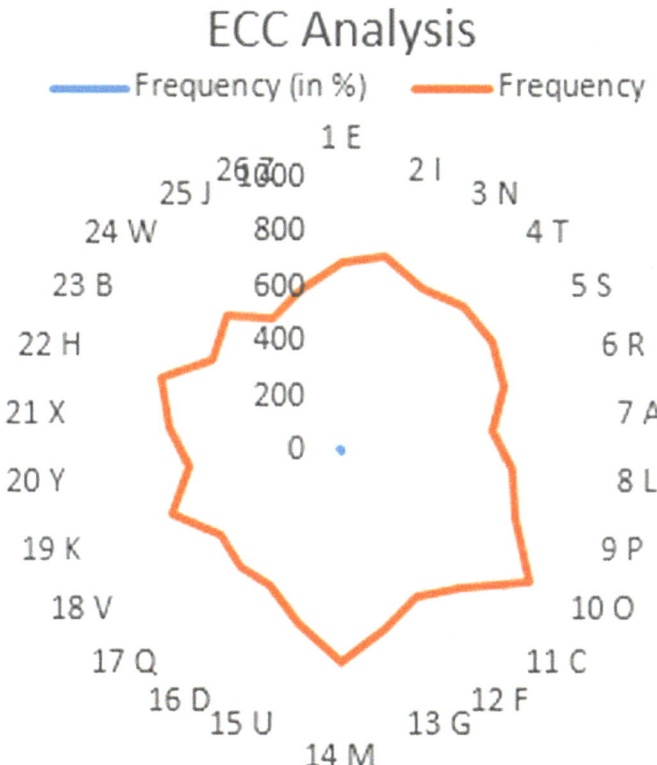

Fig. 11 Radar chart showing ECC analysis for enhanced proxy signature scheme

Table 7 ECC analysis threshold proxy signature scheme

No.	Substring	Frequency (in %)	Frequency
1	E	4.0379	682
2	I	4.257	719
3	N	3.884	656
4	T	4.1208	696
5	S	4.1089	694
6	R	3.8366	648
7	A	3.3925	573
8	L	3.8011	642
9	P	4.0971	692
10	O	5.0503	853
11	C	3.9905	674
12	F	3.5998	608
13	G	4.0024	676
14	M	4.5944	776
15	U	3.9017	659
16	D	3.3037	558
17	Q	3.3629	568
18	V	3.2564	550
19	K	4.0024	676
20	Y	3.3511	566
21	X	3.7951	641
22	H	4.2688	721
23	B	3.4162	577
24	W	3.8129	644
25	J	3.209	542
26	Z	3.5465	599

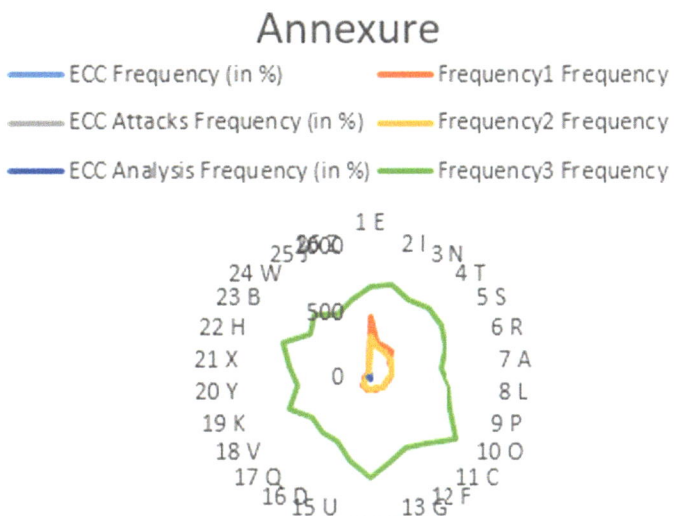

Fig. 12 Radar chart for Annexure

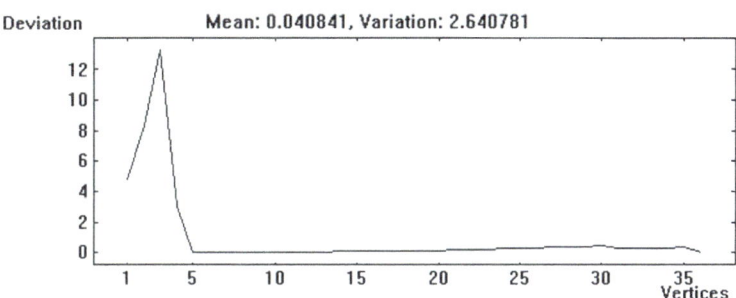

Fig. 13 Vitany diagram for ECC-enhanced proxy signature scheme

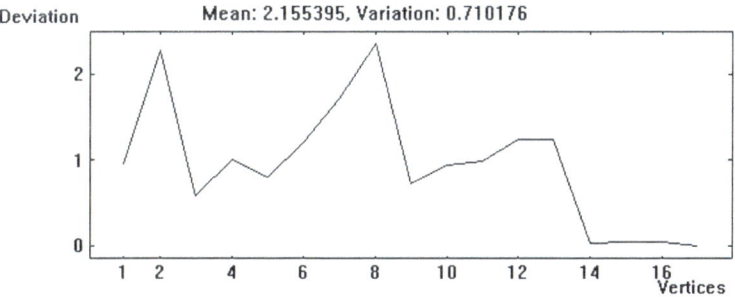

Fig. 14 Vitany diagram for attacks on ECC-enhanced proxy signature scheme

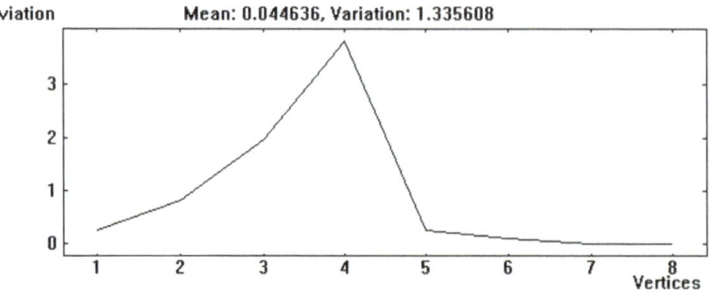

Fig. 15 Vitany diagram for analysis on ECC-enhanced proxy signature scheme

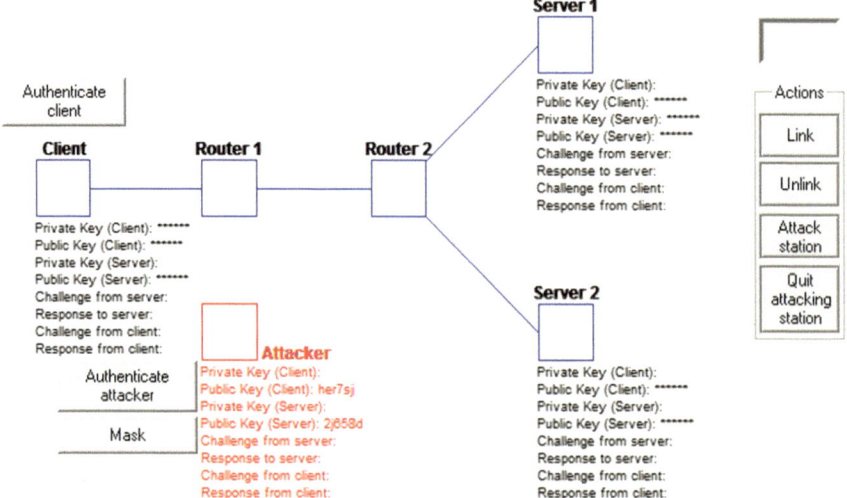

Fig. 16 Displays scenario for encryption/decryption of ECC-enhanced proxy signature scheme

5   Conclusion

In this research paper, I analyze enhanced threshold proxy signature scheme for diverse parameters like entropy, floating frequencies/intuitive synthesis, ASCII histogram, autocorrelation, histogram analysis and vitany. ECC delivers superior security and more effective performance as compared to the first-generation public key methods. As though Elliptic Curve Authentication Encryption Scheme (ECAES) bids some appreciated advantages over other cryptosystems like

RSA, the number of slightly different versions of ECAES included in the standards may hinder the adoption of ECAES. The results illustrate enhanced threshold proxy signature scheme is an efficient one for diverse real-time applications like e-commerce.

Acknowledgements The author also wishes to thank many anonymous referees for their suggestions to improve this paper.

Annexure

*	Schemes	ECC	Frequency1	ECC Attacks	Frequency2	ECC Analysis	Frequency3
No.	Substring	Frequency (in %)	Frequency	Frequency (in %)	Frequency	Frequency (in %)	Frequency
1	E	14.5136	464	11.1765	304	4.0379	682
2	I	8.3203	266	8.8971	242	4.257	719
3	N	8.2577	264	8.3088	226	3.884	656
4	T	8.0075	256	7.6103	207	4.1208	696
5	S	6.1933	198	7.5	204	4.1089	694
6	R	5.9431	190	6.9118	188	3.8366	648
7	A	5.2236	167	6.4338	175	3.3925	573
8	L	5.036	161	5.1838	141	3.8011	642
9	P	4.3791	140	5.1103	139	4.0971	692
10	O	4.3478	139	5	136	5.0503	853
11	C	3.7848	121	3.8235	104	3.9905	674
12	F	3.6597	117	3.6029	98	3.5998	608
13	G	3.3469	107	3.6029	98	4.0024	676
14	M	3.253	104	3.3824	92	4.5944	776
15	U	3.0654	98	2.9779	81	3.9017	659
16	D	2.5336	81	2.6103	71	3.3037	558
17	Q	2.4711	79	2.1691	59	3.3629	568
18	V	1.564	50	1.875	51	3.2564	550
19	K	1.2512	40	0.6618	18	4.0024	676
20	Y	1.1261	36	0.6618	18	3.3511	566
21	X	1.0322	33	0.6618	18	3.7951	641
22	H	0.9384	30	0.5515	15	4.2688	721
23	B	0.9071	29	0.4779	13	3.4162	577
24	W	0.5005	16	0.4779	13	3.8129	644
25	J	0.1877	6	0.2941	8	3.209	542
26	Z	0.1564	5	0.0368	1	3.5465	599

References

1. Koblitz, N. (1987). Elliptic curve cryptosystems. *Mathematics of Computation, 48,* 203–209.
2. BlueKrypt. (2015). *Cryptographic key length recommendation.* www.keylength.com.
3. ElGamal, T. (1985). A public-key cryptosystem and a signature scheme based on discrete logarithms. *IEEE Transactions of Information Theory, 31*(4), 469–472.
4. Miller, V. S. (1985). Uses of elliptic curves in cryptography. In *Advances in cryptology— Crypo'85.* LNCS (Vol. 218, pp. 417–426).
5. Desmedt, Y., & Frankel, Y. (1989). Threshold cryptosystems. In *The Proceedings of Advances in Cryptology. Crypto'89* (pp. 307–315).
6. Zhang, K. (1997). Threshold proxy signature schemes. In *Proceedings of Information Security Workshop* (pp. 191–197).
7. Kim, S., Park, S., & Won, D. (1997). Proxy signatures, revisited. In *The Proceedings of ICICS. ICICS'97,* LNCS (Vol. 1334, pp. 223–232).
8. Sun, H.-M. (1999). An efficient nonrepudiable threshold proxy signature scheme with known signers. *Computer Communications, 22*(8), 717–722.
9. Hwang, M.-S., IEEE Member, Lu, E. J.-L., & Lin, I.-C. (2003). A practical (t, n) threshold proxy signature scheme based on the RSA cryptosystem. *IEEE Transactions on Knowledge and Data Engineering, 15*(6), 1552–1560.
10. Wang, G., Bao, F., Zhou, J., & Deng, R. H. (2004). Comments on "A practical (t, n) threshold proxy signature scheme based on the RSA cryptosystem". *IEEE Transactions on Knowledge and Data Engineering, 16*(10), 1309–1311.
11. Kuo, W.-C., & Chen, M.-Y. (2005). A modified (t, n) threshold proxy signature scheme based on the RSA cryptosystem. In *Proceedings of the Third International Conference on Information Technology and Applications. ICITA'05* (pp. 576–579).
12. Li, F., Xue, Q., & Cao, Z. (2007). Crypanalysis of Kuo and Chen's threshold proxy signature scheme based on the RSA. In *The Proceedings of International Conference on Information Technology. ITNG'07* (pp. 815–818).
13. Geng, Y.-J., Hui, T., Fan, H. (2007). A modified and practical threshold proxy signature scheme based on RSA. In *Proceedings of the ICACT. ICACT'07* (pp. 1958–1960).
14. Lee, N. Y., Hwang, T., & Wang, C. H. (1998). On Zang's nonrepudiable proxy signature schemes. In *The Proceedings of ACISP'98,* LNCS (pp. 415–422).
15. Mambo, M., Usuda, K., & Okamoto, E. (1996). Proxy signature delegation of the power to sign message. *IEICE Transactions on Fundamentals, E-79A*(9), 1338–1353.
16. Mambo, M., Usuda, K., & Okamoto, E. (1996). Proxy signatures for delegating signing operation. In *Proceeding of Third ACM Conference of Computer and Communications Security* (pp. 48–57).
17. Sun, H.-M., Lee, N.-Y., & Hwang, T. (1999). Threshold proxy signatures. *IEEE Proceedings of Computers and Digital Techniques, 146*(5), 259–263.
18. Lee, C.-C., Lin, T.-C., Tzeng, S.-F., & Hwang, M.-S. (2011). Generalization of proxy signature based on factorization. *International Journal of Innovative Computing, Information and Control, 7*(3), 1039–1054.
19. Tzeng, S.-F., Lee, C.-C., & Hwang, M.-S. (2011). A batch verification for multiple proxy signature. *Parallel Processing Letters, 21*(1), 77–84.
20. Hwang, M.-S., Tzeng, S.-F., & Chiou, S.-F. (2009). A non-repudiable multi-proxy multi-signature scheme. *Innovative Computing, Information and Control Express Letters, 3* (3), 259–264.
21. Lu, E. J.-L., Hwang, M.-S., & Huang, C.-J. (2005). A new proxy signature scheme with revocation. *Applied Mathematics and Computation, 161*(3), 799–806.
22. Yang, C.-Y., Tzeng, S.-F., & Hwang, M.-S. (2004). On the efficiency of nonrepudiable threshold proxy signature scheme with known signers. *The Journal of Systems and Software, 73*(3), 507–514.

23. Tzeng, S.-F., Yang, C.-Y., & Hwang, M.-S. (2004). A nonrepudiable threshold multi-proxy multi-signature scheme with shared verification. *Future Generation Computer Systems, 20*(5), 887–893.

24. Tzeng, S.-F., Hwang, M.-S., & Yang, C.-Y. (2004). An improvement of nonrepudiable threshold proxy signature scheme with known signers. *Computers & Security, 23*(2), 174–178.

25. Hwang, M.-S., Tzeng, S.-F., & Tsai, C.-S. (2004). Generalization of proxy signature based on elliptic curves. *Computer Standards & Interfaces, 26*(2), 73–84.

26. Tsai, C.-S., Tzeng, S.-F., & Hwang, M.-S. (2003). Improved non-repudiable threshold proxy signature scheme with known signers. *Informatica, 14*(3), 393–402.

27. Li, L.-H., Tzeng, S.-F., & Hwang, M.-S. (2003). Generalization of proxy signature based on discrete logarithms. *Computers & Security, 22*(3), 245–255.

28. Hwang, M.-S., Lee, C.-C., & Hwang, S.-J. (2002). Cryptanalysis of the Hwang-Shi proxy signature scheme. *Fundamenta Informaticae, 53*(2), 131–134.

29. Hwang, M.-S., Lin, I.-C., & Lu, E. J.-L. (2000). A secure nonrepudiable threshold proxy signature scheme with known signers. *Informatica, 11*(2), 137–144.

30. Okamoto, T., Mitsuru, T., & Okamoto E. (1999). Extended proxy signature for smart cards. In *LNCS* (pp. 247–258). Springer.

31. Rivest, R. L., Shamir, A., & Adleman, L. M. (1978). A method for obtaining digital signatures and public-key cryptosystems. *Communications of the ACM, 21*(2), 120–126.

32. Lee, N. Y., Hwang, T., Wang, C. H., & Zhang, O. (1998). Nonrepudiable proxy signature schemes. In *Proceedings of Australasian conference on information security and privacy. ACISP'98* (pp. 415–422).

33. Katzenbeisser, S. (2001). *Recent advances in RSA cryptography* (pp. 85–90). Springer.

34. Denning, D. E. R. (1982). *Cryptography and data security* (pp. 115–265). Boston, MA, USA: Addison-Wesley Longman Publishing Co., Inc.

35. Hsu, C. L., Wu, T. S., & Wu, T. C. (2001). New nonrepudiable threshold proxy signature scheme with known signers. *The Journal of Systems and Software, 58*(5), 119–124.

36. Agrawal, M., Kayal, N., Saxena, N. (2004). PRIMES in P. *Annals of Mathematics, 160*(2), 781–793.

37. Cormen, T. H., Leiserson, C. E., Rivest, R. L., & Stein, C. (2001). *Section 31.8: Primality testing. Introduction to algorithms* (2nd ed., pp. 889–890). MIT Press, McGraw Hill. ISBN 0-262-03293-7.

38. Li, C.-T. (2008). *Multimedia foresics and security* (1st ed., pp. 73–74). IGI Global. ISBN 978-1-59904-869-7.

39. Friedman, M. (1937). The use of ranks to avoid the assumption of normality implicit in the analysis of variance. *Journal of the American Statistical Association* (*American Statistical Association*), *32*(200), 675–701.

40. Verma, H. K., Kaur, K., & Kumar, R. (2008). Comparison of threshold proxy signature schemes. In *International Conference on Security and Management. SAM'08* (pp. 227–231). USA.

41. Kumar, R., & Verma, H. K. (2010). An advanced secure (t, n) threshold proxy signature schemes based on RSA cryptosystem for known signers. In *IEEE 2nd International Advance Computing Conference. IACC'10* (pp. 293–298). India.

42. Kumar, R., & Verma, H. K. (2010). Secure threshold proxy signature scheme based on RSA for known signers. *Journal of Information Assurance and Security, USA, 5*(4), 319–326.

43. Kumar, R., Verma, H. K., & Dhir, R. (2015). Analysis and design of protocol for enhanced threshold proxy signature scheme based on RSA for known signers. *Wireless Personal Communications—An International Journal, 80*(3), 1281–1345. Springer. ISSN: 0929-6212 (Print) 1572-834X (Online).

Partial Fractional Derivative (PFD) based Texture Analysis Model for Medical Image Segmentation

S. Hemalatha and S. Margret Anouncia

Abstract The early detection of diseases such as brain tumor and lung cancer is achieved through segmenting these images. As these images are typified to contain obscure structures, a precise segmentation method needs to be evolved. The principal idea is to devise an unsupervised segmentation technique for medical images involving a partial fractional derivative (PFD)-dependent texture extraction model and an unsupervised clustering algorithm. Basically, image segmentation process allocates different tags to diverse image regions. In this chapter, the process is considered to be texture-based segmentation by representing every tag with an exclusive texture label. The textural features are extorted using PFD-based model. The ISODATA clustering is suggested for segmentation through pixel-based classification. This process is experimented on different test cases such as images of lung cancer detection and brain tumor detection and is able to produce higher accuracy than existing methods.

Keywords Partial fractional derivatives · Texture analysis · ISODATA clustering algorithm · Lung cancer detection · Brain tumor detection · Confusion matrix Classification accuracy

1 Introduction

In common, segmentation refers to partitioning an image to various localities. As it were, each one of the pixels in an image would receive a tag such that pixels having a particular tag reveal a specific feature. Particularly, when the texture-based segmentation is considered, every tag designated to a specific locality signifies a class

S. Hemalatha (✉)
School of Information Technology and Engineering, VIT University, Vellore, TN, India
e-mail: shemalatha@vit.ac.in

S. Margret Anouncia
School of Computer Science and Engineering, VIT University, Vellore, TN, India
e-mail: smargretanouncia@vit.ac.in

© Springer Nature Singapore Pte Ltd. 2018
S. Margret Anouncia and U. K. Wiil (eds.), *Knowledge Computing and Its Applications*, https://doi.org/10.1007/978-981-10-6680-1_11

213

of texture. As studying an image based on partitioned image segments is more communicative, the process of image segmentation plays an important role for accurate interpretation of images [24].

It is observed that spectral components of images are exploited for a large portion of the customary strategies of segmentation [16]. However, sometimes, these strategies might direct to incorrectness in the process. Thus, alternatively, spatial features are chosen for the intended purpose. Also, it is largely observed that texture is comprised of most of the medical images. Hence, it is decided to perform the diagnosis by texture-based segmentation. In this process, the textural features present in images are characterized numerically and segmentation is performed on these features per pixel basis.

Some of the methods are exclusively devised for texture segmentation, as it is anticipated that textural components can afford a substantial impact on segmentation [4, 7, 20, 35]. Texture feature-based segmentation is useful in diversified applications of image processing such as medical imaging [2, 6, 15, 25], remote sensing [11, 23, 30], content-based image retrieval [36, 37], and industrial applications of defect detection [12, 32]. However, the process of texture segmentation is a perplexing one as texture is defined indistinctively in the literature.

In common, the task is realized in two steps. The first step is extricating image textures and representing them as quantitative features using a texture analysis methodology. The next step is segmenting target images based on these features. Among many of the existing methods, few methods of texture analysis are studied below.

The GLCM method and the texture number method represent structural organization of pixels which are represented with a numeric value [8, 9]. A transform domain-based method was suggested by Chang and Kuo [1]. This method was well justified theoretically; however, investigational proof was missing. Yoshimura [33] attempted a texture-based segmentation process using a technique of edge detection.

Certain methods depend on concepts based on mathematics. For instance, Vese and Osher [31] proposed to use partial-differential equations and Targhi et al. [27] suggested LU transform-based method for texture analysis. Also, the concept of autocorrelation coefficients was utilized by Karthikeyan and Krishnamoorthy [14]. Of late, the researches of texture-based segmentation were accelerated with a focus on integrating it with computational methodologies. This led to improved and accurate outcomes [4, 29, 34, 35].

A fully automatic system was presented in [18] for tumor detection. The detection process is demonstrated, and the efficacy is compared between statistical features such as gray-level co-occurrence matrix (GLCM), histogram of oriented gradient (HOG), local binary pattern (LBP), gray-level run length matrix (GLRLM), and Gabor wavelet features. This study is performed by using a number of clustering algorithms, for instance k-nearest neighboring algorithm, support vector machine, nearest subspace classification, K-Means algorithm, and sparse representation clustering.

Hamouchene suggested the NBP—neighbors-based binary pattern—method in [7] by following the LBP method. The association of pixels in a neighborhood is used to descend this method. A dynamic decomposition process is adopted grouping pixels into a particular segment.

In common, the methods implied for texture analysis operate the texture characterization process by assuming that textural components and edge features are dissimilar. But, the author (Hemalatha) suggested and proved in [10] that the edge and textural features are akin by comparative analysis. Also, the textural feature is depicted as mini edges. It was also found to be successful for the segmentation of remote sensing images [11]. Hence, the fundamentals of edge detectors are used for the identification of textural components.

Usually, for edge detection in images, integer partial derivatives are applied. In PFD-based texture analysis approach, partial fractional derivatives are applied for identifying textural features, i.e., mini edges. Consequently, textural feature is typified with two descriptors using χ_n^2 statistical test. The local descriptor, LD_{tex}, exemplifies every pixel in its 3×3 neighborhood. The global descriptor, GD_{tex}, characterizes a complete image.

In order to carry on with the segmentation process, a clustering algorithm needs to be applied on texture feature vectors, i.e., local and global descriptors. In this chapter, the segmentation process is taken care of by ISODATA classification algorithm suggested [13, 17]. It is basically an unsupervised classification algorithm that follows a divide-and-combine methodology. Firstly, this algorithm picks the cluster centers at random. Consequently, clusters accumulate the nearest pixels.

2 Details of the Proposed Framework

The proposed framework is diagrammatically represented in Fig. 1. It comprises of two major steps.

 I. The proposed textural features—LD_{tex} and GD_{tex}—are extricated with the help of PFD-based model.
 II. The segmentation procedure is applied on the features extricated.

2.1 PFD-Based Texture Analysis Model

The PFD-based model for texture analysis encompasses smoother removing noise components, PFD-based filters detecting textural components, and a statistical frame typifying textural components.

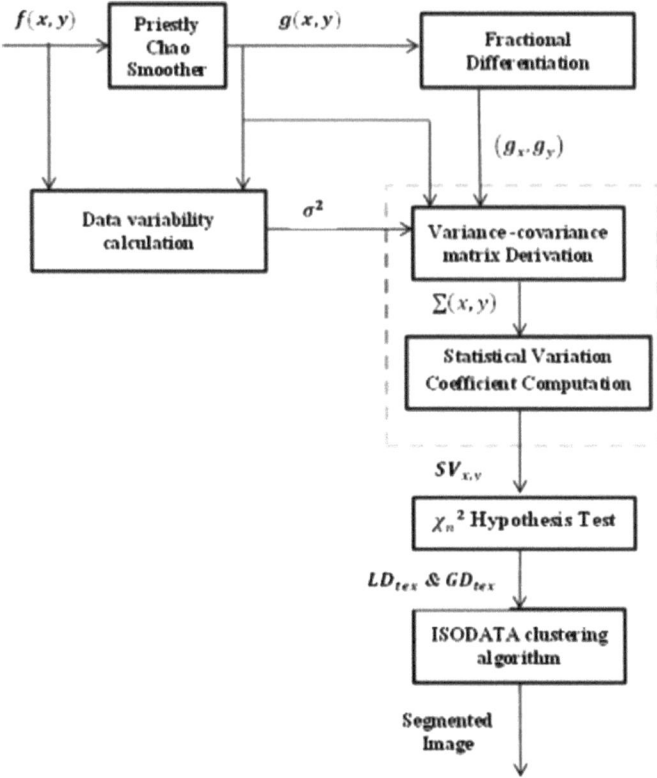

Fig. 1 Proposed framework—texture segmentation

The Priestly–Chao smoother that involves Gaussian kernel smoothing [21] is applied for the removal of noise components. This particular smoother is predominantly used for curve fitting in the case of evenly spaced and non-dependent variables. Hence, it is determined suitable for images, where the pixels are uniformly distributed. The smoother also facilitates calculating derivatives in the next step. Consequently, the Grunwald–Letnikov (GL)-based partial fractional derivatives [19, 22] are calculated at every pixel in x- and y-directions.

Finally, with the help of a statistical frame, image textures are characterized by two descriptors. As variations in small image regions better represent textures, variance–covariance coefficient is determined for the 3×3 neighborhood of every pixel. The χ_n^2 statistical test involving this coefficient is applied in order to derive LD_{tex}, the local descriptor. Then, the histogram of LD_{tex} values in an image is figured descending the global descriptor, GD_{tex}. The GD_{tex} values, after being descended in the overlying regions of size 30×30, can function as textural feature vectors for further process.

2.2 Mathematical Features of the PFD-Based Model

An $m \times n$ size grayscale image is considered, where $f(x, y)$ is the pixel value at x, y location. The Priestly–Chao smoother function SF is used to obtain the smoothed image $g(x, y)$ at each of the x, y coordinates.

$$g(x, y) = \frac{1}{2\pi mnk^2} \sum_{i=1}^{m} \sum_{j=1}^{n} \mathrm{SF}\left(\frac{x - i}{k}\right) \mathrm{SF}\left(\frac{y - j}{k}\right) f(i, j) \tag{1}$$

where k symbolizes smoothing scale and SF(t) is the smoothing function based on Gaussian kernel. SF(t) is written as

$$\mathrm{SF}(t) = \frac{1}{\sqrt{2\pi}} e^{-\left(\frac{1}{2}\right)t^2} \tag{2}$$

SF$\left(\frac{x-i}{k}\right)$ and SF$\left(\frac{y-j}{k}\right)$ are called smoothing parameters and replaced by $a_{i,k}(x) = \frac{1}{\sqrt{2\pi}} e^{-\frac{1}{2}\left(\frac{x-i}{k}\right)^2}$ and $a_{j,k}(y) = \frac{1}{\sqrt{2\pi}} e^{-\frac{1}{2}\left(\frac{y-j}{k}\right)^2}$.

As a next step, PFD of $g(x, y)$ is computed in x, y directions as shown below:

$$g_x(x, y) = \frac{1}{2\pi mnk^2} \sum_{j=1}^{n} a_{j,k}(y) \sum_{i=1}^{m} a'_{i,k}(x) f(i, j) \tag{3}$$

$$g_y(x, y) = \frac{1}{2\pi mnk^2} \sum_{i=1}^{m} a_{i,k}(x) \sum_{j=1}^{n} a'_{j,k}(y) f(i, j) \tag{4}$$

where $a'_{i,k}(x)$ and $a'_{j,k}(y)$ are GL definition-based partial fractional derivatives [19, 22] of $a_{i,k}(x)$ and $a_{j,k}(y)$ correspondingly.

$$a'_{i,k}(x) = D^v a_{i,k}(x) = \frac{1}{\sqrt{2\pi}} \frac{(-1)^P (x - i)^{2 \times (p-v)}}{2^{(p-v)} \Gamma(p - v + 1) k^{2 \times (p-v)}} \tag{5}$$

$$a'_{j,k}(y) = D^v a_{j,k}(y) = \frac{1}{\sqrt{2\pi}} \sum_{p=0}^{\infty} \frac{(-1)^P (y - j)^{2 \times (p-v)}}{2^{(p-v)} \Gamma(p - v + 1) k^{2 \times (p-v)}} \tag{6}$$

$a'_{i,k}(x)$ and $a'_{j,k}(y)$ are devised as per the derivation shown in [10]. In Eqs. (3) and (4), $\left(g_x(x, y), g_y(x, y)\right)$ signify the gradient vector of $g(x, y)$.

The measure of data fluctuation in the input image is computed as

$$\sigma^2 = \frac{1}{mn} \sum_{i=1}^{m} \sum_{j=1}^{n} (f(i, j) - g(i, j))^2 \tag{7}$$

At every (x, y), the variance–covariance matrix $\sum(x, y)$ is determined as follows:

$$\sum(x, y) = \begin{bmatrix} \sigma_{11}(x, y) & \sigma_{12}(x, y) \\ \sigma_{12}(x, y) & \sigma_{22}(x, y) \end{bmatrix} \tag{8}$$

where $\sigma_{11}(x, y)$, $\sigma_{12}(x, y)$, $\sigma_{12}(x, y)$, and $\sigma_{22}(x, y)$ are estimated as per Eqs. (9), (10), and (11).

$$\sigma_{11}(x, y) = \tau \sum_{i=1}^{m} \sum_{j=1}^{n} a_{j,k}^2(y) \, a_{i,k}'^2(x) \tag{9}$$

$$\sigma_{22}(x, y) = \tau \sum_{i=1}^{m} \sum_{j=1}^{n} a_{i,k}^2(x) \, a_{j,k}'^2(y) \tag{10}$$

$$\sigma_{12}(x, y) = \tau \sum_{i=1}^{m} \sum_{j=1}^{n} a_{i,k}(x) a_{j,k} k(y) a_{i,k}'(x) a_{j,k}'(y) \tag{11}$$

It is noted that $\tau = \frac{\sigma^2}{(4\pi^2 m^2 n^2 k^4)}$.

In order to identify and describe textural features, the χ_n^2 statistical test [3] is used. The feature detection in [3] is basically accomplished using the following statistic

$$S(Z) = (Z - \mu_z)^T \sum\nolimits^{-1}(Z - \mu_z) \tag{12}$$

and the following definition for critical region:

$$C(\gamma) = \{y : K(Z) > C_\gamma\} \tag{13}$$

In Eqs. (12) and (13), Z is a random variable with $\chi_n{}^2$ distribution with n degrees of freedom. And the significance level of the test is depicted by the size of critical region (γ). The term $(Z - \mu_z)$ in Eq. (12) and (g_x, g_y) correspond to variation. Hence, $(Z - \mu_z)$ can be replaced with (g_x, g_y). Thus, the statistical variation coefficient, $SV_{x,y}$, is obtained as with Eq. (14)

$$SV_{x,y} = (g_x, g_y) \sum\nolimits^{-1}(x, y)(g_x, g_y)^T \tag{14}$$

When Eq. (14) is shortened, the following equation is obtained.

$$SV_{x,y} = \frac{g_x^2 \sigma_{22}(x, y) + g_y^2 \sigma_{11}(x, y) - 2\sigma_{12}(x, y) g_x g_y}{\sigma_{11}(x, y) \sigma_{22}(x, y) - \sigma_{12}^2(x, y)} \tag{15}$$

Fig. 2 χ_n^2 statistical test
result

$t_{x-1,y-1}$	$t_{x,y-1}$	$t_{x+1,y-1}$
$t_{x-1,y}$	$t_{x,y}$	$t_{x+1,y}$
$t_{x-1,y+1}$	$t_{x,y+1}$	$t_{x+1,y+1}$

In each of the overlying 3×3 regions, the texture identification process is executed by a significance test on $SV_{x,y}$ with the help of Eq. (14). Considering a significance level (γ), the values of $SV_{x,y}$ at (x, y) and $C(\gamma)$ from statistical table are compared. Based on whether $SV_{x,y}$ lies outside or inside the region marked by $C(\gamma)$, the significance of a pixel for texture is encoded as either 1 or 0. The testing process is applied for the eight neighbors in each of the 3×3 regions and the outcomes are interpreted as in Fig. 2

The coefficients in Fig. 2 are put in an array similar to the one shown in Eq. (16), and LD_{tex} is computed with Eq. (17).

$$T(i) = \left\{t_{x,y-1}, t_{x+1,y-1}, t_{x+1,y}, t_{x+1,y+1}, t_{x,y+1}, t_{x-1,y+1}, t_{x-1,y}, t_{x-1,y-1}\right\} \quad (16)$$

$$LD_{tex} = \sum_{i=1}^{7} T(i) * 2^i \quad (17)$$

Finally, the input image is converted into a two-dimensional array containing LD_{tex} values. The global descriptor, GD_{tex}, is devised by estimating the histogram of local descriptors.

2.3 Texture Segmentation Using ISODATA

In common, the Interactive Self-Organizing Data Analysis Technique (ISODATA) is utilized for segmenting images. There are three basic stages in this algorithm. Initially, centers for clustering are picked arbitrarily. Consequently, all the pixels are dispersed to the clusters very adjacent to them. Finally, each of the clusters is revised and so their centers. The second and third stages are repetitively executed while waiting for the transition between two repetitions to become modest.

With each stage, the following criteria need to be satisfied by the algorithm.

1. The center of every cluster is computed as the feature-based mean of the samples that reside in it.

2. Every cluster is assigned with its nearest samples. The nearness is confirmed with the help of feature-based distance measure.
3. If the feature-based distance between two clusters is less significant, then the clusters are combined.
4. If the variance of a cluster is big, then it is divided into two.

ISODATA, in essence, is an unsupervising one and was developed from K-Means algorithm. The difference between the two is about the value of K, the number of clusters. K is invariable in case of K-Means and is varying in ISODATA clustering. ISODATA comprises of several benefits, and a few are minimal computations and maximal speed. This chapter uses ISODATA for the segmentation stage and employs GD_{tex} assuming the role of feature vectors. The texture-based feature vector, GD_{tex}, is descended for every pixel in the input image via overlying neighborhoods of size 30×30.

3 Results and Discussion

The PFD-based texture segmentation method was intended to be used for the detection of lung cancer and brain tumor by segmentation. In common, medical images possess intricate structures, and hence their clinical analysis needs them to be segmented accurately. In this work, two sets of medical images from MedPix [28] are considered for segmentation. They are images of brain tumor detection and lung cancer detection.

Early diagnosis of lung cancer is the inspiring problem due to the unusual forming of cells that may sometimes intersect. A small attempt is made for detecting early lung cancer by applying the developed PFD-based descriptors to carry out segmentation process to segment the image followed by a binarization process to highlight the affected region. The original and the segmented images are depicted in Fig. 3.

Thus, the efficacy of the PFD-based texture analysis model for identifying the cancer cells in the lung image is exhibited.

Similarly, another image from medical domain is subjected to the same procedure. Brain tumor detection is a frenetic process to find the tumor appearance clearer in MRI images. Thus, the tumor detection in MRI images of brain is experimented through segmentation by the developed methodology. The original and outcome images are depicted in Fig. 4.

Therefore, the PFD-based framework of texture segmentation for identifying the tumor cells in the lung image is demonstrated.

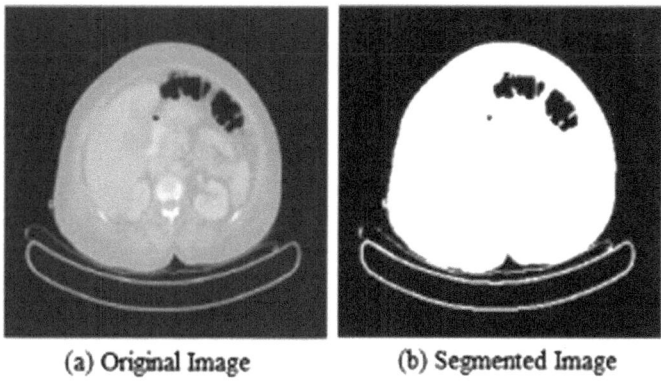

(a) Original Image   (b) Segmented Image

Fig. 3 Lung cancer detection

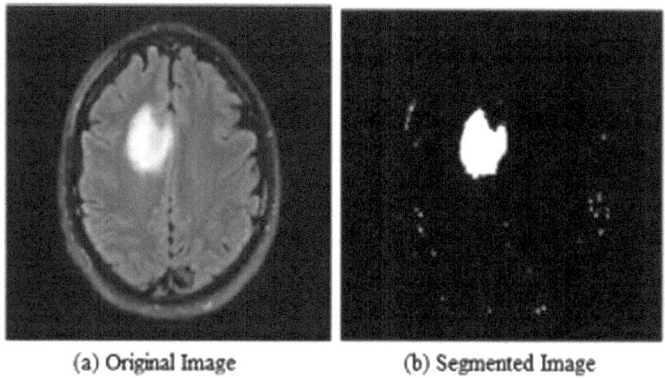

(a) Original Image   (b) Segmented Image

Fig. 4 Brain tumor detection

3.1 Performance Assessment

The PFD-based texture segmentation for medical images is validated with the help of classification devised using confusion matrix approach [5, 26]. The proposed segmentation is considered as the process of multiclass classification. That is, each pixel is exclusively designated with one of the texture labels. Thus, the segmentation correctness is measured with confusion matrix approach.

The PFD-based method is benchmarked with statistical feature-based method [18] and NBP method [7] by evaluating through the confusion matrix measures, ACC, ERR, SN, PR, and F_m. As far as the detection of lung cancer and brain tumor is concerned, confusion matrix approach corresponding to binary classification is followed.

The measures ACC, ERR, SN, PR, and F_m calculated for the lung cancer detected image with the help of PFD method, statistical feature-based method [18], and NBP method [7] are exhibited in Table 1.

Table 1 Classification measures for lung cancer detection in Fig. 3

Method being used	ACC (%)	ERR (%)	SN (%)	PR (%)	F_m (%)
Statistical feature-based method	94.27	5.73	94.13	93.89	93.92
NBP method	95.05	4.95	94.93	94.71	94.32
PFD-based method	96.86	2.61	98.59	98.71	98.65

Table 2 Classification measures for brain tumor detection in Fig. 4

Method being used	ACC (%)	ERR (%)	SN (%)	PR (%)	F_m (%)
Statistical feature-based method	94.72	5.28	94.65	94.42	94.57
NBP method	95.75	4.25	95.53	95.31	94.93
PFD-based method	97.75	2.25	98.98	98.68	98.84

Similarly, these measures are computed for brain tumor detection of Fig. 4 and displayed in Table 2.

The segmentation accuracy is evaluated with the measures ACC (Accuracy), SN (Sensitivity), PR (Precision), and F_m (F measure). And the error rate is evaluated with ERR (Error Rate). These measures are compared between the PFD-based method, the statistical feature-based method, and NBP method. From Tables 1 and 2, it is observed that the accuracy measure (ACC) is coherent with other measures (SN), (PR), and (F_m).

The statistical-feature based method and NBP method provided 94–95% accuracy, whereas, the PFD-based method provided 96–97% accuracy. Therefore, it is proved that the PFD-based texture segmentation approach has enhanced the accuracy upto 3%. Correspondingly, the error measure is also diminished. Thus, it is demonstrated that our PFD-based method outperforms the statistical-feature based method and the NBP method.

The PFD-based approach for texture segmentation is built based on fractional differential filters that are principally developed for enhancing image textures [19, 22]. Thus, in the proposed approach for texture segmentation, the textural components are first enriched and then characterized. Therefore, better segmentation accuracy is attained than the other methods like statistical feature-based method and NBP method.

4   Conclusion

This chapter presented an unsupervised texture segmentation approach for medical diagnosis, i.e., the detection of brain tumor and lung cancer. This is accomplished with the PFD-based texture model and ISODATA clustering algorithm. The PFD-based model employs Priestly–Chao smoother, GL FD filters, and a statistical frame. The PFD-based model is applied on images to be diagnosed for lung cancer and brain tumor. Firstly, the texture feature vectors in terms of LD_{tex} and GD_{tex} are derived. Consequently, segmentation is implemented with ISODATA grouping on LD_{tex} and GD_{tex}. Finally, binarization procedure is applied for diagnosis. The developed methodology is found to perform reasonably for a satisfactory diagnosis.

By the numerical evaluation of the segmentation performance through confusion matrix measures, it is justified that the PFD-based texture segmentation methodology provided considerable values for the accuracy measures ACC, SN, PR, and F_m. The error rate ERR is also reduced considerably. It is interesting to note that the results obtained from the developed methodology are superior than the statistical feature-based method and NBP method. By comparing these three classification methods, a higher percentage of accuracy is achieved by the developed methodology for the brain tumor classification and lung cancer detection.

References

1. Chang, T., & Kuo, C. C. (1993). Texture analysis and classification with tree-structured wavelet transform. *IEEE Transactions on Image Processing, 2*(4), 429–441.
2. Chen, Y. T. (2017). Medical image segmentation using independent component analysis-based kernelized fuzzy-means clustering. *Mathematical Problems in Engineering, 2017.*
3. Chuang, E. R., & Sher, D. (1993). χ^2 test for feature detection. *Pattern Recognition, 26*(11), 1673–1681.
4. Cimpoi, M., Maji, S., Kokkinos, I., & Vedaldi, A. (2015). Deep filter banks for texture recognition, description, and segmentation. *International Journal of Computer Vision, 1–30.*
5. Costa, E., Lorena, A., Carvalho, A. C. P. L. F., & Freitas, A. (2007). A review of performance evaluation measures for hierarchical classifiers. In *Evaluation methods for machine learning II: Papers from the AAAI-2007 workshop* (pp. 1–6). Vancouver, Canada: AAAI.
6. Danesh, H., Kafieh, R., Rabbani, H., & Hajizadeh, F. (2014). Segmentation of choroidal boundary in enhanced depth imaging OCTs using a multiresolution texture based modeling in graph cuts. *Computational and Mathematical Methods in Medicine, 2014.* Article ID. 479268.
7. Hamouchene, I., & Aouat, S. (2016). A new approach for texture segmentation based on NBP method. *Multimedia Tools and Applications, 1–20.*
8. Haralick, R. M., & Shanmugam, K. (1973). Textural features for image classification. *IEEE Transactions on Systems, Man, and Cybernetics, 6,* 610–621.
9. He, D. C., & Wang, L. (1990). Texture unit, texture spectrum, and texture analysis. *IEEE Transactions on Geoscience and Remote Sensing, 28*(4), 509–512.
10. Hemalatha, S., & Anouncia, S. M. (2016). A computational model for texture analysis in images with fractional differential filter for texture detection. *International Journal of Ambient Computing and Intelligence (IJACI), 7*(2), 93–113.
11. Hemalatha, S., & Anouncia, S. M. (2017). Unsupervised segmentation of remote sensing images using FD based texture analysis model and ISODATA. *International Journal of Ambient Computing and Intelligence (IJACI), 8*(3), 58–75.
12. Iyer, M. (2014). Defect detection in pattern texture analysis. In *Proceedings of the 2014 International Conference on Communications and Signal Processing (ICCSP)* (pp. 172–175). Melmaruvathur, TN, India: IEEE.
13. Jain, A. K., Murty, M. N., & Flynn, P. J. (1999). Data clustering: A review. *ACM Computing Surveys, 31*(3), 264–323.

14. Karthikeyan, T., & Krishnamoorthy, R. (2012). Autoregressive model based on Bayesian approach for texture representation. *ICTACT Journal on Image and Video Processing, 3*(1), 485–491.
15. Korchiyne, R., Sbihi, A., Farssi, S. M., Touahni, R., & Alaoui, M. T. (2012, May). Medical image texture segmentation using multifractal analysis. In *2012 International Conference on Multimedia Computing and Systems (ICMCS)* (pp. 422–425). IEEE.
16. Lu, D., & Weng, Q. (2007). A survey of image classification methods and techniques for improving classification performance. *International Journal of Remote Sensing, 28*(5), 823–870.
17. Memarsadeghi, N., Mount, D. M., Netanyahu, N. S., & Le Moigne, J. (2007). A fast implementation of the ISODATA clustering algorithm. *International Journal of Computational Geometry & Applications, 17*(01), 71–103.
18. Nabizadeh, N., & Kubat, M. (2015). Brain tumors detection and segmentation in MR images: Gabor wavelet vs. statistical features. *Computers & Electrical Engineering, 45,* 286–301.
19. Neel, M. C., & Joelson, M. (2011). Generalizing Grünwald-Letnikov's formulas for fractional derivatives. In *Proceedings of the 6th EUROMECH Nonlinear Dynamics Conference.* Saint Petersburg, RUSSIA: IPACS Electronic Library.
20. Ojala, T., Valkealahti, K., Oja, E., & Pietikäinen, M. (2001). Texture discrimination with multidimensional distributions of signed gray-level differences. *Pattern Recognition, 34*(3), 727–739.
21. Priestley, M. B., & Chao, M. T. (1972). Non-parametric function fitting. *Journal of the Royal Statistical Society. Series B (Methodological),* 385–392.
22. Pu, Y. F. (2010). Fractional differential mask: A fractional differential-based approach for multiscale texture enhancement. *IEEE Transactions on Image Processing, 19*(2), 491–511.
23. Roy, M., Ghosh, S., & Ghosh, A. (2014). A novel approach for change detection of remotely sensed images using semi-supervised multiple classifier system. *Information Sciences, 269,* 35–47.
24. Sezgin, M. (2004). Survey over image thresholding techniques and quantitative performance evaluation. *Journal of Electronic Imaging, 13*(1), 146–168.
25. Sharma, N., Ray, A. K., Sharma, S., Shukla, K. K., Pradhan, S., & Aggarwal, L. M. (2008). Segmentation and classification of medical images using texture-primitive features: Application of BAM-type artificial neural network. *Journal of Medical Physics, 33*(3), 119.
26. Sokolova, M., & Lapalme, G. (2009). A systematic analysis of performance measures for classification tasks. *Information Processing and Management, 45*(4), 427–437.
27. Targhi, A. T., Hayman, E., & Olof Eklundh, J. (2006, May). Real-time texture detection using the LU-transform. In *Proceeding of the Workshop on Computation Intensive Methods for Computer Vision in Conjunction with ECCV* (Vol. 713). Graz, Austria: Springer-Verlag Berlin Heidelberg.
28. The National Library of Medicine. (2010). MedPix, Online. Accessed December 15, 2016. Also available as https://medpix.nlm.nih.gov/home.
29. Tirandaz, Z., & Akbarizadeh, G. (2016). A two-phase algorithm based on kurtosis curvelet energy and unsupervised spectral regression for segmentation of SAR images. *IEEE Journal of Selected Topics in Applied Earth Observations and Remote Sensing, 9*(3), 1244–1264.
30. Tsai, F., Chou, M. J., & Wang, H. H. (2005). Texture analysis of high resolution satellite imagery for mapping invasive plants. In *Proceedings of the IEEE International Geoscience and Remote Sensing Symposium* (Vol. 4, pp. 3024–3027). Seoul: IEEE.
31. Vese, L. A., & Osher, S. J. (2003). Modeling textures with total variation minimization and oscillating patterns in image processing. *Journal of Scientific Computing, 19*(1–3), 553–572.
32. Xie, X. (2008). A review of recent advances in surface defect detection using texture analysis techniques. *ELCVIA Electronic Letters on Computer Vision and Image Analysis, 7*(3).
33. Yoshimura, M. (1997, July). Edge detection of texture image using genetic algorithms. In I. SICE'97 (Ed.), *Proceedings of the 36th SICE Annual Conference* (pp. 1261–1266). Tokushima, Japan: IEEE.

34. Yu, H., Yang, W., Xia, G. S., & Liu, G. (2016). A color-texture-structure descriptor for high-resolution satellite image classification. *Remote Sensing, 8*(3), 259.
35. Yuan, J., Wang, D., & Cheriyadat, A. M. (2015). Factorization-based texture segmentation. *IEEE Transactions on Image Processing, 24*(11), 3488–3497.
36. Yue, J., Li, Z., Liu, L., & Fu, Z. (2011). Content-based image retrieval using color and texture fused features. *Mathematical and Computer Modelling, 54*(3), 1121–1127.
37. Zhang, D. W. (2000). Content-based image retrieval using Gabor texture features. In *Proceeding of the IEEE Pacific-Rim Conference on Multimedia* (pp. 392–395). Shanghai, China: IEEE.

Breast Cancer Classification Using Deep Neural Networks

S. Karthik, R. Srinivasa Perumal and P. V. S. S. R. Chandra Mouli

Abstract Early diagnosis of any disease can be curable with a little amount of human effort. Most of the people fail to detect their disease before it becomes chronic. It leads to increase in death rate around the world. Breast cancer is one of the diseases that could be cured when the disease identified at earlier stages before it is spreading across all the parts of the body. The medical practitioner may diagnose the diseases mistakenly due to misinterpretation. The computer-aided diagnosis (CAD) is an automated assistance for practitioners that will produce accurate results to analyze the criticality of the diseases. This chapter presents a CAD system to perform automated diagnosis for breast cancer. This method employed deep neural network (DNN) as classifier model and recursive feature elimination (RFE) for feature selection. DNN with multiple layers of processing attained higher classification rate than SVM. So, the researchers used deep learning method for hyper-spectral data classification. This chapter used DNN to learn deep features of data. The DNN with multiple layers of processing is applied to classify the breast cancer data. The system was experimented on Wisconsin Breast Cancer Dataset (WBCD) from UCI repository. The dataset partitioned into different sets of train-test split. The performance of the system is measured based on accuracy, sensitivity, specificity, precision, and recall. From the results, the accuracy obtained 98.62%, which is better than other state-of-the-art methods. The results show that the system is comparatively outperformed than the existing system.

Keywords Classification · Deep learning · Healthcare system · Feature selection
Breast cancer diagnosis

S. Karthik (✉) · R. Srinivasa Perumal
School of Information Technology and Engineering, VIT University, Vellore, India
e-mail: karthik.s2012@vit.ac.in

R. Srinivasa Perumal
e-mail: r.srinivasaperumal@vit.ac.in

P. V. S. S. R. Chandra Mouli
School of Computer Science and Engineering, VIT University, Vellore, India
e-mail: chandramouli@vit.ac.in

© Springer Nature Singapore Pte Ltd. 2018
S. Margret Anouncia and U. K. Wiil (eds.), *Knowledge Computing
and Its Applications*, https://doi.org/10.1007/978-981-10-6680-1_12

1 Introduction

Nowadays, there is a drastic level of increase in the amount of women around the world, getting affected by breast cancer and it is raising over time. Earlier detection of breast cancer reduces death rate and avoids it till reaches the chronic level. In 2012, almost 1.7 million women are affected by breast cancer [20]. The impact of breast cancer is increasing day by day, due to that the healthcare professionals are not in a condition to state the people affected at different stages earlier to save their lives. Digital mammography is a major diagnosis model used throughout the world for breast cancer detection. Computer-aided diagnosis (CAD) is widely used in detecting numerous diseases with accurate decision. It assists the healthcare professionals to analyze and conclude the stages of the various diseases. CAD systems are developed in a way to provide promising results and perfect decisions on patient's condition that helps medical practitioners to diagnose the stages of the diseases. It supports radiologists to avoid misconceptions and wrong diagnosis due to inaccurate data, lack of focus, or inexperience, who uses visually screening mammogram of patients. The aim of the system is to develop a novel CAD model to diagnose a breast cancer in earlier stages with more accurate results to save their precious lives by using DNN.

DNN followed by a similar artificial neural network with a complex network structure which has 'n' hidden layers, which can process the input data from the previous layer. The error rate of the input data will be consistently reduced by adjusting the weights of every node, which leads to achieve an accurate result. It helps to create a model and define its complex hierarchies in a simple form. It supports all kinds of algorithms, namely supervised, unsupervised, semi-supervised, and reinforcement. So, the system did not define any specific algorithm. The DNN generates a better model themselves to train the given data. The architecture of the DNN is shown in Fig. 1.

The rest of the chapter is summarized as follows: Section 2 describes the related work, Sect. 3 explains the proposed method, Sect. 4 demonstrates the experimental results, and Sect. 5 concludes the work.

2 The Literature Review

Neural Network is inspired by the working principle of biological neural networks, which has its own input and output channels called as dendrites and axons, respectively. A typical ANN will have millions of processing units or elements, which forms a highly interconnected network that processes a huge amount of information based on the response, given by the external input of a computing system. Every single neuron in a typical neural network is called as unit. A layer in a neural network is considered as a set of neurons in a stack. A layer may have n number of nodes in it. A typical neural network system has single input layer and

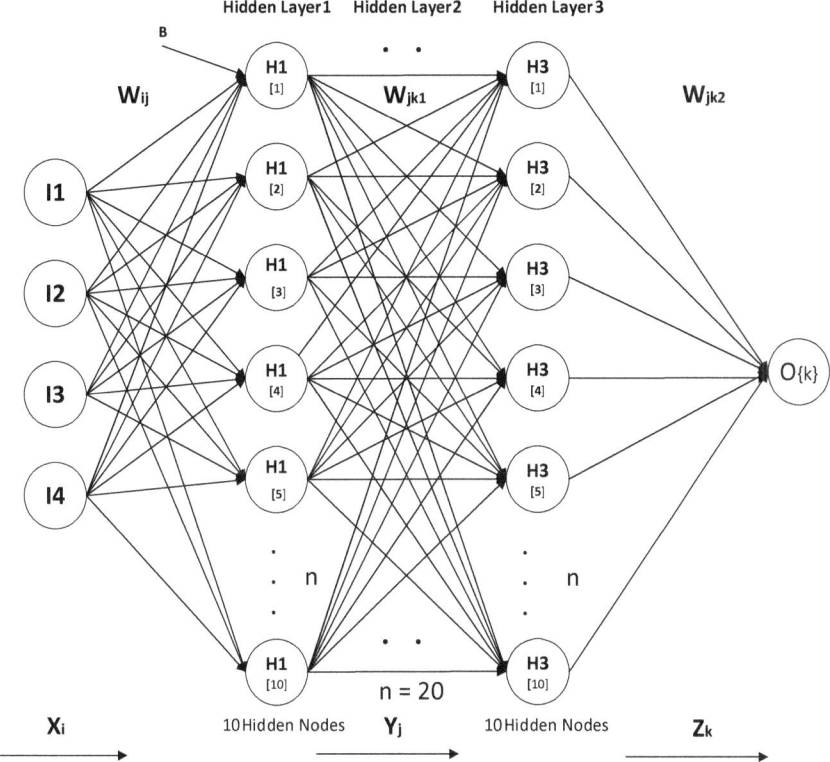

Fig. 1 Deep neural network architecture

may have a single or two hidden layers, which is directly connected to the output layer, which receives input from the input layer, i.e., previous layer of the current node.

Abbass developed a system with pareto-differential evaluation algorithm with local search scheme, called memetic pareto-artificial neural network (MPANN) [1]. MPANN analyzes the data effectively than other models. The method achieved 98.1% accuracy on random split. Tuba and Tulay proposed the statistical neural network-based breast cancer diagnosis system [13]. In the diagnosis system, they used RBF, general regression neural network (GRNN), and statistical neural network structures on WDBC dataset. The system obtained 98.8% on 50–50 partitioning split.

Paulin and Santhakumaran [19] developed a system with back-propagation neural network (BPNN) and obtained 99.28% accuracy with Levenberg–Marquardt algorithm. They used median filter for preprocessing and normalized the data using min–max technique. None of the features are eliminated from the dataset. The accurate result attained from 80:20 partition scheme. Karabatak and Ince [12] developed an expert system for breast cancer detection. Association rules (AR) are used to reduce the dimensions of the dataset. In the system, AR1 and AR2 are developed to reduce the features. AR1 reduces one feature from 9, and AR2 reduces 5 features out of 9. The conventional neural network is used for classification in both AR1 and AR2. The method attained 95.6% accuracy on AR1, 97.4% on AR2, and 95.2% on all 9 features with threefold cross-validation scheme.

Mert et al. [14] used radial basis function neural network (RBFNN) for medical data classification and independent component analysis for feature selection. The method selects the one feature vector randomly from 30 features. The method obtained the accuracy in the average of 86%. Bhattacherjee et al. [6] used BPNN for classification. The method achieved 99.27% accuracy. An intelligent medical decision model was developed based on evolutionary strategy [8]. They validated the performance of the method by testing on different datasets. Neural network (NN), genetic algorithm (GA), support vector machine (SVM), K-nearest neighbor (KNN), multilayer perceptron (MLP), radial basis function (RBF), probabilistic neural network (PNN), self-organizing map (SOM), and Naive Bayes (nB) are used as classifiers. Crossover and mutation techniques are applied between different algorithms. The method proves that the SVM classifier on WBC dataset attained better recognition rate than other classifiers.

Ahmad and Ahmed investigated three classification methods, namely RBF, MLP, and PNN and applied to WBC dataset [4]. Among these, PNN shows better result of 97.66% and outcome of MLP is 96.34%. RBF outperformed than MLP, but validation error is little bit higher. Jhajharia et al. [10] proposed a model for breast cancer classification using PCA, to extract features from the dataset and a feed-forward neural network to classify the data. The result is obtained by dividing data by percentage split for training and testing data. Kemal Polat and Salih Gne examined WBCD dataset with least square support vector machine (LSSVM). A set of linear equations is used for training process in LSSVM, but quadratic optimization problem is used in SVM for the same process. This system achieved 98.53% accuracy with k-fold cross-validation method.

Jouni et al. [11] proposed a model based on artificial neural network with multilayered perceptron networks and BPNN. This model learns to classify whether the result of the simulation will be malignant or benign. It also includes weight adjustment factors and bias values. Bewal et al. [5] used multilayered perceptron network with four back-propagation training algorithms such as quasi-Newton, gradient descend with momentum and adaptive learning, Levenberg–Marquardt, and resilient back propagation to train the network. Steepest descent back propagation is used to measure the performance of other neural networks. Levenberg–Marquardt algorithm with MLP achieved best accuracy rate of 94.11%.

SVM with recursive feature elimination (RFE) applied on Wisconsin Diagnostic Breast Cancer (WDBC) Dataset. Principal component analysis (PCA) applied separately for the same dataset for dimensionality reduction process, and SVM is used to classify the dataset. After PCA applied, 25 features are selected. It achieved 98.58% which outperforms SVM and SVM-RFE techniques [22]. Huang et al. [9] investigated a new classification system to develop a robust system using SVM ensembles. Two ensemble learning methods are called bagging and boosting.

Any kernel function in SVM applied to dataset without feature selection process will increase computational time of the system. GA is used to select best features from dataset. The results show that, for small scale datasets, linear kernel with bagging ensembles and RBF with boosting ensembles outperforms than other classifiers. The dataset is divided into 90–10% splits based on k-fold cross-validation. GA+RBFSVM achieved accuracy of 98.00 and 99.52%, respectively, for small and large datasets. Nayak et al. [16] proposed a system which includes adaptive resonance theory (ART-1) network for classification, and it is compared with PSO-MLP and PSO-BBO algorithms which prove that ART is best than other two classifiers. They split the dataset into 70–30 for training and testing the dataset.

Anooj [3] used weighed fuzzy rules to develop a clinical decision support system (CDSS). This system mainly involves two processes, namely generation of fuzzy rules and developing a fuzzy-rule-based decision support system. To ensure better predictions, fuzzy-rule-based decision support system is taken decisions based on the fuzzy rules generated with decision support system. Fuzzy rules are generated based on historical data for better learning. It applied weighted fuzzy rule based on the importance of attributes. Onan [18] proposed a novel classification model based on fuzzy-rough nearest neighbor method. This system consists of three phases, namely instance selection, feature selection, and classification. Fuzzy-rough instance selection method is used for instance selection with weak gamma evaluator to remove erroneous and ambiguous instances. Consistency-based feature selection method is used in conjunction with re-ranking algorithm to efficiently search for possible enumerations in search space. Fuzzy-rough nearest neighbor method is used for the classification process. The method outperformed than other fuzzy approaches.

Ghosh et al. [7] introduced a neuro-fuzzy-based breast cancer classification system. They used WBC, WDBC, and mammographic mass (MM) datasets to apply and evaluate the method. The dataset fuzzified using sigmoidal membership functions and computes degree of membership for individual patterns to various classes. Multilayer perceptron model is used for classification process. At last, defuzzification is applied to generate the results. This method achieved 97.8% accuracy rate. Nilashi et al. [17] proposed knowledge-based system with fuzzy logic for breast cancer. Maximization technique is applied for grouping (clustering) of data. To overcome the problem of multicollinearity, PCA was used. Classification and regression technique (CART) is applied to dataset to generate fuzzy rules for knowledge-based system. Finally, fuzzy-rule-based reasoning system is used the fuzzy rules for classification. Kindie developed breast CDSS with rough set and BPNN-based knowledge mining process [15]. Rough set indiscernibility relation method is applied on dataset to extract minimal set of attributes. Also missing of data is handled in this process. Then, BPNN is used for classification of the dataset. The dataset was divided into 80–20% splits, and it attained 98.6% accuracy.

Schmidhuber [21] provides an overview of deep learning in neural networks. The method proves that the deep learning algorithms reduced the error rate and increase the accuracy with respect to training of algorithm. Abdel and Eldeib [2] applied deep belief network (DBN) for WBC dataset and achieved 99.68% accuracy. The dataset was divided into train-test split of 54.945.1%. DBN follows unsupervised path and back-propagation network to follow supervised path. This system was constructed by BPNN with Levenberg–Marquardt learning function. Here, the weights are initialized with DBN path. This system provides the promising result and outperforms than other classifiers. This motivated to use deep learning concepts for medical data classification. Deep learning reduces the error rate, and it will improve the accuracy rate.

3 Proposed System

The proposed method is used DNN model for classification process and RFE system for selecting a subset of features from all given features. The steps involved in the proposed method are described below, and it is represented in Fig. 2.

1. The WBC dataset is preprocessed to remove the noise from the instances.
2. Select the best four features from the dataset by applying logistic regression model.
3. For iteration process, recursive feature elimination is used.
4. RFE ranked and extracted the best feature and eliminates other features.
5. Classify the dataset using deep neural network.

3.1 Preprocessing

In a machine learning model, to normalize and eliminate redundant, ambiguous data from the dataset, preprocessing techniques are applied. In breast cancer dataset, it consists of 699 instances with 9 feature variables. This dataset supports binary classification models since it has only two class labels called benign and malignant.

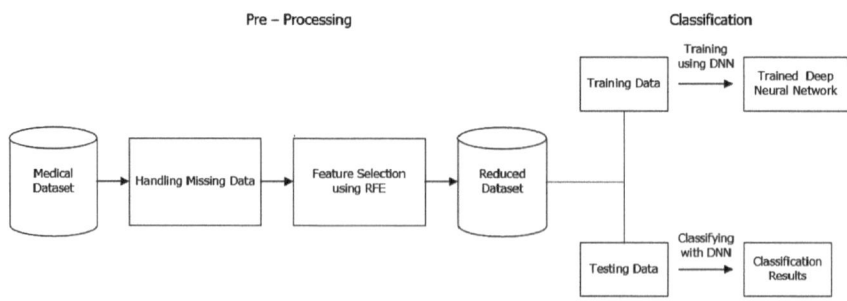

Fig. 2 Architecture of the system

Benign is to identify patients without cancer, and malignant is to identify patients with cancerous tumors. In the 699 instances, 16 of them have missing values. To handle those values, the system removed all the 16 instances from the dataset before feature selection to improve the stability of the system. This system selects the best features from the feature variables to improve the performance and accuracy of the system.

3.2 Feature Selection

The importance of feature selection in a machine learning model is inevitable. It turns the data to be free from ambiguity and reduces the complexity of the data. Also, it reduces the size of the data, so it is easy to train the model and reduces the training time. It avoids over fitting of data. Selecting the best feature subset from all the features increases the accuracy. Some feature selection methods are wrapper methods, filter methods, and embedded methods.

3.2.1 Recursive Feature Elimination

RFE is one of the best feature selection techniques, which adopts greedy optimization technique that comes under wrapper methods. It selects a feature subset from all the given features. In every iteration, the subset is selected, using logistic regression model to train the features. It will be decided whether to keep the feature or to remove it from the subset of features. The model will be constructed by removing unnecessary features. This process continues until all the best features are selected as a subset. Finally, those selected features are ranked based on their elimination order that can calculate every iteration. The RFE process is illustrated in Fig. 3.

In this system, RFE is applied on the pre-processed data to rank the features for next level. These features are chosen based on the outcome of the result from the logistic regression model. At the end of all the iterations, remaining features will be available for selecting those specific features for reducing the dataset. The system

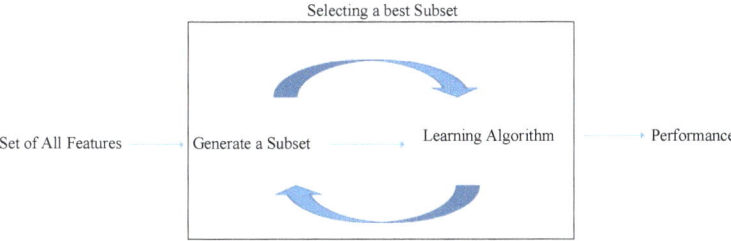

Fig. 3 Process of recursive feature elimination

Table 1 Ranking of
attributes using RFE

Attributes	Ranking
Clump thickness	1
Uniformity of cell size	1
Uniformity of cell shape	1
Marginal adhesion	5
Single epithelial cell size	6
Bare nucleoli	1
Bland chromatin	2
Normal nucleoli	4
Mitosis	3

selected the four best ranked features to provide input for DNN classifier. Those features are clump thickness, uniformity of cell size, uniformity of cell shape, and bare nucleoli. The ranking of attributes are shown in Table 1.

Algorithm of Recursive Feature Elimination

1. Initially, select all the features for feature selection

$$f_n = f_1, f_2, f_3 \cdots f_{n-1}$$

2. Define a model to train the selected features during every iteration

$$\text{Logistic regression: } y = f(x) = \begin{cases} 1 & \beta_0 + \beta_1 x + \varepsilon > 0 \\ 0 & \text{Otherwise} \end{cases}$$

3. At every iteration, some features will be removed from the set of all features, based on the inference of the training model.
4. A final subset of features will be generated based on the limitation of the model or selections of best features are done.
5. Ranking of selected features is done by following the sequence order of elimination of features during every iteration.

3.3 Classification

The system divides the dataset into 80–20, 70–30, and 60–40% train-test split for experimental purpose. Splitting of the dataset is done randomly without following any sequences. After partitioning, a training set of data is initially applied to the classifier. This deep neural network classifier has an input layer with four input nodes, three hidden layers with 10, 20, and 10 hidden nodes, and an output layer with a single node. Since this network has multiple layers with a huge amount of inner nodes, computationally it is expensive but provides promising results after training the model.

3.3.1 Deep Neural Network

Deep neural networks follow the structure of a typical artificial neural network with a complex network model. It helps us to create a model and define its complex hierarchies in a simple form. It has 'n' hidden layers and processes the data from the previous layer called as the input layer, and after every epoch, error rate of the input data will be gradually reduced by adjusting the weights of every node, back propagating the network and continues till reaches better results. Any number of inputs can be assigned as input nodes in input layer. Normally, number of nodes in DNN will be more than the input layer to increase the learning process intensively. Number of outputs can be defined individually as unique output nodes in output layer. The parameters used in DNN are said to be number of nodes in input and output layer, bias, learning rate, initial weights for adjustment, number of hidden layers, number of nodes in every hidden layers, and stop condition for terminating the epochs while execution. In this model, bias value is assigned as 1, which is usually assigned to be 1 in any neural network to avoid nullified network results. Also, learning rate is assigned as 0.15 by default and randomly changed later by trial and error for obtaining varying outcomes from the model. The initial weight of the nodes can be assigned randomly and changed by the network during back propagation by calculating error rate and updated periodically after every epoch. Number of hidden layers and number of nodes in every hidden layer are decided based on the number of inputs and size of the data. Termination condition of the network is said to be either the number of epochs are reached or the expected result is achieved from the learning model. If the number of layers and nodes in a network is more, it will take more time and resource to train the model.

Algorithm of Deep Neural Network

1. Define a neural network with an input layer having n input nodes.
2. Initialize the number of hidden layers needed to train the data.
3. Define the learning rate and bias value for every node. The weight will be randomly selected in intial forward propagation.
4. Define the activation function

Rectified linear unit (ReLU): $f(x) = \max(0, x)$.

5. Define the number of epochs to back propagate the value from the output node.
6. Train the network for given set of training data.
7. After the network is trained, pass the test data to the trained network to find the classification rate of the model.
8. Train the network until the number of epochs is completed (or) expected output is achieved.
9. Calculate the accuracy of the model using evaluation metrics.

4 Experimental Results

WBC dataset consists of 699 instances and 9 feature variables. The dataset supports for binary classification models, since it has only binary values as class label values, i.e., 0 for benign and 1 for malignant. But, the actual values given in the dataset for benign and malignant would be 2 and 4, respectively. To keep the system more stable, convert all the values of the class label from 2 and 4 to 0 and 1. Out of 699 instances, 16 instances contain missing value. The missing instances are removed to reduce the error rate of the system. Finally, 683 instances are used for feature selection. The description of WBC dataset is shown in Table 2.

In recursive feature elimination process, initially 9 features are given as an input to logistic regression model, which is used as learning algorithm. The objective of RFE is to select best possible subset of features from all given inputs. Initially, all the features are trained using learning algorithm and the performance of each feature is separately maintained. Then, the features with least coefficient will be omitted during every epoch and retrained after removing each feature from the input set and continued until required number of features are retained. As an outcome of RFE, single epithelial cell size, marginal adhesion, normal nucleoli, mitosis, and bland chromatin are eliminated individually during every iteration of training process. These features are removed from dataset, and selected four features are used as inputs to DNN for classification process.

The performance of a model is estimated through confusion matrix. The confusion matrix helps to find the classified and misclassified rate of the system. Effectiveness and performance of a system can be measured by calculating the accuracy. The accuracy is defined in Eq. (1).

$$\text{Accuracy} = \frac{\text{TP} + \text{TN}}{\text{TP} + \text{TN} + \text{FP} + \text{FN}} \tag{1}$$

Table 2 Description of Wisconsin Breast Cancer Dataset

Number	Attribute name	Domain	Missing values
1	Clump thickness	1–10	0
2	Uniformity of cell size	1–10	0
3	Uniformity of cell shape	1–10	0
4	Marginal adhesion	1–10	0
5	Epithelial cell size	1–10	16
6	Bare nucleoli	1–10	0
7	Bland chromatin	1–10	0
8	Normal nucleoli	1–10	0
9	Mitosis	1–10	0
10	Class	2, 4	0

where TP represents true positive, TN represents true negative, FP represents false positive, and FN represents false negative.

Sensitivity can be measured to calculate the proportion of correctly identified instances with actual positives.

$$\text{Sensitivity} = \frac{TP}{TP + FN} \tag{2}$$

Specificity is a measure to find the proportion of correctly identified instances with actual negatives.

$$\text{Specificity} = \frac{TN}{TN + FP} \tag{3}$$

F-score is calculated to find the test accuracy of the model. To compute *f*-score, precision and recall are also calculated.

$$\text{Precision} = \frac{TP}{TP + FP} \tag{4}$$

$$\text{Recall} = \frac{TP}{TP + FN} \tag{5}$$

$$F - \text{Score} = \frac{\left(\beta^2 + 1\right) * \text{precision} * \text{recall})}{\beta^2 * \text{precision} + \text{recall}} \tag{6}$$

While calculating *F*-score, it is balanced with $\beta = 1$ and here, bias value is mentioned as β. Also, when $\beta < 1$, it is favor for precision and when $\beta > 1$, it favor for recall.

Once trained the network model with training set, then test set with 20% of instances is applied to explore the accuracy rate of the classifier. As expected, this model outperforms and gives promising result of 98.62% for 80–20% partition split. Also, this system produces a better result of 97.66 and 96.88% for 70–30% and 60–40% splits, respectively. Table 3 shows the accuracy rate, and it is compared with other existing methods.

Table 3 Analysis of result with existing methods

Classification model	Accuracy	Model evaluation
EM-PCA-CART	93.2	10-fold
ART-1	93.2	70–30
LP-SVM	97.33	75–25
SVM-RBF	93.47	10-fold
Proposed method	98.62	80–20
Proposed method	97.66	70–30
Proposed method	96.88	60–40

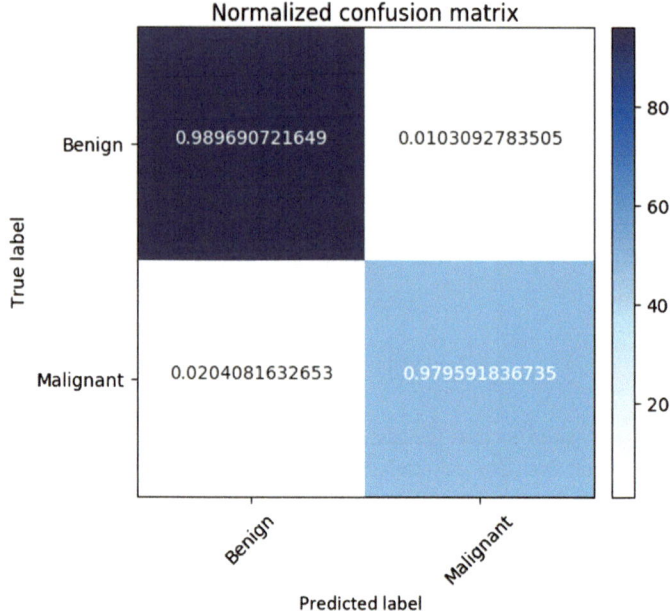

Fig. 4 Confusion matrix for the proposed system

Fig. 5 Performance for the proposed system

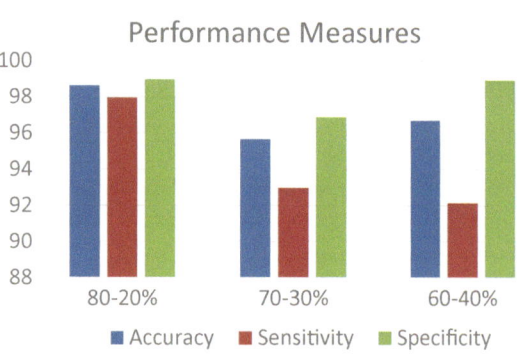

This system achieved better accuracy of 98.62% for 80–20% partitioning. Figure 4 helps to visualize the result of the proposed method. Figure 5 shows the performance of the method in terms of accuracy, sensitivity, and specificity with different train-test split partitioning. Figure 6 displays the ROC curve to analyze the performance of the system.

Fig. 6 Receiver operational
characteristic curve

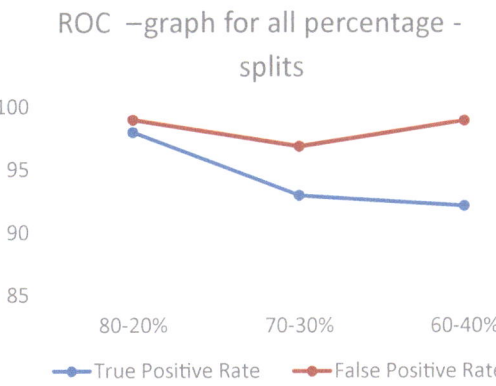

ROC −graph for all percentage -
splits

5 Conclusion

In this modern era, lots of people are facing many problems with modern age diseases. Breast cancer is one of the most common types of deadliest disease raising over time among different countries. Lack of awareness and post-identification of disease will be the major reason for more death rates. Computer-aided diagnosis will be a perfect solution for all kind of peoples to diagnose with accurate results. CAD system will not be a perfect replacement for professional doctors, but this aid will help them a lot, by assisting practitioners, to make a perfect decision by analyzing patient reports. Sometimes, practitioners may do some mistake due to lack of experience or poor analysis of reports. So, it will act as a better remedy for the current medical environment. More accurate decisions are taken, only if the model used to train the system will be unambiguous.

This system provides better results, compared with previous models, and needs a little improvement. The limitation of this system is training time of the algorithm since it has deeply trained the neural network. In GPU-incorporated systems, this system will take less amount of time compared with commercial hardware. So it is expected that the user of this system will have a computationally better system to test and process their data.

6 Future Work

The proposed system used RFE for feature selection and deep neural network system for classifying the data. It provides 98.62% of accuracy in DNN-RFE model. This system has complex multiple hidden layers with a lot of neurons in every layer, which can propagate the input and parse through multiple layers. So, in every epoch, during back-propagation process, the error rate of the system is gradually reduced by adjusting the weights of nodes and network fine tuning the values of nodes in every layer. Since this network architecture has very complex structure, increase in training time is inevitable. In future, to improve the classification rate the researchers can perform particle swarm optimization technique or genetic algorithm for feature selection, which may increase the accuracy of the overall model, by selecting the features based on its fitness values. Deep learning algorithms are more powerful and need high-end computing resources to run for both training and testing phases of learning models, which may lack in its performance when executed in local machines due to less amount of computational power. Cloud-assisted virtual machines or parallel processing systems may be used to optimize the computational efficiency of the model. It will reduce the time to train the system and makes the system to be computationally inexpensive.

References

1. Abbass, H. A. (2002). An evolutionary artificial neural networks approach for breast cancer diagnosis. *Artificial Intelligence in Medicine, 25*(3), 265–281.
2. Abdel-Zaher, A. M., & Eldeib, A. M. (2016). Breast cancer classification using deep belief networks. *Expert Systems with Applications, 46,* 139–144.
3. Anooj, P. (2012). Clinical decision support system: Risk level prediction of heart disease using weighted fuzzy rules. *Journal of King Saud University-Computer and Information Sciences, 24*(1), 27–40.
4. Azar, A. T., & El-Said, S. A. (2014). Performance analysis of support vector machines classifiers in breast cancer mammography recognition. *Neural Computing and Applications, 24*(5), 1163–1177.
5. Bewal, R., Ghosh, A., & Chaudhary, A. (2015). *Journal of Clinical and Biomedical Sciences, 5*(4), 143–148.
6. Bhattacherjee, A., Roy, S., Paul, S., Roy, P., Kausar, N. & Dey, N. (2015). Classification approach for breast cancer detection using back propagation neural network: a study. *Biomedical Image Analysis and Mining Techniques for Improved Health Outcomes,* p. 210.
7. Ghosh, S., Biswas, S., Sarkar, D. C., & Sarkar, P. P. (2016). Breast cancer detection using a neuro-fuzzy based classification method. *Indian Journal of Science and Technology, 9*(14).
8. Gorunescu, F., & Belciug, S. (2014). Evolutionary strategy to develop learning-based decision systems. Application to breast cancer and liver fibrosis stadialization. *Journal of Biomedical Informatics, 49,* 112–118.
9. Huang, M. W., Chen, C. W., Lin, W. C., Ke, S. W., & Tsai, C. F. (2017). Svm and svm ensembles in breast cancer prediction. *PloS one, 12*(1), e0161501.
10. Jhajharia, S., Varshney, H. K., Verma, S., & Kumar, R. (2016). A neural network based breast cancer prognosis model with pca processed features. In: *2016 International Conference on Advances in Computing, Communications and Informatics (ICACCI),* pp. 1896–1901.
11. Jouni, H., Issa, M., Harb, A., Jacquemod, G., & Leduc, Y. (2016). Neural network architecture for breast cancer detection and classification. In: *IEEE International Multidisciplinary Conference on Engineering Technology (IMCET),* pp. 37–41.
12. Karabatak, M., & Ince, M. C. (2009). An expert system for detection of breast cancer based on association rules and neural network. *Expert Systems with Applications, 36*(2), 3465–3469.
13. Kiyan, T., & Yildirim, T. (2004). Breast cancer diagnosis using statistical neural networks. *Journal of Electrical and Electronics Engineering, 4*(2), 1149–1153.
14. Mert, A., Kılıç,, N., Bilgili, E., & Akan, A. (2015). Breast cancer detection with reduced feature set. *Computational and Mathematical Methods in Medicine.*

15. Nahato, K. B., Harichandran, K. N., & Arputharaj, K. (2015). Knowledge mining from clinical datasets using rough sets and backpropagation neural network. *Computational and Mathematical Methods in Medicine*.
16. Nayak, T., Dash, T., Rao, D. C., & Sahu, P. K. (2016). Evolutionary neural networks versus adaptive resonance theory net for breast cancer diagnosis. In: *Proceedings of the International Conference on Informatics and Analytics* (ACM), p. 97.
17. Nilashi, M., Ibrahim, O., Ahmadi, H., & Shahmoradi, L. (2017). A knowledge-based system for breast cancer classification using fuzzy logic method. *Telematics and Informatics, 34*(4), 133–144.
18. Onan, A. (2015). A fuzzy-rough nearest neighbor classifier combined with consistency-based subset evaluation and instance selection for automated diagnosis of breast cancer. *Expert Systems with Applications, 42*(20), 6844–6852.
19. Paulin, F., & Santhakumaran, A. (2011). Classification of breast cancer by comparing back propagation training algorithms. *International Journal on Computer Science and Engineering, 3*(1), 327–332.
20. Prevention Control: Center for Diseases Control and Prevention (2014). URL https://www.cdc.gov/cancer/breast/index.htm.
21. Schmidhuber, J. (2015). Deep learning in neural networks: An overview. *Neural Networks, 61*, 85–117.
22. Yin, Z., Fei, Z., Yang, C., & Chen, A. (2016). A novel svm-rfe based biomedical data processing approach: Basic and beyond. In: *IECON 2016-42nd Annual Conference of the IEEE Industrial Electronics Society*, pp. 7143–7148.

Part III
Knowledge Computing and Competency Development

Sense Disambiguation of English Simple Prepositions in the Context of English–Hindi Machine Translation System

D. Jyothi Ratnam, M. Anand Kumar, B. Premjith, K. P. Soman and S. Rajendran

Abstract In the context of developing a Machine Translation System, the identification of the correct sense of each and every word in the document to be translated is extremely important. Adpositons play a vital role in the determination of the sense of a particular word in a sentence as they link NPs with the VPs. In the context of developing English to Hindi Machine Translation system, the transfer of the senses of each Preposition into the target langue needs done with much attention. The linguistic and grammatical role of a preposition is to express a variety of syntactic and semantic relationships between nouns, verbs, adjectives, and adverbs. Here we have selected the most important and most frequently used English simple prepositions such as '*at*', '*by*', '*from*', '*for*', '*in*', '*of*', '*on*', '*to*' and '*with*' for the sake of contrast. A supervised machine learning approach called Support Vector Machine (SVM) is used for disambiguating the senses of the simple preposition '*at*' in contrast with Hindi postpositions.

Keywords Prepositions · Postpositions · Support Vector Machine
Word embedding

D. J. Ratnam (✉) · M. A. Kumar · B. Premjith · K. P. Soman · S. Rajendran
Center for Computational Engineering and Networking (CEN),
Amrita School of Engineering, Coimbatore, India
e-mail: d_jyothiratnam@cb.amrita.edu

M. A. Kumar
e-mail: m_anandkumar@cb.amrita.edu

B. Premjith
e-mail: b_premjith@cb.amrita.edu

K. P. Soman
e-mail: kp_soman@amrita.edu

S. Rajendran
e-mail: s_rajendran@cb.amrita.edu

D. J. Ratnam · M. A. Kumar · B. Premjith · K. P. Soman · S. Rajendran
Amrita Vishwa Vidyapeetham, Coimbatore, India

© Springer Nature Singapore Pte Ltd. 2018
S. Margret Anouncia and U. K. Wiil (eds.), *Knowledge Computing and Its Applications*, https://doi.org/10.1007/978-981-10-6680-1_13

1   Introduction

Word sense disambiguation (WSD) [16] is one of the major tasks in Natural Language Processing (NLP). For getting the correct and quality MT output, it is indispensable to realize the correct sense of each and every word in a particular sentence. In WSD, we try to detect the actual meaning of a particular word/word group (phrase) in a given context. At the time of translation a text containing English sentences are mainly shady or ambiguous because, [2] it exhibit different sense interpretations. The ambiguity can arise due to the presence of simple prepositions like 'at', 'by', 'for', 'from', 'in', 'of', 'on', to, and 'with'.

Prepositions are most frequently used words in English [1]. According to British National Corpus, four out of the top-ten most frequent words in English is prepositions (of, to, in, and for). Prepositions belong to a grammatical category of word class with defined grammatical and linguistic functions [4]. They possess extraordinary features to bring syntactic and semantic variations. Most of the prepositions are proficient to mark the features of time, space, direction, manner, and a galaxy of abstract senses [6]. In a particular sentence the preposition/Prepositional Phrase (PP) establishes a unique grammatical relationship with its Prepositional object (P-object) in a specific context [13]. It also establishes the semantic relationships of spatial, temporal or logical in nature. The sense of a preposition and the semantic role of a prepositional phrase (PP) are different [9]. The reorganization of the correct semantic role of a PP will helpful to determine the correct sense of a preposition in a given context [11]. Similarly the correct sense reorganization of a preposition will be helpful to us to determine the semantic role of the prepositional phrase (PP). The mapping of English preposition into Hindi postpositions depends on the sense and the semantic role of preposition [20]. The Hindi equivalent will determine the main verb of the verb phrase (VP) or the preceding or following noun (P-object) of the prepositional phrase (PP) [26]. In Machine learning approach, the linguistically analyzed feature of adpositions of both Source Language (SL) and Target Language (TL) are the main components for the sense disambiguation as well as the errors corrections of adpositions [3].

2   Task Determinations

In this section, we will examine the sense disambiguation and the semantic role identification [5] of English simple preposition Phrase (SPP). Same English simple prepositional Phrase (PP) and Hindi Postpositional Phrase (POP) can have the ability to bring a lot of sense with same syntactic construction [20]. Word Sense Disambiguation (WSD) is considered as one of the major tasks in all types of Machine Translation systems (MT systems) [18].

Locative prepositions of all languages express the topological relations between objects in the sense of *area, volume, goal, line of contact, proximity, superiority, contiguity, direction*. The English simple prepositions 'at', 'in', and 'on' are used to mention the topological relation between the primary object/'trajector' and reference object/'landmark' [4]. Sentences like,

(a) The toy car at the gate is mine.
(b) The toy car on the step is mine.
(c) The toy car in the basket is mine.

In all the above sentences, the NP_1 'the toy car' is the primary object or landmark. Here the prepositions reveal the physical alliance between the trajector and the landmark. Each preposition expresses how it differs from each other in their sense with respect to physical alliance of trajector or NP_1 and landmark or NP_2. By the changing of the preposition we can easily bring sense variation without affecting the syntax of the sentence [10]. In other words, prepositions combine with nouns flexibly when describing concrete locative relations. The preceding NP/VP and following NP of a preposition in a sentence play a vital role to determine the semantic role and sense of a Prepositional Phrase (PP) [9]. English simple locative prepositions [15] 'at', 'in' and 'on' are highly flexible, because they show same kind of topological relationship between different objects (trajector and landmark) and the same way they show different kinds of topological relationship between the same object (trajector and landmark). For the determination of the sense and semantic role of English simple spatial preposition/Prepositional Phrase (SLPP) we extracted the linguistic features of the preceding NP/VP and the following NP of the preposition in a sentence [30].

3 Various Relations or Meanings Denoted by Prepositions

Linguistically prepositions are closed set of words show various syntactic and semantic functions [28]. Preposition combine with nouns to form prepositional phrase (PP) [23], alike postposition combine with nouns to form postpositional Phrase (POP). Preposition is the head word of a PP and the noun with which it combines is called prepositional object (P-object); it may be a single noun/a noun group (NP) and NP can have one or more modifiers also. Prepositions and postpositions together can be called Adpositons. Adpositons generally show a relation between nouns and verbs and sometimes between noun and noun (for example, possessive relations in Hindi *rAm kA* beta *'Ram's boy'*). Adpositons show various spatial and temporal relations apart from some abstract relations. The spatial relations include interiority, exteriority, superiority, inferiority, anteriority, posteriority, proximity, contiguity, direction, separation, oppositeness, verticality, transversality, horizontality, circularity, and separation [4]. The temporal relations include point in time, relative to a point, period of time, relative to a period, anteriority, posteriority, and frequency. Abstract relations include cause, reason, etc [17]. The most important fact is that adpositons are able to produce vital semantic variations without any re-arrangement of other lexical items at the sentence level. By simply changing the preposition, we can change the semantics of a sentence [4].

Table 1 Functions of prepositions and examples

Sense	Examples
position near	*I sat by my secretary*
movement near	*I walked* by *the post office*
agency	*The letter will be typed by the secretary*
relative to a point in time	*She will finish them by tomorrow*
Means	*It will be sent by registered post*
Extent	*The envelop measures 9 cm by 6 cm*

4 Polysemy and Synonymy in Prepositions

Prepositions exhibit polysemy and synonymy. A single preposition can have a number of functions or meanings and a number of prepositions can denote a single function or meaning [19]. The preposition '*by*' [4]. Like a single function or meaning can be shared by more than one preposition as shown below (Table 1) [4].

1. I felt sick in the stomach - I felt sick at the stomach
2. I dreamt of you last night - I dreamt about you last night
3. He died of a heart attack - He died from a heart attack
4. In consequence of this... - As a consequence of this...
5. In reference to your letter - With reference to your letter

5 Mapping English Prepositions with Hindi Postpositions

English Prepositions show higher degree of ambiguity. This particular nature of English prepositions increases the complexity of translation [31]. Hindi postpositions too show different senses in different contexts with the same syntactic structure [26]. So Hindi postpositions are also highly complex and the selection of postpositional equivalents for English prepositions is not an easy job, especially in the context of possessive postpositions. Possessive postposition in Hindi has six variant forms: three variant forms (kA/ke/ki) combine with proper and personal nouns and three variant forms (rA/re/rI) combine with pronouns of first and second persons. The postpositional equivalents of Hindi for English prepositions are not parallel [20]. The matching of English preposition with Hindi postposition is complex [17].

5.1 Mapping of preposition 'by' with Hindi

English prepositions 'by' denotes different senses [24] depending on the context.

1) Did you come by car?

   क्या आप कार से आये ?
   (kyA Ap kAr se aye?)

2) Most of the people cook by firewood.

   अधिकतर लोग लकडी से भोजन पकाते हैं।
   (adlhikatar log lakaDI se bhojan pakAte hain)

3) The meal was served by his servant.

   खाना उसके नौकर के द्वारा परोसा गया था।
   (khAnA uske nOkar ke dwArA parosA gaya thA)

4) I sat by my secretary

   मैं अपने सचिव के बगल में बैठा।
   (main apane saciv ke bagal men baithA)

5) I walked by the post office.

   मैं डाक घर के नज़दीक से चला।
   (main DAk ghar ke nazdIk se clA)

6) He stopped our bus by the hotel for breakfast.

   उन्होंने नाश्ते केलिए होट्ल के पास गाडी रोकी।
   (unhone nASte keliye hottal ke pAs gADI rokI)

7) The referee ran up and caught Graham by the shoulder.

   दौड़ते हुए रफरी ने ग्हाम के कन्धे पर पकडा।
   (daudte huye rafarI ne graham ke kandhe par pakadA)

8) By 11.p.m. all the students were back.

   रात ग्यारह बजे तक सभी छात्र लौटे।
   (rAt gyArh bje tk sbhI CAtr laute)

9) He will complete this work by next month.

   वह यह काम कम से कम अगले महीने में पूरा करेगा।
   (vah yah kAm km se km agle mahIne men pUra karegA)

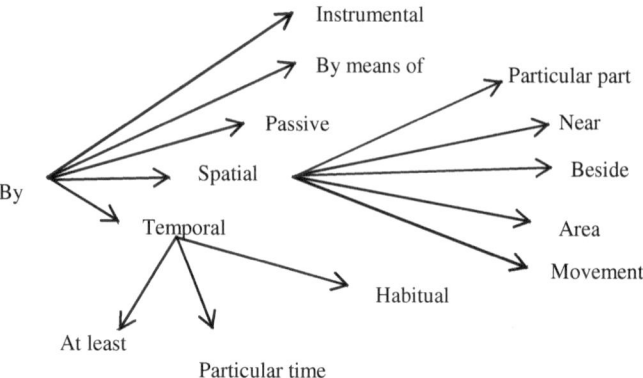

Fig. 1 Different senses denoted by 'by'

10) Young man who goes about his work by day as peaceable civilian turns into Soldier by night.

शान्तिप्रिय जवान नागरिक जो दिन में अपने काम में व्यस्त रहता है वह रात में सैनिक बन जाता है।

(SAntipriy javAn nAgrik jo din meM apane kAm men vyast rhtA hai vah men sainik bnjAtA hai)

'By' denotes different functions such as 'spatial', 'by means of', 'instrumental', 'passive', 'area' and temporal'. 'Spatial' function has different senses such as 'position beside', 'position near', 'position of a particular part of something' and 'movement' [25]. 'Temporal' function has different senses such as 'at least', 'particular time' and 'an action done in a habitual manner and performed during day or night. 'The passive voice of a verb is always denoted by the preposition by. It is also used to indicate 'area'. The polysemy of by is shown in the following diagram (Fig. 1).

The comparison of English by-prepositional phrases with the equivalent postpositional phrases in Hindi discloses the following information,

- '*by-prepositional phrases*' denoting different spatial senses are mapped into different postpositions in Hindi.
- '*by-prepositional phrase*' denoting the spatial sense 'beside' is mapped with 'ke bagal men' in Hindi [14].
- '*by-prepositional phrase*' denoting the spatial sense 'near' is mapped with '*ke pAs*' in Hindi.
- '*by-prepositional phrase*' denoting the spatial sense particular part' 'is mapped with '*par*' in Hindi.
- '*by-prepositional phrase*' denoting the spatial sense 'movement' is mapped with 'ke nzdIk *se*' in Hindi [14].
- *by-prepositional phrase* denoting the sense 'by means' is mapped with '*se*' in Hindi.

Table 2 Equivalents of Hindi Postpositions against '*by*'

English preposition 'by'	Hindi postpositions
Spatial	ke bagal meM, ke nazdIk se,ke pAs
by means of temporal	se
Instrument	se
Passive	ke dwara
Area	par
Temporal	tk, meM, kam se kam

- '*by-prepositional phrase*' denoting the sense 'instrumental' is mapped with '*se*' in Hindi.
- '*by-prepositional phrase*' denoting the sense 'passive voice' is mapped with 'ke dwArA' in Hindi.
- '*by-prepositional phrase*' denoting the temporal sense 'at least' is mapped with '*kam se kam*' in Hindi [14].
- '*by-prepositional phrase*' denoting the temporal sense 'particular time' is mapped with '*tk*' in Hindi.
- '*by-prepositional phrase*' denoting the temporal sense 'habitual' is mapped with 'men' in Hindi.
- '*by-prepositional phrase*' denoting the sense 'area' is mapped with '*par*' in Hindi.

Table 2 shows the mapping of English preposition '*by*' with different postpositions in Hindi according to the different contextual senses denoted by it.

5.2 Mapping of Simple Preposition 'to' into Hindi

The English preposition to also denote different senses depending upon the context of its occurrence. The following examples will illustrate this fact (Fig. 2).

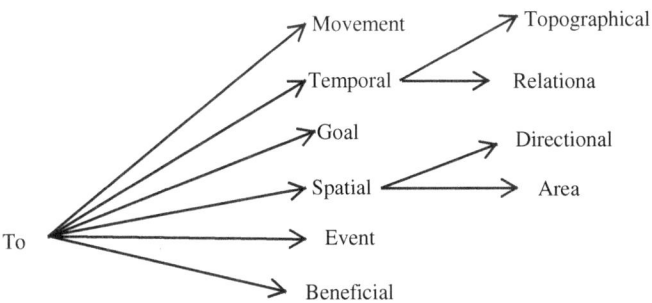

Fig. 2 Different sense produced by '*to*'

11) He went to Delhi.

वह दिल्ली गया।
(vah diLLI gyA)

12) He went to his grandfather.

वह अपने दादाजी के पास गया।
(vah apane dAdAjI ke pAs gyA)

13) He gave some sweets to the children.

उसने बच्चों को मिठाईयां दीं।
(usne baccOM ko mithAyiyAM dIM)

14) Please send it back to me.

कृपया इसे मुझ को वापस भेज दो।
(krpayA ise mujh ko vaps bhej dO)

15) He worked from Monday to Friday.

उसने सोम वार से शुक्र वार तक काम किया।
(usne somavAr se SukrvAr tak kAm kiya)

16) The court was adjourned to coming Monday.

अगले सोमवार तक अदालत स्थगित की गयी।
(agle somavAr tak adAlat sthgith kI gayI)

17) You must come to dinner.

आप को डिन्नर केलिये ज़रूर आना चाहिए।
(Ap ko dinner keliye zaruur AnA chAhie)

18) The entire wall should be painted from top to bottom.

ऊपर से नीचे तक पूरी दीवार की पेयिन्ट करनी चाहिए।
(Upr se nIce tk pUrI dIvAr peyinT karnI cAhie)

19) The killer whale broke the surface no more than twenty yards to the north of me.

कातिल व्हेल मेरे दायीं ओर पानी की सतह पर कम से कम बीस गज दूरी पर उभरी।

(kAtil vhail apane daylmor pAnI kI stah ljbhj bIs gj dUrI pr ubhrI)

The comparison of English prepositional phrase (PP) with simple preposition 'to' and its equivalent Hindi postpositional phrases reveals the following information:

Table 3 Equivalents of Hindi Postpositions against 'to'	English preposition 'to'	Hindi postposition
	movement	*null, kepAs*
	Beneficial	*ko*
	Goal	*ko*
	Spatial	*tk, IMor*
	Temporal	*tk*
	Event	*ke liye*

- *'to-prepositional phrase'* denoting the sense 'moving towards a topographical region' is mapped with *'null unit'* in Hindi [14].
- *'to-prepositional phrase'* denoting the sense of 'moving to the P-object' is mapped with *'ke pAs'* in Hindi.
- *'to-prepositional phrase'* denoting the sense 'area' is mapped with *'tak'* in Hindi.
- *'to-prepositional phrase'* denoting the spatial sense 'direction' is mapped with *'IM or'* in Hindi.
- *'to-prepositional phrase'* denoting the temporal sense 'up to' is mapped with *'tak'* in Hindi.
- *'to- prepositional phrase'* denoting the sense 'beneficial' is mapped with *'ko'* in Hindi.
- *'to-prepositional phrase'* denoting the sense 'goal' is mapped with *'ko'* in Hindi.
- *'to-prepositional phrase'* denoting the sense 'event' is mapped with *'keliye'* in Hindi.

As shown in the examples given above *'to'* be can equaled by *'null'*, *'ke pAs'*, *'ko'*, *'tak'*, *'IMor'* and *'ke liye'* in Hindi. Table 3 illustrates this fact.

5.3   Mapping of Simple Preposition 'with' with Hindi

The preposition 'with' in English is also capable of expressing different senses depending on the context as illustrated below (Fig. 3).

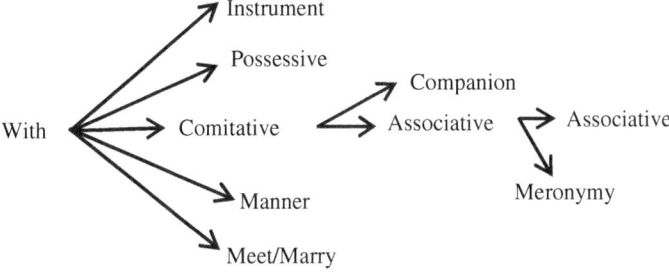

Fig. 3 Different senses expressed by *'with'*

20) I wrote the poem with this pen.

मैंने इस कलम से कविता लिखी।

(meMne is kalam se kavita likhI)

21) People celebrated the victory with great joy.

लोगों ने अति उत्साह से जीत को मनाया।

(logoM ne ati utsAh se jIt ko manAyA)

22) Yesterday I met with his family.

कल मैं उनके परिवार से मिला।

(kal maine unke parivar se milA)

23) He attended her wedding with our friends.

उन्होंने हमारे मित्रों के साथ उसकी शादी में भाग लिया।

(unhonne hamAre mitroM ke sAth uskI SAdI meM bhAg liyA)

24) Grandfather ate apple with grapes.

दादाजी ने सेब के साथ अंगुर खाये।

(dAdAjI ne seb ke sAth anjur khAye)

25) She bought a silk sari with golden brocade.

उसने सुनहले जरीवाली साडी खरीदी।

(usne sunhale jarIvAlI sAdI kharIdI)

26) I saw a girl with blue eyes.

मैंने नीली आंखोंवाली लड़की को देखा।

(mainne nIlI aMkhOm vAlIaDkI ko dekhA)

In the first sentence, '*with*' denotes instrumental sense; in the second, manner sense; and in the third and fourth, commutative sense which can be divided into two: companion sense and associative sense; in the fifth and sixth sentences, '*with*' denotes associative sense; and finally in the seventh sentence, '*with*' denotes possessive sense (Table 4).

- '*with-prepositional phrase*' denoting the sense 'instrumental' matches with '*se*' in Hindi.
- '*with-prepositional phrase*' denoting the sense 'manner' matches with '*se*' in Hindi [14].

Table 4 Equivalents of Hindi Postpositions against '*with*'

English preposition 'with'	Hindi postpositions
Instrumental	*se*
Comitative	*ke sAth, vAlA-forms*
Possessive	*vAlA-forms*

- *'with-prepositional phrase'* denoting the sense 'meeting' matches with *'se'* in Hindi.
- *'with-prepositional phrase'* denoting the sense 'companion (together)' matches with *'ke sAth'* in Hindi.
- *'with-prepositional phrase'* denoting the associative sense (along with) matches with *'ke sAth'* in Hindi.
- *'with-prepositional phrase'* denoting the associate sense (with meronymy relation) matches with *'vAlA-forms'* in Hindi.
- *'with-prepositional phrase'* denoting possessive sense matches with *'vAlA-forms'* in Hindi.

Thus English preposition *with* can be equated with three postpositions Hindi: '*se*','*ke sAth*', and '*vAlA*-forms'. The 'meet and marry verbs' classified by Levin [7] which takes with-prepositional phrase is equaled by '*se*'. But in the context of social interaction '*ke sAth*' is preferred for '*se*' in Hindi [14] as in the following examples.

27) Hon. Minister interacted with foreign delegates.

माननीय मन्त्री जी ने विदेशी प्रतिनिधियों के साथ बातचीत की।
(mAnanIy mntrI jI ne videSI prtinidhiyoM ke sAth bAtcIt kI)

5.4   Mapping of the Preposition 'in' with Hindi

The use of the preposition '*in*' in English with two different senses depending on the context.

28) Who sleep in the other wing?

उस विंग में कौन सो रहा है?
(us ving meM kOn so rahaahaI?)

29) Deva stayed in America for four weeks.

देवा चार हफ्ते अमेरिका में बीता है।
(deva chAr hafte amerikaa meM bItA hai)

30) The assurance was given in Parliament.

संसद में आश्वासन दिया गया था।
(samsad meM ASvAsan diyA gyA thA)

31) Tender lymph nodes in your neck and armpits.

आप की टोंटी और बग़ल में स्थित मृदु लिंफ़ गांठ।
(Ap kI ToMTI Or bajal men sthit mrudu liMf gAnd)

Table 5 Equivalents of Hindi Postpositions against 'in'	English preposition 'in'	Hindi postpositions
	Spatial	mem
	Temporal	mem

32) What did you want to become in the future?

आप भविष्य में क्या बनना चाहते थे?

(Ap bhavishya meM kon bananA chAhate thae?)

33) In the days that followed her operation, she spent a lot of time in bed.

उसके ऑपरेशन के बाद के दिनों में, वह बिस्तर पर बहुत समय बितायी।

(uske aupreSn ke baad ke dinoM meM, vha bistar par bahut samay bitAyI)

The preposition *in* with spatial sense indicates the relation between the trajector and the landmark; here the primary object which is clearly bounded by the landmark or the landmark act as a container. The temporal sense of '*in*' indicates a long period of time. The comparison of 'in-prepositional phrases' with its equivalents from Hindi will reveal the following information (Table 5).

- '*in-prepositional phrase*' in the spatial sense matches with '*men*' in Hindi.
- '*in-prepositional phrase*' in the temporal sense matches with 'men' in Hindi.

5.5 Mapping of the Preposition 'on' with Hindi

The use of preposition '*on*' in English with two different senses depending on the context is illustrated below (Table 6).

34) I was on that hill.

मैं उस पहाड़ पर था।

(main us pahAD par thA)

35) He himself slept on the floor.

वह खुद ही फर्श पर सोया।

(vah khud hI frS par soyA)

36) Dr. Watson standing on the moor.

ऊसर मैदान पर डाकटर वाट्सन खडा है।

(Usar maidAn pra Dr.vATsane khadA hai)

37) He will be here on 10 May.

वह दस मई को यहां होगा।

(vah das mai ko yahAn hogA)

Table 6 Equivalents of Hindi postpositions against *'on'*	English preposition 'on'	Hindi postpositions
	Spatial	*par*
	Temporal	*ko*

The comparison of 'on-prepositional phrases' with its equivalents from Hindi will reveal the following information:

- 'on-*prepositional phrase*' in spatial sense matches with *'par'* in Hindi.
- 'on-*prepositional phrase*' in temporal sense matches with *'ko'* in Hindi.

5.6 Mapping of the Preposition 'from' with Hindi

The simple spatial preposition *'from'* has three different senses; all these three senses can be mapped into Hindi with the postposition *'se'*. The following examples will illustrate the translation equivalents of *'from- prepositional phrase'* in Hindi.

38) Remove the crusts from the cake,

केक से पापडी हटाओ,
(kek se papadI haTAo)

39) He has learnt from professionals.

वह अनुभवी व्यक्तियों से सीख चूका है।
(vah anubhavI vyaktiyon se sIkh chukA hai)

40) The light rays emitted from the stars.

तारों से उत्सर्जित प्रकाश किरण।
(tAron se utsarjit prakAS kiraN)

41) She looked at it from a distance.

उसने दूर से इसे देखा।
(usne dUr se ise dekhA)

42) He says he is from the Times of India.

वह कहता वह टैंस आंफ इडिया से आता है।
(vah kahatA vah times Anf idiyA se AtA hai)

The comparison of 'from-prepositional phrases' with its equivalents from the Hindi sentences reveals the following information:

- *'from-prepositional phrase'* denoting the sense of starting point of a particular action is mapped into Hindi with the postposition *'se'*.
- *'from-prepositional phrase'* denoting sense of standing position of an observer is mapped into Hindi with the postposition *'se'*.
- *'from-prepositional phrase'* denoting sense of separation is mapped into Hindi with the postposition *'se'*.
- *'from-prepositional phrase'* denoting sense of someone's origin or place of living is mapped into Hindi with the postposition *'se'*.

5.7 Mapping of the Preposition 'Of' with Hindi

The *'of-prepositional phrase'* shows the relation between the two Noun Phrases (NPs) in a particular sentence. The *'possessive sense'* of *'of-prepositional* phrase' is equated into Hindi with the same *'possessive postpositional phrase'*. But the forms of Hindi possessive postpositional phrase depends on the Number and Gender (N &G) of the preceding and the following nouns of the *'of'* preposition. The examples given below will illustrate the various forms of Hindi 'possessive postposition' against the English preposition *'of'*.

43) A piece of cake.

केक का टुकड़ा।
(*kek kA tukadA*)

44) The plays of Shakespeare are very interesting for us.

शेक्सपियर के नाटक हमारे लिए बहुत दिलचस्प हैं।
(*sheksapiyar ke nAtak hamAre lie bahut dilachasp hai*)

45) Manasa Devi the wife of Lord Siva

भगवान शिव की पत्नी मनसा देवी।
(*bhagavAn shiv kI patnI manasa devI*)

46) The wives of King George,

महाराजा जार्ज की पत्नियां,
(*mahArAjA jArj kI patniyan*)

47) King Dasaratha's three wives and the princess of Mithila

महाराजा दशरथ के तीन रानियां और मिथिला की राजकुमारी
(*maharAjA daSarath ke tIn rAniyan aur mithilA kI rAjakumArI*)

Table 7 Equivalents of Hindi Postpositions against '*of*'

English Preposition 'of'	Hindi Postpositions
Noun masculine singular +of	kA
Noun masculine plural +of	ke
Noun feminine singular +of	kI
Noun feminine plural +of	kI

The English simple preposition '*of*' have three forms of translation equivalents '*kA/ke/kI*' is possible in Hindi. The following information can be inferred by comparing the '*of-prepositional phrase*' with their equivalents from Hindi (Table 7).

- If the noun in the Noun phrase2 (NP2) is 'masculine singular, '*of*'-prepositional phrase' must be translated into Hindi with the postposition '*kA*'.
- If the noun in the Noun Phrase2 (NP2) is 'masculine plural', '*of*'-prepositional phrase' must be translated into Hindi with the postposition '*ke*'.
- If the noun in the Noun Phrase2 (NP2) is 'feminine singular or plural' the '*of*'-prepositional phrase' must be translated into Hindi with the postposition '*kI*'.
- If an adjective-numeral which precedes the noun of the '*of-prepositional phrase*' it must be translated into Hindi with postposition '*ke*'. (Example sentence 47) (Table 7)

5.8 Mapping of the Preposition 'at' with Hindi

English simple spatial preposition 'at' has many senses depending on the context of its use in spite of occurring in the same syntactic construction. All these senses are mapped by different postpositions in Hindi as illustrated below (Fig. 4).

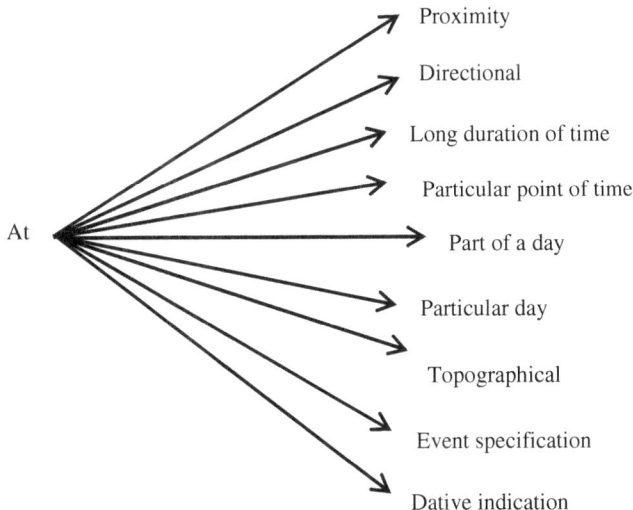

Fig. 4 Different senses expressed by 'at'

48) Yes we'll meet at the school gate.

हां हम स्कूल के प्रवेश द्वार पर मिलेंगे।

(*hAM ham skUl ke praveS dvAr par mileM ge*)

49) I heard someone at the door.

फाटका पर किसी को मैंने सूना।

(*fATak par kisI ko meM ne sunA*)

50) Yesterday we met at the bus stop.

कल हम बस स्टाप पर मिले थे।

(*kal ham bas stop par mile the*)

51) We stopped at KFC on our way to Chennai.

चेन्नै जाते समय हम केफसी के फाटक पर रुके।

(*Chennai jAte samay ham KFC ke dvAr par ruke*)

52) Who is the man standing at the bus stop?

बस स्टाप पर खड़ा आदमी कौन है ?

(*bas stop par khadA AdmI kaun hai?*)

53) He saw a police van at the back gate.

उसने पिछवाड़े के द्वार पर एक पुलिस वैन को देखा।

(*us ne piChvADe ke darvAze par ek pulIs vAN ko deghA*)

54) We had stopped at Milan, Florence, and Pisa on our way to Rome.

रोम जाते समय हम मिलन,फ्लोरंस और पिसा में रुके थे।

(*rom jAte samay ham Milan,florans aur pisa meM ruke the*)

55) Devan had come to the airport to meet me at Los Angles.

लोस आन्जल्स हवाई अड्डे में मुझे देखने केलिए देवन आया था।

(*los enjal havAE adde meM mujhe dekhne keliye devan AyA thA*)

56) I'm sure we stopped at York during our trip.

मुझे यकीन है कि अपनी यात्रा के दौरान हम यॉर्क में हम रुके थे।

(*mujhe yakIn hai ki apanI yAthra ke daurAn ham yArk meM ruke the*)

57) We went to a festival at Royal Festival Hall.

हम रॉयल फेस्टिवल हॉल में एक त्योहार देखने गए थे।

(*ham royal festival hAnl men ek tyohAr dekhne gaye the*)

58) The meeting took place at the company's headquarters.

कंपनी के मुख्यालय में बैठक होगी।
(kampanI ke mukhyAlay meM baiDahak hogI)

59) He is at Manchester studying linguistics.

वह मैनचेस्टर में भाषाविज्ञान का अध्ययन कर रहा है।
(vah mAncastra men bhAshA vijyAn kA adhyayan kar rahA hai)

60) Were there many people at the meeting?

क्या बैठक में बहुत सारे लोग थे?
(kyA baithk meM bahut saare log the?)

61) Mrs. Thomas expressed her idea at the meeting.

श्रीमती थॉमस ने अपने विचार को बैठक पर व्यक्त किया।
(shrImatI thomas ne apane vichaar ko baithak par vyakt kiya)

62) I saw Jack at a football match.

मैंने एक फुटबॉल मैच में जैक को देखा।
(menne ek fudbAl maIch men jAk ko dekhA thA)

63) I saw him at a concert last Sunday.

गत इतवार को एक संगीत कार्यक्रम में मैंने उसे देखा।
(gata etavAr ke sangIta karyakram men meMne use dekhA thA)

64) They were great success at Edinburgh Festival.

एडिनबर्ग समारोह में उन्होंने बहुत बड़ी सफलता पाई।
(edinbarg festival meM unhonne bahut badI safalatA pAI)

65) He shoots at them.

वह उन कीओर गोली मारता है।
(vah unkIor golI mArtA hai)

66) He pointed at me.

उसने मेरी ओर इशारा की।
(usne merIor iSArA kI)

67) He threw the ball at me.

उसने मेरी तरफ गेंद फेंकी।
(usne merI tarf gend feMkI)

68) He aimed the gun at them.

उसने उन कीओर बंदूक तानी।
(usne unkIor bandUktAnI)

69) Shayam tossed the ball at Ann.

श्याम ने अन्ना कीओर गेंद फेंकी।
(shayAM ne annA kI or gend fenkI)

70) He splashed at me.

उसने मेरी ओर छिडकाया।
(uasne merI aur ChidakAyA)

71) The stranger stared at him.

उस अजनबी आदमी ने मुझे घूरकर देखा।
(us ajnabi AdmaI ne mujhai ghUr kar dekhA)

72) The stray dog stared at the child.

भगेड़ू कुत्ता बच्चे को घूर कर देखा।
(bhageDU kuttA bacce ko ghUr kar dehgA)

73) Sita gazed at the moon.

सीता ने टकटकी लगाकर चाँद को देखा।
(sItA ne takatakee lagaakar chAnd ko dekha)

74) My obedient dog Jimmy jumped at them.

मेरा आज्ञाकारी कुत्ता जिम्मी ने उनपर छलांग मारा।
(merA AghyakArI kuttA jimmI ne unpar chalAng mArA)

75) Tom looked at the dog.

टॉम ने कुत्ते को देखा।
(tAM ne kutte ko dekhA)

76) Linda winked at the audience.

लिंडा ने दर्शकों को देख कर आँख मारी।
(lindA ne drSkoM ko dekh kar AMkh mArI)

77) I have a meeting at 9.a.m.

मुझे सुबह नौ बजे पर एक बैठक है।
(mujhe subah nau baje par ek baithak hai)

78) The show will start at 6.30.

प्रदर्शन छह बज कर तीस मिनिट पर शुरू होगा।

(pradarshan chhah baj kar tIs minit par shurU hogA)

79) We'll start our journey at 10.30 a.m.

हम अपनी यात्रा सुबह दस बज कर तीस मिनट पर शुरू करेंगे।

(ham apanI yAtrA subah das bajakar tIs minat par shurU karenge)

80) At daytime, we collected firewood from forest.

हम दिन में जंगल से लकडी एकत्रित करते थे।

(ham din ko jangal se lakadI ekatrit karate the)

81) At midnight, I heard a sob of a woman.

आधी रात को मैंने एक महिला की सिसकियां सुनी।

(Adhee rAt ko mainne ek mahilA kee sisakiyAn sunI)

82) She makes cakes only at Christmas.

वह सिर्फ क्रिसमस को केक बनाती है।

(vah sirf krisamas ko kek banAtI hai)

83) We get paid at the end of the month.

हमें महीने के अंत में भुगतान मिलता है।

(hameM mahIne ke ant meiM bhugatAn milatA hai)

84) He is going to Varanasi at the end of this year.

वह इस साल के अंत में वाराणसी जायेगा।

(vah is sAl ke ant mein vArAnasI jAyegA)

The following information can be inferred by comparing the *'at-prepositional Phrases'* with their equivalents form Hindi (Table 8).

Table 8 Equivalents of Hindi postpositions against 'at'

English preposition 'at'	Hindi postpositions
Proximity	*par*
Topographical relation	*meM*
Event specification	*meM*
Dative indication	*meM*
Direction	*kI aur/kI tarf*
Particular point of time	*par*
Part a day	*ko*
Particular day	*ko*
Indication of long duration of time	*men*

- If the trajector is a particular point or very closer to the landmark, '*at*' must be translated into Hindi with the postposition '*par*'.

- If the '*at -prepositional phrase*' expresses the topographical relation between the trajector and the landmark indicating a wider geographical or topographical area, it must be translated into Hindi with the postposition '*men*'.

- If the noun in the preceding NP denotes a particular event, the preceding NP will determine the sense of 'at'; 'at-prepositional phrase' in this context must be translated into Hindi using the postposition 'men'.

- When '*at*' collocates with certain types of verbs such as verbs of gestures/signs involving body parts, throw verbs (projectile) [7] the trajector becomes the target of the action. In that context, '*at- prepositional phrase*' denotes the sense of direction and so it must be translated into Hindi using the postpositional phrase '*kIour/kItarf*'.

- When prepositional phrases with '*at*' collocate with the certain verbs like 'laugh', 'gaze', 'stare', 'bark', 'shout', (bark verbs/wink verbs, verbs of gestures/signs involving body parts) [7] '*at-prepositional phrase*' indicates the object of the preposition. So in that context, '*at-prepositional phrase*' must be translated into Hindi with the postposition '*ko*'.

- If '*at-prepositional phrase*' gives the sense of Particular point of time, it must be translated into Hindi with the postposition '*par*'.

- If '*at-prepositional phrase*' gives the sense of a part of the day, it must be translated into Hindi with the postposition '*ko*'.

- If '*at-prepositional phrase*' gives the sense of a particular day, it must be translated into Hindi with the postposition '*ko*'.

- If '*at-prepositional phrase*' gives the sense of long duration of time, it must be translated into Hindi with the postposition '*men*'.

6   Machine Learning Approach to the Disambiguation of '*At*'

Support Vector Machine (SVM) [27, 29] is a supervised machine learning algorithm which is mainly used for classification. Basically SVM is a binary classifier. Let $X = \{x_1, x_2, \ldots, x_N\}$ be the training data and $y_i \in \{-1, +1\}$ be the set of class labels. Now, SVM classifier finds a hyper-plane which classifies data points in two classes with maximum separation. Data points that lie on the boundary of each class are called support vectors. Since SVM [12] is a linear classifier, algorithm always tries to achieve a linear classification between the classes. This is easy when the data are linearly separable. But when the data are non-linearly separable, kernel trick comes into play to obtain the linear classification. With kernel trick, input data is mapped to a higher dimensional feature space where a linearly separating hyper-plane can be drawn.

Formulation for the SVM algorithm is given as,

$$\min_{w,b} \frac{1}{2} w^T w + c \sum_{i=1}^{N} \zeta_i$$

Subject to the constraints $y_i(w^T \phi(x_i) + b) \geq 1 - \zeta_i$, $\zeta_i \geq 0$, $i = 1, \ldots, N$

Where w is the coefficient vector, c is the regularization parameter, ζ_i is the slack variable for handling non-separable data, y_i is the class label and $\phi(.)$ is the kernel function for mapping the input data into a higher dimensional space. When a new data point comes, the decision function $f(x) = sign(w^T x + b)$ finds target class.

6.1 Random Kitchen Sink (RKS)

A. Rahimi and B. Recht used an explicit random feature mapping algorithm, Random Kitchen Sink (RKS) [21] to handle non-linearly separable large datasets. RKS algorithm is a random Fourier approximation for the kernel function.

A kernel function can be written as,

$$k(x, y) = \langle \phi(x), \phi(y) \rangle$$

If we choose an appropriate function, one could approximate the kernel function as very well. But for general kernels, this mapping may yield infinite dimensional feature representation. So in order to obtain a finite dimensional higher order feature mapping, we use RKS algorithm. So RKS algorithm is a mapping function which maps input data into a higher dimensional space. The main idea behind RKS algorithm is Bochner's theorem [22]. Based on Bochner's theorem, Rahimi and Recht proposed a method to sample from Fourier transform of shift invariant kernels to obtain the approximations.

Radial Basis Function (RBF) is one of the most popular shift invariant, positive definite kernel functions. It can be mathematically written as,

$$k(x, y) = k(x - y) = e^{-\sigma \|x-y\|^2}$$

So according to Bochner's theorem,

$$k(x, y) = \int_{\mathbb{R}^N} p(\omega) e^{-i\langle \omega, (x-y) \rangle} d\omega$$

So to obtain an approximate mapping $z(x)$, RKS algorithm computes

$$z(x) = \frac{1}{\sqrt{d}} \left[\cos(\omega^T)x, \sin(\omega^T)x \right]$$

6.2 Experiments and Results

Features are the major ingredient of machine learning algorithms. So in order to perform the classification, the sentences in English must be represented numerically and we used fastText toolbox [8], which is a toolbox developed by Facebook Research team for learning vector representation.

Using fastText, each sentence is represented using a vector of length 100. Initially, the model was trained using our corpus to capture the syntactic and semantic information in the corpus and then using this model; vector representation for each sentence in the corpus is derived. While training the model, the minimum number of word occurrence is set to 1 so that presence of all the words in the corpus can be taken into consideration. Size of the context window is set to 5 to take account of the effect of phrases such as '*at the end of*', '*at the beginning of*', '*look at*'.

After obtaining the sentences as vectors, SVM was used for classification of different senses. An explicit feature mapping algorithm was used to map the input features to a higher dimensional feature space. The input data is highly non-linearly separable and hence the features were projected to a dimension 40 times the input dimension to obtain relatively better classification. We obtained a classification accuracy of 64% for disambiguating the sense of '*at*'. The model was generated with 100 sentences in each class and also tested with another 100 sentences in each class. The confusion matrix for this sense disambiguation is shown in Table 10 and its graphical representation is shown in Fig. 5. X-axis of the figure represents the class labels (1, 2, 3 and 4) and the Y-axis represents the number of senses of English preposition '*at*' classified into each of the classes.

The class labels for each sense is shown in Table 9. From Table 10, we can infer that the senses are classified with reasonable accuracy. The data in class 1 and 2 (*at_par*, *at_men*) are classified with good accuracy, where as, for data in class 3 and 4 (*at_ko*, *at_klaur/kltraf*) showed poor results. This is mainly due to the complicated structure of the given sentences.

Fig. 5 Graph showing the distributions of confusion matrix

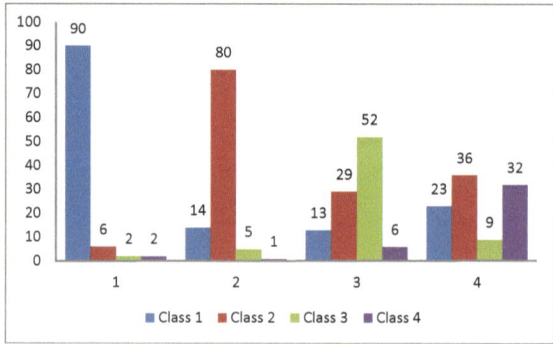

Table 9 Senses and respective class labels

Sense	Label
at_par	1
at_meM	2
at_ko	3
at_kI tarf/kI aur	4

Table 10 Confusion Matrix

Class label	1	2	3	4
1	90	14	13	23
2	6	80	29	36
3	2	5	52	9
4	2	1	6	32

7　Conclusion

Sense disambiguation of prepositions is a difficult task as they denote different senses depending on the contexts of their use. This polysemy in prepositions leads to complexity in the cross lingual transfer, i.e. transfer from English to Hindi. In this paper, we have elaborately discussed about the transfer of English prepositional phrases into Hindi postpositional phrases pointing out the parallels between them and tabulating them. As a sample we have shown how to disambiguate the different senses of '*at*' using Support Vector Machine algorithm with Random Kitchen Sink.

References

1. Alam Y (2004). Decision Trees for Sense Disambiguation of Preposition. Case over. In HLT-NAACL, Computational Lexical Semantics Workshop, Boston: MA, pp 52–59.
2. Anand Kumar, M., Rajendran, S., & Soman, K. P. (2015). Cross-lingual preposition disambiguation for machine translation. *Procedia Computer Science, 54,* 291–300.
3. Aravind, A., & Anand Kumar, M. (2014). Machine Learning approach for correcting preposition errors using SVD features. *2014 International conference on Advances in Computing, Communications and informatics (ICACCI),* New Delhi, India, 359–376.
4. Downing, A., & Locke, P. (2002). A university course in English grammar. *New Fetter Lane, London: Routledge, 11,* 590–601.
5. Baldwin, T. V., & Kordoni, A. Villavicencio. (2009). Prepositions in applications. A survey and introduction to the special issue. *Computational Linguistics, 35*(2), 119–149.
6. Bannard, C., & Baldwin, T. (2003). Distributional Models of Preposition Semantics. ACL-SIGSEM, Workshop on the Linguistic Dimensions of Prepositions and Their Use in Computational Linguistics Formalism and Applications, Toulouse: France, pp 169–180.
7. Beth, L. (1993). *English verb classes and alternations a preliminary investigation* (p. 201). Chicago, IL: University of Chicago press.
8. Bojanowski, P., et al. (2016). Enriching word vectors with sub word information arXiv preprint arXiv: 1607.0460.
9. Dorr, B. (1992). The use of lexical semantics in interlingual machine translation. *Machine Translation, 7*(3), 135–193.

10. Harabagiu, S. (1996). *An application of word net to prepositional attachment* (pp. 360–363). Santa Cruz: ACL.
11. Hovy, D., Tratz, S., & Hovy, E. (2010). What's in a preposition? Dimensions of sense disambiguation for an interesting word class. In *Coling 2010: Posters*, (pp. 454–462). Beijing, China, August. Coling 2010 Organizing Committee.
12. Joachims, T. (1999). Transductive inference for text classification using support vector machines. In *International Conference on Machine Learning (ICML)*.
13. Kamakshi, S., & Rajendrarn, S. (2008). *Preliminaries to the Preparation of a machine aid to translate linguistic texts written in English into Tamil*. DLA publications.
14. Kamata Prasad Guru (1992). Hindi Vyakaran. Nagaripracharinisabha, Varanasi, India, 359-376.
15. Litkowski, K. (2002). Digraph analysis of dictionary preposition definition. In *ACL-SIGLEX, SENSEVAL Workshop on Word Sense Disambiguation: Recent Success and Future Directions, Philadelphia: PA*, pp 9–16.
16. Litkowski, K., & Hargraves, O. (2007). SemEval-2007 task 06: word-sense disambiguation of prepositions. In *Proceedings of the 4th International Workshop on Semantic Evaluations (SemEval-2007), Prague, Czech Republic*.
17. O'Hara, T. J. Wiebe. (2002). *Classifying preposition semantic roles using class-based Lexical Associations Technical Report NMSU-CS-2002- 013*. Computer Science Department: New Mexico State University.
18. O'Hara, T., & Wiebe, J. (2003). Preposition semantic classification via Penn Treebank and FrameNet. In *Proceedings of CoNLL* (pp. 79–86).
19. O'Hara, T. J. Wiebe. (2009). Exploiting semantic role resources for preposition disambiguation. *Computational Linguistics, 35*(2), 151–184.
20. Mamidi, R. (2004). Disambiguating prepositions for machine translation using lexical semantical resources. In *National Seminar on Theoretical and Applied Aspect of Lexical Semantics. Center of Advanced Study in Linguistics, Osmania University, Hyderabad*.
21. Rahimi, A., & Recht, B. (2007). Random features for large-scale kernel machines. In *NIPS* (Vol. 4).
22. Rudin, Walter (2011). Fourier Analysis On Groups. John Wiley& Sons.
23. Rudzicz, F., & Mokhov, S. A. (2003). *Towards a heuristic categorization of prepositional phrases in English with word net*. Technical report, Cornell University.
24. Saint-Dizier, P., & Vazquez, G. (2001). A compositional framework for prepositions. In *ACLSIGSEM, International Workshop on Computational Semantic*. Tilburg: Netherlands.
25. Sablayrolles, P. (1995). The semantics of motion. *EACL* (pp. 281–283). France: Toulouse.
26. Husain, S., Sharma, D. M., Reddy, M. (2007). Simple Preposition Correspondence: A problem in English to Indian language machine Translation. Language Technologies research Centre, IIIT, Hydrabed, India.
27. Schölkopf, B., & Smola, A. J. (2002). Learning with kernels: support vector machines, regularization, optimization, and beyond. Cambridge, MA: MIT press.
28. Stephen, T., & Dirk, H. (2009). Disambiguation of preposition sense using linguistically motivated features. In *Proceedings of the NAACL HLT Student Research Workshop and Doctroal Consortium* (pp. 96–100). Boulder, Colarado.
29. Soman, K. P., Loganathan, R., & Ajay, V. (2009). *Machine learning with SVM and other kernel methods*. New Delhi: PHI Learning Pvt. Ltd.
30. Sopena, J. A. L., & Loberas, J. Moliner. (1998). A connectionist approach to prepositional phrase attachment for real world text. *ACL, Montreal* (pp. 1233–1237). Canada: Quebec.
31. Tratz, S., & Hovy, D. (2009). Disambiguation of preposition sense using linguistically motivated features. In *Proceedings of Human Language Technologies: The 2009 Annual Conference of the North American Chapter of the Association for Computational Linguistics* (pp. 96–100).

User Interface Design Recommendations Through Multi-Criteria Decision Analysis

Subbiah Vairamuthu, Amalanathan Anthoniraj, S. Margret Anouncia and Uffe Kock Wiil

Abstract End users encounter difficulties while spending time and effort in learning and using some software products. Ultimately, the foremost goal of all the organizations and companies developing product lies in the improvement of user's satisfaction and interest toward the product. One of the methods by which the improvement could be performed is by designing a simple and guiding user interface that is suitable for all the users of an application. Especially, in the domain of information systems, the impediment raises due to the lack of effective user interfaces. The context of use and the end user's knowledge about user interface pose a strenuous demand in usability. Thus, there is a demand to design interface with utmost care considering the end user's motor skills and their abilities. Though several laws and principles exist for the stated purpose, effective design involves creativity and some amount of craft which may involve series of procedures, techniques, and tools. However, the decision making in dynamic environments to choose the techniques and tools is seemingly intractable and is complex because of natural trade-offs. Apparently, choosing from existing alternatives will frequently involve making trade-offs that may fail to satisfy requirements. Hence, the selection of alternate solution in lieu of multiple criteria has been challenging. One of the methods that were successful for this objective is multi-criteria decision analysis (MCDA). This technique defines several approaches for decision making and tries to converge toward the optimal solution. The work attempts to design user interface for any information systems with minimal number of interactions for the different categories of users.

Keywords Recommender systems · MCDA · TOPSIS · User interface design

S. Vairamuthu (✉) · A. Anthoniraj · S. Margret Anouncia
VIT University, Vellore, India
e-mail: svairamuthu@vit.ac.in

U. K. Wiil
University of Southern Denmark, Odense, Denmark

© Springer Nature Singapore Pte Ltd. 2018
S. Margret Anouncia and U. K. Wiil (eds.), *Knowledge Computing and Its Applications*, https://doi.org/10.1007/978-981-10-6680-1_14

1 Introduction

Traditionally, recommender system is considered to be a system that produces personalized individual recommendations as output for any product or a system. It involves learning from the behavior of end users dynamically with traditional products or a system through which a machine can be made to learn and perform various mining tasks for recommending the same based on a quantification scheme. Most of the Web information systems follow this technique to attract the users, and it facilitates the users to choose the right options with less searching time. Ultimately, it improves the usability of a system with minimum number of interactions.

Considering the qualifications of the recommendation system, a recommendation model is attempted for creating a case based repository that helps in decision making toward using the appropriate user interface design. An elaboration of the accomplished procedure in design of the case based repository is provided in the subsequent section. The recommender model is created with a focus on three subtasks such as user interface concept extraction, multi-criteria decision model, and creation of case based repository. The objective of the model as a whole is to generate cases that would help in the process of designing alternate user interface design that improves user satisfaction level.

2 Review of Literature

The section highlights the major contributions in the domain related to this contribution.

2.1 Automatic Generation of UI

Researchers have acknowledged that usability is normally viewed in terms of relationship across users and their desirability of interfaces that direct the repeated use of the system for a specific purpose. Existing language-based tools use special purpose languages and specifications for designing user interface effectively. Yet, to improve the process, automation has become a major concern and number of trials had been tried to address this concern. Initially, the progress toward it was not given much importance. Few of the researchers had given a thrust on this aspect [12, 13, 23]. An automation of system that incorporates adaptation to perform interaction selection in user interface design was proposed by Eisenstein and Puerta [13]. Yet, the system worked out only for smaller number of examples and had limitations on dialog layout and application structure. An adaptable explicit and user customization of User Interface through rendering was carried out in their work [16, 17].

The process had a limitation of undemanding evaluation. And hence, the reach of the process was partial. A formal approach and methodology for analysis and generation of human–machine interfaces, with special emphasis on human–automation interaction, was suggested by Heymann and Degani [23]. However, the system could not gain enough strength of usage as the formal descriptions were less encouraged for software development. An automated user interface specific to pharma industry is developed to show the usefulness of user interface automation to reduce the unnecessary elapse of time duration [12].

With the idea of automation that is carried forward, it is observed that neither the existing guidelines nor the standards of user interface design have been precisely followed, and hence, it has led to the conflict of interest in the product and other external applications. To avoid such issues, several user interface tools have been generated. These tools follow the guidelines and standards of user interface and help process to become automated or semiautomated. Number of tools had been evolving with definite focus on certain aspects of user interface design. The following subsection 2.4 provides a highlight of such tools which are for automatic generation of user interfaces.

2.2 UI Generation Tools

In the present scenario of human–computer interaction, variety of tools and techniques are available for improving the productivity and quality of any system. Especially, for the improved productivity, one of the factors considered is the usability. The usability usually depends on number of factors, and one of the most influencing factors is the user interface design. To obtain an elegant and usable user interface design, well-established tools and techniques have been followed. Yet, the accommodation of user interface standards and guidelines was very much limited. Hence, different themes focusing on the user interface guidelines had been formulated by different researchers for specific purposes. Some of the notable contribution on building such tools and techniques is highlighted in this section.

Among the available tools, three different categories such as language-based tools, graphics-based specification tools, and model-based user interface generation tools play a significant role in user interface design. Earlier, a very strong background for research in generation and management of user interfaces was laid. A User Interface Management Systems (UIMS) called COUSIN had been proposed as a model-based tool to provide I/O specification of the user interface design [21]. The research continued to progress, and as a result, automatic generation of user interfaces to make the novice designers more comfortable began to evolve.

According to Olsen [35], automatic generation of menus and dialog boxes using the comments and function signatures had been attempted. It is language-based tool which had a tight binding toward a particular programming language. Tools like ITS [59], JADE [62], HUMANOID [52, 53], and UIDE [54] provided substantial impact in the user interface design. In these systems, the user interfaces are generated with the models specified by the designers and the specification of models appeared to be a complex process as it was too abstract and time-consuming. Though it appeared to be time-consuming, it contributed a valuable share toward usability.

In the meanwhile, a model-based system JADE [62] insisted on generating layouts for dialog boxes using text specifications. The system permitted complex relationship as a simpler representation to create a syntactic ally correct specification. In a similar view, a system called ITS [59] had been formulated by IBM research team to focus on explicit separation of concerns from specification. Unfortunately, group labeling and graphical objects were not addressed completely. An application model mapping application specifications to the desired interface had been attempted by [54]. The proposed UIDE model helped the designers to focus on modeling and helped to defer the interface design issues. A tool called HUMANOID supported generating entire applications apart from menus and dialog boxes [52, 53]. The tool lacked in visualization of complex data and validation of generated UI. A project called MASTERMIND was attempted by [54] to propose a hybrid model combining goodness of UIDE and HUMANOID. It was appealing as a model-based UI development environment which supported grids and guides for complex layouts.

However, expressiveness, extensibility, semantics, and task modeling schemes were less addressed. TRIDENT [56], a system for the similar use, has evolved in addition to MASTERMIND. The system evolved as knowledge-based one by combining an automated design assistant and an automatic interface design generator. The system failed to address the issue of flexibility toward the tailored demands and local needs. A model-based user interface design environment that is capable of generating large-scale interfaces for declarative model has been proposed by [41]. Though the method explored on Internet-based UI, the extendibility toward the component and distributed interfaces were limited.

A compact specification of dynamic task behavior had been modeled as ConcurTaskTrees [38] to support graphical constructs and operators that are derived from formal specifications. The approach had been applied to various business applications, and it left an opportunity toward the industrial applications. A multiplatform user interface design had been attempted by Farooq Ali et al. [14]. A single language UIML (User Interface Markup Language) is extended to support vocabularies and transformational algorithms that would allow specifications for a family of devices. However, the system had a limitation of handling difficult platforms related to voice and digital assisting devices.

An XIML (eXtensible Interface Markup Language) had been proposed by Puerta and Eisenstein [42, 43] to provide a representation and manipulation of interaction data that is helpful to interoperate within the user interface. TERESA, a semi-automated model-based system helped User Interface design of heterogeneous devices facilitated transforming the User Interfaces between different contexts as proposed by Morri et al. [29]. The system supported only few modalities. User Interface eXtensible Markup Language (USIXML) allowed multiple path development of user interfaces. Yet the model had less attention toward the translation of task from different generation systems [26]. A similar tool called SUPPLE was proposed by Gajos et al. [16]. The tool was intended to generate user interfaces and had a limitation of proper evaluation. Similar approaches viz OLEX2 [11], CHARMM-GUI [25], Gabedit [1], Gmsh [19], etc., had been progressing; unfortunately, the attempts were confined to specific domains and were not generic. Table 1 summarizes the most promising contributions for specifying user interface as introduced at the beginning of this section viz. language-based specification, model-based generation, and graphical specification which actually laid the foundation for several other tools.

Thus, the study revealed the importance of user interface design for improving usability through user satisfaction. And it is noted from the investigation that most of these tools and techniques had been proposed in consideration to particular

Table 1 Summary of early contributions for specification

Nature of specification	Specification format	Examples	Contributors
Language-based generation	Application frameworks	MacApp, UniDraw	[60, 57]
	Constraint languages	ThingLab, C32	[4, 31]
	Context-free grammars	YACC, LEX, SYNCGRAPH	[36]
	Declarative languages	COUSIN, Open Dialog, and MotifUIL	[21, 46]
	Event languages	ALGAE, Sassafras	[15, 24]
	Screen scrapers	Easel	[31]
	Visual programming	LabVIEW, Prograph	[58, 20]
Model-based generation	Models	UIDE, ITS, HUMANOID	[54, 59, 52, 53]
Graphical specification	Cards	Menulay and HyperCard	[5]
	Data visualization tools	DataViews	[28]
	Graphical editors	Peridot, Lapidary, Marquise	[30, 33, 32]
	Interface builders	DialogEditor, NeXT Interface Builder, Prototyper, UIMX	[6, 18, 51, 31]
	Prototypers	Bricklin's Demo, Director	[31, 22]

domain and targeted toward addressing number of utility criteria. Still, not all the criteria contributing for user satisfaction have been addressed. In addition, the influence of the selected criteria is less known and it is mostly chosen at random. Therefore, it is required to have a mechanism that would determine the key criteria influencing the user satisfaction of the system. Of course, the key criteria are not generic for any system, and they depend on the domain of interest. Therefore, considering the objective of improving user satisfaction through usability of the selected domain of investigation—information system, the criteria are analyzed and a proper set of criteria having impact toward the user satisfaction level is pulled out. In order to proceed, it is preferred to perform a criteria decision analysis, through which a ranking of influencing criteria can be expedited.

2.3 Criteria Analysis for the Domain of Interest

One of the major elements of effective software development is design of the user interface. In any system, the interfaces are viewed to replicate the mental model of functionality flow of the system. A number of well-contrived and robust methods have been evolving in addressing the issue of user interface design. Unfortunately, all the impressing criteria of user interface design may not be possible to model due to several characteristics pertaining to the application development. Therefore, certain criteria are left out unmolded while few are attempted partially. In some cases, the entire process is completed as a random walk. This approach of user interface may help to sustain the workability of the model. Yet, the satisfaction of the users and the usability of the model are questionable. To avoid this issue, it is highly essential to understand the impact of the different criteria contributing toward user satisfaction of the system [61]. Mostly, these criteria create an association between context and the interface. Hence, it becomes difficult to adopt single design for all kinds of interaction. Ultimately, the adoption of inappropriate interface design or user interface with inappropriate controls would reduce the user satisfaction. And hence, adaptive user interfaces are needed. Several milestones have been reached in this context of user interface research. Especially in the last decade, the research toward the adaptive user interface gained more confidence for improving the usability of a system and it is viewed to be in terms of prime performance characteristics such as understandability, learnability, memorability, ease of use, efficiency, navigation, interaction, and speed. Initially in the earlier years of the decade, understandability issues about the process such as ill-formed error messages, unclear instructions, lack of system helpfulness were a major focus [55].

A model-to-model transformation for reducing the number of manual changes in an information system is addressed through the design of beautification operations of user interfaces which add limitations toward the final user satisfaction and usability [39]. Eventually, availability of context-aware systems for concurrent running operations was a major focus of interface design [2]. On the other hand, efficiency in the runtime environment through distributed user interface had been focused as a major concern as proposed [3]. Extensibility and controllability of user interface design had been approached through a novel architectural style called COMET. The architectural style had a less focus toward the visualization and transformation at run time [10]. In the meantime, Akash Singh and Janet Wesson explored that navigation, learnability, customization, task support and presentation were the predominent performance characteristics of information systems User Interface Design for improved usability [48].

Bringing runtime efficiency in the model of business logic was a greater concern in the research of HCI, and hence, manipulability was appearing to be puzzling task in the user interface design [8]. Ultimately, improving the speed and accuracy for all the users especially users with the motor impairments had been greater challenges concluded by Gajos et al. [17]. Consequently, behavioral measures such as response time, accuracy with respect to keystrokes, and screen contents were found to be performance deciding factors of a user interface design [49]. In another dimension, Reinecke and Bernstein [44] revealed that usability is closely connected to cultural aesthetics of user interface with less clicks and fewer errors. In due course, addressing users with respect to their experience, age, and physical abilities was found to be challenging. Especially, satisfying older people with appropriate design was a cumbersome task as quantified in several studies [27, 45, 50]. This idea is motivated toward utilizing multimodal human–computer interaction (MMHCI) as a prospective avenue for the HCI [47].

According to the elaborative observations made from the study, it is noted that criteria such as learnability, understandability, efficiency, response time, speed of use, accuracy were the leading concerns of user interface design. In addition, minimizing the level of interaction and reducing the errors are the other aspects to be emphasized in user interface design. In another perspective, users with different challenging abilities and older users were found to be prominent cases of interest for investigation. Thus, the criteria such as age, experience, abilities, and the level of interactions appear to play a vital role in designing user interface of a system. Yet, choosing the appropriate criteria for the application is a perplexing task for any designer. Hence, an initiative toward the multi-criteria decision analysis (MCDA) is found to be essential.

2.4 Recommender System

In general, recommender system is a system that collects information on the preferences for a group of items from the users and facilitates them to make decisions from the existing alternatives. A tremendous number of recommendation systems are evolving today in parallel with the growth of information in Web. Nowadays, this kind of recommendation systems is applied in several domains from simple things to more complex items as exemplified by various contributions. Some of the notable recommendation systems for varied categories of domains were analyzed by Park et al. [37]. Due to the availability of enormous volume of data and information, several recommendation systems and recommendations were evolved. Out of the existing recommendation systems, the following contributions for recommending movies, ebooks, music and many others were considered to be vital [7, 9, 34, 40]. In the forthcoming era of information technology (IT), the recommender systems will use information from Internet of Things (IoT). Though many information systems have been recommended for diverse purpose by recommendation systems, none of them advises decisions with respect to user interface design. Considering this as an aid, this investigation attempted to apply this recommendation system concept in this study to recommend alternative user interface designs with the primary objective of improving usability.

3 User Interface Concept Extraction Model

The user interface extraction model is designed to arrive at a good understanding of the user interaction style and design charted for any information system. A pattern of controls that are usually associated with each of the interaction styles is discovered using association mining process. Thus, a set of controls that are possible for modeling user interactions are discovered into pattern. Using the discovered pattern, suggestive rules are formulated.

3.1 Association Rule Mining for Case Base Representation

Association rule learning is one of the popular and most widely accepted methods for identifying interesting relationship between items in a larger database. Association rules are created after analyzing the frequent patterns together with the criteria support and confidence to ensure durability of the derived rules. The support of an item set is the proportion of transactions in the data set which contain the item set, while confidence of a rule can be interpreted as an estimate in probability of finding right-hand side (RHS) of a rule in transaction under the condition that the same transaction also contains the left-hand side (LHS). In this study, to identify the interesting patterns of user interface controls used during user interface design, this association mining concept is used. It attempted to discover the most commonly used user interface controls for various interaction styles. For identifying this interestingness measure, six different academic portals are visited and analyzed for its user interface design.

Most of the existing information systems use menu, form, command line, direct manipulation, Windows Icons Menus & Pointers (WIMP) and natural language interfaces for modeling their interactions. Out of this, the natural language interface is less likely used as its complication on the request requires cumbersome processing. The rest of the interaction styles have standard controls that are being used repeatedly. In order to confirm the controls and its standard usage, the association mining process is performed to claim the support and confidence of the user interface control patterns.

According to the observation of the six selected information system, a menu interaction style used controls like buttons, text box, date and time pickers, toggles. To ascertain the pattern of controls shadowed in the system, the computation of support and confidence of the pattern occurring in these information systems is performed. The pattern of controls used for menu interaction style for these six systems are considered, and the support and confidence of the same are computed using the theory of association mining.

According to association mining, a transaction t contains X, a set of items (item set) in I, if $X \subseteq t$. An association rule is an implication of the form:

$$X \rightarrow Y, \text{ where } X, Y \subset I, \text{ and } X \cap Y = \varnothing \tag{1}$$

An item set is a set of items. In this case, the item set is the set of controls used for menu interaction and transaction is assumed to be different information system that uses the similar controls for the same interactions. The support measure is defined to be the rule holds with support sup in T (the transaction data set) if sup% of transactions contain $X \cup Y$, and hence, $sup = \Pr(X \cup Y)$. Similarly, the confidence is defined to be the rule holding in T with confidence $conf$ if $conf$ % of transactions that contain X also contain Y. So, $conf = \Pr(Y \mid X)$. An association rule is a pattern that states when X occurs, Y occurs with a certain probability. Therefore, the support count is determined through the expression,

$$\text{support} = \frac{(X \cup Y).\text{count}}{n}$$

$$\text{confidence} = \frac{(X \cup Y).\text{count}}{X.\text{count}}$$

Table 2 Support and confidence of menu-based interaction styles

Web site	Controls	Support	Confidence	Support (%)	Confidence (%)
1	{Text box, Buttons}	3/6	3/5	50	60
2	{Text box, buttons, Date and Time Pickers}	2/6	3/5	33	60
3	{Text box, Toggles}	2/6	2/3	33	67
4	{Text box, Buttons, Date and Time Pickers}	2/6	3/5	33	60
5	{Text box, Toggles, Date and Time Pickers}	1/6	1/3	17	33
6	{Buttons, Toggles}	1/6	1/3	17	33

Using the expressions of support and confidence, the possible UI control patterns are derived and the % of support and confidence of user interface controls is given in Table 2.

Observing the support and confidence of the submitted controls for the interaction style, it is seen the controls set of four information systems gained the confidence level of around 60%, and hence, it is evident that this combination of controls would be appropriate for the menu interaction style. Similarly, for the form-based interaction, the controls that are being used are check box, radio buttons, list, drop down list box, buttons, text box, date and time pickers, toggles. The support and confidence measure for the controls are tabulated in Table 3.

According to the chosen information systems, the controls text box and toggles have claimed strong conclusion with good confidence level in command language interface interaction style as indicated in Table 4.

The direct interaction style strongly clinched to have association of controls like list, button, and toggles. Mining for the frequent control patterns in the chosen information systems, it is clear that the listed user interface controls are mostly used in interface design. Also, looking into the % of support and confidence values in Table 5, the identified controls will best suit for this interaction style.

For the WIMP interaction style, the support and confidence level of the mined interaction patterns justifies that the controls like check box, radio buttons, list box, drop down list box, buttons, text box, date and time pickers and toggles influence the user interface design with decent support and confidence percentage as shown in Table 6.

Considering the support and confidence measure of the different patterns of user interface controls, the interaction styles such as menu, forms, command language interface, direct manipulation, and WIMP had visible pattern of user interface controls. The support and confidence level of natural language interface is found to lack significant pattern of user interface controls. The computation of support and

Table 3 Support and confidence for form-based interaction

Web site	Controls	Support	Confidence	Support (%)	Confidence (%)
1	{Check box, List, Text box, Buttons}	4/6	4/5	67	80
2	{Check box, List, Text box, Buttons}	4/6	4/5	67	80
3	{Radio buttons, Text box, List, Buttons}	1/6	1/1	17	100
4	{Text box, List, Toggles, Buttons}	2/6	1/1	33	100
5	{Buttons, Text box, List, Drop Down List}	1/6	1/1	17	100
6	{Check box, List, Text box, Buttons, Toggles}	1/6	1/2	17	50

Table 4 Command language interface support and confidence

Web site	Controls	Support	Confidence	Support (%)	Confidence (%)
1	{Text Box}	4/6	1/6	67	17
2	{Text Box}	4/6	4/6	67	67
3	{Text Box, Toggles}	2/6	2/2	33	100
4	{Text box, toggles}	2/6	2/2	33	100
5	{Text box}	4/6	4/6	67	67
6	{Text Box}	4/6	4/6	67	67

Table 5 Support and confidence for direct manipulation style

Web site	Controls	Support	Confidence	Support (%)	Confidence (%)
1	{List, Buttons}	3/6	3/6	50	50
2	{List, Buttons, Toggles}	3/6	3/6	50	50
3	{List, Buttons}	3/6	3/6	50	50
4	{List, Buttons, Toggles}	3/6	3/6	50	50
5	{List, Buttons, Toggles}	3/6	3/6	50	50
6	{List, Buttons}	3/6	3/6	50	50

Table 6 Support and confidence for WIMP interaction style

Web site	Controls	Support	Confidence	Support (%)	Confidence (%)
1	{Buttons, Text box, Check Box, Radio buttons}	3/6	3/6	50	50
2	{Radio buttons, Check box, Date and Time Pickers, Toggles}	2/6	2/2	33	100
3	{Radio buttons, Check box, Date and Time Pickers, Toggles}	2/6	2/2	33	100
4	{Buttons, Text box, Check Box, Radio buttons}	3/6	3/6	50	50
5	{Radio buttons, Check box, Toggles}	1/6	1/3	17	33
6	{Buttons, Text box, Check Box, Radio buttons}	3/6	3/6	50	50

confidence for the interface as depicted in Table 7 indicates the minimum support and good confidence to state that the coverage is less; however, the usable controls are completely noteworthy.

The frequent control patterns thus identified ensure that the support and confidence level for the proposed patterns precisely matches the need of information system. Yet, the inherent difficulty to hold a specific style of unambiguous

Table 7 Support and confidence level for natural language interaction style

Web site	Controls	Support	Confidence	Support (%)	Confidence (%)
1	{Text Box, Button}	2/6	2/2	33	100
2	–	0	0	0	0
3	–	0	0	0	0
4	{Text Box, Button}	2/6	2/2	33	100
5	–	0	0	0	0
6	–	0	0	0	0

interaction by each user creates a greater limitation of using natural language interface for an information system. With the established evidence for each of the patterns and interaction styles, the preliminary cases for case based repository are constructed.

3.2 Design of Multi-Criterion Decision Model

The recommendation model is aimed to generate recommendations on the user interface design so as to improve the user satisfaction. While performing the process, it is desired to arrive at recommendations that are acceptable for all types of end users. Also, the recommendations should favor all the end users by providing optimality on the suggestive factors. Considering these concerns in the framework that is designed for recommending optimal UI design, initially, a multi-criteria analysis involving UI parameters such as intensity of interaction (II), user experience (UE), and end user's age (AF) is performed. In order to carry out the process, the widely used TOPSIS model is used. The steps involved in carrying out the MCDA are depicted in Fig. 1.

3.3 Insight into MCDM and TOPSIS

The MCDA (multi-criteria decision analysis) approach can be considered as one of the complex tools for decision making (DM) process which involves qualitative and quantitative factors. Various MCDA approaches and techniques were introduced in the literature to choose the possible optimal solution. Many researchers proposed various studies to introduce changes in MCDA so that the method can be employed in their related domain. MCDA technique had been devised to elect a preferred alternative or classify the alternatives from a number of choices or to rank them based on a subjective preference. The method was considered to be a general method to help users facilitating decision making from more than one available preference where the chances of conflicts were high.

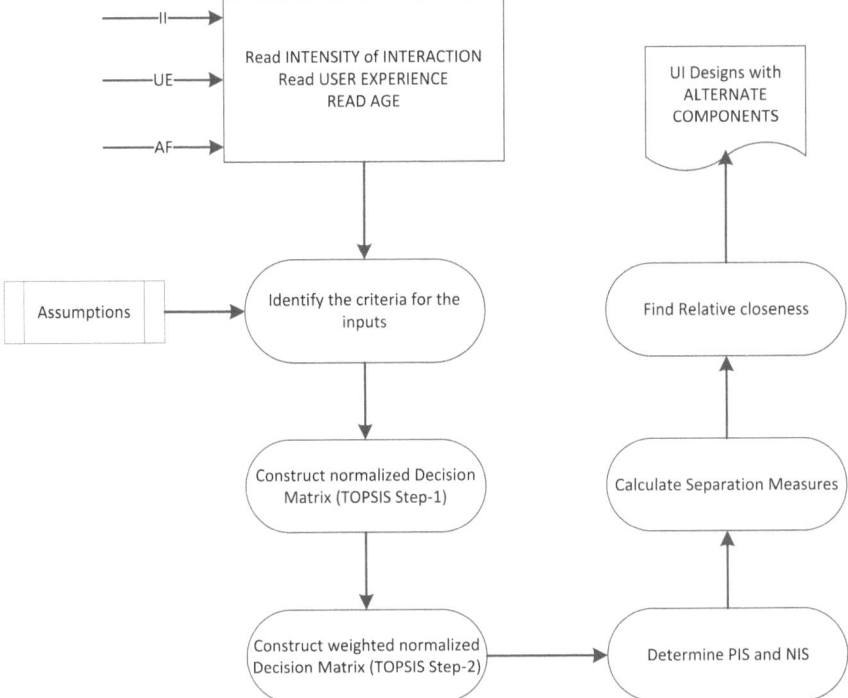

Fig. 1 TOPSIS for UI design

3.3.1 TOPSIS Illustrated for Problem-Solving

According to the TOPSIS method, two artificial alternatives such as positive ideal alternative and negative ideal alternative are hypothesized. The positive ideal alternative indicates the best level for all chosen attributes of decision, while the negative ideal alternatives possess the worst values for the attributes. TOPSIS selects the alternative that is the closest to the ideal solution and farthest from negative ideal alternative. The procedure of TOPSIS with respect to the multiple criteria is as listed.

Consider that the chosen domain has 'm' alternatives (options) and 'n' attributes/criteria and each alternative is assigned with a score in accordance with the impact of criterion.

Let x_{ij} be the score of option i with respect to criterion j.

Hence, a decision matrix $X = (x_{ij})$.

Let J be the set of benefit attributes or criteria (more is better).

Let J' be the set of negative attributes or criteria (less is better).

Step 1: Construct normalized decision matrix to transform various attribute dimensions into non-dimensional attributes allowing comparisons across criteria using Expression 2.

$$r_{ij} = x_{ij} / \sqrt{\sum x_{ij}^2} \text{ for } i = 1, 2, \ldots, m; j = 1, 2, \ldots, n \qquad (2)$$

Step 2: Construct the weighted normalized decision matrix using Expression 3

$$v_{ij} = w_i r_{ij} \qquad (3)$$

Step 3: Determine the ideal and negative ideal solutions. Positive ideal solution (PIS)

$$A^* = \left\{ v_1^*, v_2^*, \ldots, v_n^* \right\} \qquad (4)$$

where

$$v_j^* = \left\{ \max_i (v_{ij}) \text{ if } j \in J; \min_i (v_{ij}) \text{ if } j \in J' \right\} \qquad (5)$$

Negative ideal solution (NIS)

$$A' = \left\{ v_1', v_2', \ldots, v_n' \right\} \qquad (6)$$

where

$$V' = \left\{ \min_i (v_{ij}) \text{ if } j \in J; \max_i (v_{ij}) \text{ if } j \in J' \right\} \qquad (7)$$

Step 4: Calculate the separation measures for each alternative. The separation from the ideal alternative is

$$S_i^* = \left[\sum_j \left(v_j^* - v_{ij} \right)^2 \right]^{1/2} i = 1, 2, \ldots, m \qquad (8)$$

Similarly, the separation from the negative ideal alternative is

$$S_i' = \left[\sum_j \left(v_j' - v_{ij} \right)^2 \right]^{1/2} i = 1, 2, \ldots, m \qquad (9)$$

Step 5: Calculate the relative closeness to the ideal solution C_i^*

$$C_i^* = S_i'/\left(S_i^* + S_i'\right), 0 < C_i^* < 1 \tag{10}$$

Step 6: Select the alternative with C_i^* closest to 1.

The outcome obtained as a result of executing the mentioned steps projects the ranked alternate solutions to help in formulating the acceptable alternate solutions. Thus, the idea of TOPSIS is applied in improving user satisfaction of services/product through a recommendation of an optimal user interface design that satisfies most of the end users.

3.3.2 TOPSIS for GUI Design

The purpose of this work is to utilize an optimum number of user interface controls or elements in user interface of an application so as to minimize the number of interactions. The implicit requirement of this core objective is to identify the suitable interaction style that is applicable to any category of users irrespective of their age, ability, and experience in the chosen domain. The process flow of applying TOPSIS in GUI design is depicted in Fig. 1.

As given in Fig. 1, the system needs three basic inputs from the user who wants to make design decision. Based on the experience in the relevant field, certain assumptions are considered initially for further analysis. Three basic inputs required for this process are as follows:

Intensity of interaction (II),
User experience (UE), and
End user's age (AF).

Basically, interaction in any information system would be decided by the combination of multiple factors viz. keystrokes, mouse clicks, and navigation. Considering these factors, the intensity of interaction is determined. Usually, in traditional interaction, keystrokes and mouse clicks dominate in all the levels though there are other factors like navigation, icons, and windows. Hence, these two factors are primarily considered for modeling intensity of interaction. Initially, the percentage of intensity of interaction is computed as

Intensity of Interaction $(\text{II}) = (A + B)/(A + B + C + D + E) * 100$

where

A   Number of mouse clicks,
B   Number of keystrokes,
C   Number of navigations,
D   Number of windows, and
E   Number of icons/images.

Based on the calculated value, this criterion is categorized into three levels namely low, medium, and high and it is given below in Expression 11.

$$\text{Intensity of Interaction (II)} = \begin{cases} \text{Low,} & 0 < II \le 30 \\ \text{Medium,} & 31 < II \le 60 \\ \text{High,} & 61 < II \le 100 \end{cases} \tag{11}$$

Similarly, the user experience (UE) is computed based on the two aspects namely years of familiarity (YE) with computers and adequacy of devices for interaction (IDA). Due to the unavailability of direct measure for the UE, it is computed through other known aspects. Hence, by considering the impact of the mentioned aspects of UE, it is assumed that these factors YE and IDA share a composition of 20:80. To commence the computation of user experience, users are expected to rate their proficiency with respect to these factors on an ordinal scale ranging between 1 and 5. The scaling is grouped as 1 for <5 years, 2 for YE between 6 and 10 years, 3 for YE between 10 and 15 years, 4 for YE between 16 and 20 years, and 5 if YE greater than 20. Similarly, for IDA, an input of ordinal range between 1 and 5 is obtained. The scale 1–5 is interpreted as follows:

1: Novice,
2: Manageable,
3: Adequate,
4: Skilled, and
5: Expert.

With the interpreted values, the UE is computed using the Expression 3.

$$UE = YE * 0.2 + IDA * 0.8 \tag{12}$$

And categorization is made as per the crisp class assignment represented using Expression 4:

$$\text{User Experience (UE)} = \begin{cases} \text{Novice,} & 0 < UE \le 5 \\ \text{Manageable,} & 6 < UE \le 20 \\ \text{Adequate,} & 21 < UE \le 40 \\ \text{Skilled,} & 41 < UE \le 60 \\ \text{Expert,} & 61 < UE \le 100 \end{cases} \tag{13}$$

Conversely, end user's age (AF) is categorized as stated below using 5:

$$\text{Age (AF)} = \begin{cases} \text{Young,} & 17 < \text{Age} \le 30 \\ \text{Middle,} & 31 < \text{Age} \le 50 \\ \text{Old,} & 51 < \text{age} \le 80 \end{cases} \tag{14}$$

Thus, the assumed crisp values for performing evidential reasoning are formulated and are represented in Table 8.

Table 8 Mapping crisp values for inputs

Main criteria	Sub-criteria	Value (on a 10-point scale)
Intensity of interaction (II)	Low	1–3
	Medium	4–6
	High	7–10
User experience (UE)	Novice	0, 1
	Manageable	2, 3
	Adequate	4, 5
	Skilled	6–8
	Expert	9, 10
Age factor (AF)	Young	1–3
	Middle	4–7
	Old	8–10

Based on the values and procedure discussed above, different categories are determined to apply in TOPSIS model. According to the model, three factors such as intensity of interaction (II), user experience (UE), and age factor (AF) are mapped against six interaction styles and the suitability of the interaction style with preference is ranked. An illustration of the same is outlined in the ensuing section.

Illustration of TOPSIS

To apply TOPSIS, the weightage for each factors confining to decision making were initialized with some random values which sums up to 100 on some assumptions. With the collected inputs from the literature, it is observed that for any information system, the UE has more impact than II and AF. Therefore, a weightage of 0.3, 0.4, and 0.3 is assumed with both II and AF with equal weights while UE is assumed to have elevated value.

To better understand the TOPSIS procedure, consider the following inputs:

Intensity of interaction (II)—LOW
User experience (UE)—Manageable
Age (AF)—Young

Step 1: Frame initial decision matrix and normalize it.
A decision matrix using decision parameters interaction style and usability parameter is drawn as shown in Table 9.
Further, a normalized decision matrix is arrived as shown in Table 10 using Expression (1).

Step 2: Construction of weighted decision matrix by applying Eq. (3). Table 11 shows the result of Step 2.

Table 9 Initial decision matrix for the chosen input category

Alternative versus criteria	II (0.3)	UE (0.4)	AF (0.3)
DIRECT	3	2	1
MENU	1	3	3
FORM	1	2	1
CLI	1	2	1
WIMP	1	2	3
NATURAL	3	3	1

Table 10 Normalized decision matrix

Alternative versus criteria	II	UE	AF
DIRECT	0.136	0.058	0.045
MENU	0.045	0.088	0.136
FORM	0.045	0.058	0.045
CLI	0.045	0.058	0.045
WIMP	0.045	0.058	0.136
NATURAL	0.136	0.088	0.045

Table 11 Weighted decision matrix

Alternative versus criteria	II	UE	AF
DIRECT	0.040	0.023	0.013
MENU	0.013	0.035	0.040
FORM	0.013	0.023	0.013
CLI	0.013	0.023	0.013
WIMP	0.013	0.023	0.040
NATURAL	0.040	0.035	0.013

Step 3: Find PIS and NIS using Eqs. (4) through (7). The result obtained is given in Table 12.

Step 4: Calculate separation measures using Eq. (8) for PIS and Eq. (9) for NIS. The result is given in Table 13 for PIS, and Table 14 gives the same for NIS.

Step 5: Find the relative CLOSENESS to PIS by applying Eq. (10). The result is given in Table 15.

From the final relative closeness of the PIS in Table 15, it is observed that any user between 17 and 30, who is dynamic to have own previous experience, is recommended with the MENU or NATURAL language-based interaction styles. Also, it is noted for these kinds of users, interaction styles like FORM fill-in and command line interface (CLI) would be least appropriate.

Table 12 PIS and NIS values

Solution versus criteria	II	UE	AF
PIS	0.040	0.035	0.040
NIS	0.013	0.023	0.013

Table 13 Separation measure for PIS

Alternative versus criteria	II	UE	AF
DIRECT	0	0.0001	0.0007
MENU	0.0007	0	0
FORM	0.0007	0.0001	0.0007
CLI	0.0007	0.0001	0.0007
WIMP	0.0007	0.0001	0
NATURAL	0	0	0.0007

Table 14 Separation measure of NIS

Alternative versus criteria	II	UE	AF
DIRECT	0.0007	0	0
MENU	0	0.0001	0.0007
FORM	0	0	0
CLI	0	0	0
WIMP	0	0	0.0007
NATURAL	0.0007	0.0001	0

Table 15 Relative closeness matrix

Alternatives	Closeness	Rank
DIRECT	0.457	2
MENU	0.542	1
FORM	0	NA
CLI	0	NA
WIMP	0.457	2
NATURAL	0.542	1

In a similar way, the procedure is repeated for different possible combinations to complete the mapping process and hence to derive the recommendation cases. The summary of the arrived recommendations with respect to these input parameters is listed in Table 16.

Thus, for any category of users with varying levels of experience with any depth of interactions, the interaction styles are reached with the help of TOPSIS method. Subsequently, a mapping of possible user interface elements is required to propose the suitable alternatives.

Table 16 Recommendation cases for user interaction styles

Intensity of interaction	User experience	Age factor	Preferred interaction style
LOW	NOVICE	YOUNG	DIRECT
		MIDDLE	NATURAL
		OLD	DIRECT
	MANAGEABLE	YOUNG	MENU
		MIDDLE	FORM
		OLD	WIMP
	ADEQUATE	YOUNG	FORM
		MIDDLE	WIMP
		OLD	NATURAL
	SKILLED	YOUNG	MENU
		MIDDLE	FORM
		OLD	DIRECT
	EXPERT	YOUNG	FORM
		MIDDLE	WIMP
		OLD	DIRECT
MEDIUM	NOVICE	YOUNG	DIRECT
		MIDDLE	MENU
		OLD	MENU
	MANAGEABLE	YOUNG	DIRECT
		MIDDLE	FORM
		OLD	MENU
	ADEQUATE	YOUNG	MENU
		MIDDLE	MENU
		OLD	WIMP
	SKILLED	YOUNG	MENU
		MIDDLE	WIMP
		OLD	CLI
	EXPERT	YOUNG	MENU
		MIDDLE	CLI
		OLD	DIRECT
HIGH	NOVICE	YOUNG	MENU
		MIDDLE	FORM
		OLD	MENU
	MANAGEABLE	YOUNG	WIMP
		MIDDLE	FORM
		OLD	FORM
	ADEQUATE	YOUNG	WIMP
		MIDDLE	FORM
		OLD	NATURAL
	SKILLED	YOUNG	CLI
		MIDDLE	DIRECT
		OLD	DIRECT
	EXPERT	YOUNG	MENU
		MIDDLE	FORM
		OLD	MENU

3.4   Case Based Repository

A case based repository is created to store the rules in the form of cases. These cases imply the different levels of association between interaction style, user interface controls, and the type of modality. The rules generated and evaluated though the association mining and MCDA were formalized into cases. The cases are pruned for similarity with respect to the input parameters, and the consequences of the extracted cases are learned through transitivity for producing the final conclusion. This characteristic of case base is helpful for the chosen area of investigation in suggesting a better alternate UI design that would emphasize on the improved user satisfaction.

3.4.1   Design of Case Based Repository

Case-based reasoning (CBR) is a kind of problem-solving approach which aims at reusing the previously memorized and solved problem situations called cases. Cases are concrete experiences in CBR approach. Learning can be simple as memorizing new cases or can entail refining the memory organization of existing cases. Based on these premises, CBR has been identified as a suitable approach for the area of investigation. In general, the CBR approaches follow primarily the four steps such as:

Retrieve: Most similar cases are retrieved from the existing case bases.
Reuse: From the previous experiences, use the cases and attempt to solve the new problem.
Revise: The proposed solution can be revised for better solution if necessary.
Retain: The new solution can become the part of case bases for future use.

The CBR model is created to store the validated recommendations that can be employed for suggesting new recommendation toward the UI design with a consideration of the interaction style, user experience, and age. Given the category of users with their competency levels, the recommendations for UI design can be retrieved using query-based search from the constructed case based repository.

4   Conclusion

Having established the case based repository, the recommendations are retrieved based on the keyword constituted from interaction style, user interface controls, and interactive devices. Thus, a recommendation model is attempted to enclose the user interface extraction that is studied to associate a user interaction style, user interface controls. A multi-criteria decision analysis is performed through TOPSIS, and a ranking of the possible recommendation is established. The ranking helps to project all possible cases that can be utilized for choosing a better alternative design that is optimal for all categories of end users. The developed recommendations are stored in a case based repository which is further used for retrieving recommendations for any new request. The outcome expected out of this model is a recommendation toward constructing user interface with appropriate interaction style and with minimal user interface controls so as to facilitate users' interaction easier through which usability can be improved.

References

1. Allouche, A. R. (2011). Gabedit—a graphical user interface for computational chemistry softwares. *Journal of Computational Chemistry, 32*(1), 174–182.
2. Bihler, P., & Mügge, H. (2007). Supporting cross-application contexts with dynamic user interface fusion. In *GI Jahrestagung (1)* (pp. 459–464).
3. Blumendorf, M., Feuerstack, S., & Albayrak, S. (2007). Multimodal user interaction in smart environments: Delivering distributed user interfaces. In *European Conference on Ambient Intelligence* (pp. 113–120).
4. Borning, A. (1981). The programming language aspects of ThingLab, a constraint-oriented simulation laboratory. *ACM Transactions on Programming Languages and Systems (TOPLAS), 3*(4), 353–387.
5. Buxton, W., Lamb, M. R., Sherman, D., & Smith, K. C. (1983). Towards a comprehensive user interface management system. In *ACM SIGGRAPH Computer Graphics* (Vol. 17, pp. 35–42).
6. Cardelli, L. (1988). Building user interfaces by direct manipulation. In *Proceedings of the 1st Annual ACM SIGGRAPH Symposium on User Interface Software* (pp. 152–166).
7. Carrer-Neto, W., Hernández-Alcaraz, M. L., Valencia-García, R., & García-Sánchez, F. (2012). Social knowledge-based recommender system. Application to the movies domain. *Expert Systems with Applications, 39*(12), 10990–11000. https://doi.org/10.1016/j.eswa.2012.03.025.
8. Coutaz, J. (2010). User interface plasticity: Model driven engineering to the limit! In *EICS'10 —Proceedings of the 2010 ACM SIGCHI Symposium on Engineering Interactive Computing Systems* (pp. 1–8). Retrieved from http://www.scopus.com/inward/record.url?eid=2-s2.0-77955106658&partnerID=tZOtx3y1.
9. Crespo, R. G., Martínez, O. S., Lovelle, J. M. C., García-Bustelo, B. C. P., Gayo, J. E. L., & De Pablos, P. O. (2011). Recommendation system based on user interaction data applied to intelligent electronic books. *Computers in Human Behavior, 27*(4), 1445–1449.
10. Demeure, A., Calvary, G., & Coninx, K. (2008). COMET (s), a software architecture style and an interactors toolkit for plastic user interfaces. In *International Workshop on Design, Specification, and Verification of Interactive Systems* (pp. 225–237).
11. Dolomanov, O. V., Bourhis, L. J., Gildea, R. J., Howard, J. A. K., & Puschmann, H. (2009). OLEX2: A complete structure solution, refinement and analysis program. *Journal of Applied Crystallography, 42*(2), 339–341.
12. Edeki, C. (2015). Automated user interface design for HEPA filter recertification. *British Journal of Applied Science & Technology, 6*(6), 652.
13. Eisenstein, J., & Puerta, A. (2000). Adaptation in automated user-interface design. In *Proceedings of the 5th International Conference on Intelligent User Interfaces* (pp. 74–81).
14. Farooq Ali, M., Pérez-quiñones, M. A., & Abrams, M. (2005). Building multi-platform user interfaces with UIML. *Multiple User Interfaces: Cross-Platform Applications and Context-Aware Interfaces,* March (pp. 93–118). https://doi.org/10.1002/0470091703.ch6.

15. Flecchia, M. A., & Bergeron, R. D. (1987). Specifying complex dialogs in ALGAE. In *ACM SIGCHI Bulletin* (Vol. 18, pp. 229–234).
16. Gajos, K., & Weld, D. S. (2004). SUPPLE: Automatically generating user interfaces. In *Proceedings of the 9th International Conference on Intelligent User Interfaces* (pp. 93–100).
17. Gajos, K. Z., Weld, D. S., & Wobbrock, J. O. (2010). Automatically generating personalized user interfaces with supple. *Artificial Intelligence, 174*(12), 910–950.
18. Garfinkel, S. L., Mahoney, M. K., Silbar, R. R., Mallinckrodt, A. J., McKay, S., & others. (1993). NeXTSTEP programming, step one: Object-oriented applications. *Computers in Physics, 7*(3), 287–288.
19. Geuzaine, C., & Remacle, J.-F. (2009). Gmsh: A 3-D finite element mesh generator with built-in pre-and post-processing facilities. *International Journal for Numerical Methods in Engineering, 79*(11), 1309–1331.
20. Golin, E. J. (1991). Tool review: Prograph 2{\textperiodcentered} 0 from TGS systems. *Journal of Visual Languages & Computing, 2*(2), 189–194.
21. Hayes, P. J., Szekely, P. A., & Lerner, R. A. (1985). Design alternatives for user interface management systems based on experience with COUSIN. In *ACM SIGCHI Bulletin* (Vol. 16, pp. 169–175).
22. Hendratman, H. (2008). *The magic of macromedia director*. Bandung: Informatika Bandung.
23. Heymann, M., & Degani, A. (2007). Formal analysis and automatic generation of user interfaces: Approach, methodology, and an algorithm. *Human Factors: The Journal of the Human Factors and Ergonomics Society, 49*(2), 311–330.
24. Hill, R. D. (1986). Supporting concurrency, communication, and synchronization in human–computer interaction—the Sassafras UIMS. *ACM Transactions on Graphics (TOG), 5*(3), 179–210.
25. Jo, S., Kim, T., Iyer, V. G., & Im, W. (2008). CHARMM-GUI: A web-based graphical user interface for CHARMM. *Journal of Computational Chemistry, 29*(11), 1859–1865.
26. Limbourg, Q., Vanderdonckt, J., Michotte, B., Bouillon, L., & López-Jaquero, V. (2004). USIXML: A language supporting multi-path development of user interfaces. In *International Workshop on Design, Specification, and Verification of Interactive Systems* (pp. 200–220).
27. Lindsay, S., Jackson, D., Schofield, G., & Olivier, P. (2012). Engaging older people using participatory design. In *Proceedings of the SIGCHI Conference On Human Factors in Computing Systems* (pp. 1199–1208).
28. McCrory, E. S. (1993). Easy and effective application programs using DataViews. In *Particle Accelerator Conference, 1993, Proceedings of the 1993* (pp. 1952–1954).
29. Mori, G., Paterno, F., & Santoro, C. (2004). Design and development of multidevice user interfaces through multiple logical descriptions. *IEEE Transactions on Software Engineering, 30*(8), 507–520. https://doi.org/10.1109/TSE.2004.40.
30. Myers, B. A. (1988). *Creating user interfaces by demonstration*. Boston, MA: Academic Press Professional, Inc.
31. Myers, B. A. (1992). *State of the art in user interface software tools*. Carnegie-Mellon University. Department of Computer Science.
32. Myers, B. A., McDaniel, R. G., & Kosbie, D. S. (1993). Marquise: Creating complete user interfaces by demonstration. In *Proceedings of the INTERACT'93 and CHI'93 Conference on Human Factors in Computing Systems* (pp. 293–300).
33. Myers, B. A., Zanden, B. V., & Dannenberg, R. B. (1989). Creating graphical interactive application objects by demonstration. In *Proceedings of the 2nd annual ACM SIGGRAPH symposium on User interface software and technology* (pp. 95–104).
34. Nanopoulos, A., Rafailidis, D., Symeonidis, P., & Manolopoulos, Y. (2010). Musicbox: Personalized music recommendation based on cubic analysis of social tags. *IEEE Transactions on Audio, Speech and Language Processing, 18*(2), 407–412.
35. Olsen D. R. Jr. (1989). A programming language basis for user interface. In *ACM SIGCHI Bulletin* (Vol. 20, pp. 171–176).
36. Olsen, D. R., Jr., & Dempsey, E. P. (1983). SYNGRAPH: A graphical user interface generator. *ACM SIGGRAPH Computer Graphics, 17*(3), 43–50.

37. Park, D. H., Kim, H. K., Choi, I. Y., & Kim, J. K. (2012). A literature review and classification of recommender systems research. *Expert Systems with Applications, 39*(11), 10059–10072.
38. Paternò, F., Mancini, C., & Meniconi, S. (1997). ConcurTaskTrees: A diagrammatic notation for specifying task models. In *Human–Computer Interaction INTERACT'97* (pp. 362–369).
39. Pederiva, I., Vanderdonckt, J., España, S., & Panach, I. (2007). The beautification process in model-driven engineering of user interfaces. *Ifip International Federation For Information Processing, 4662*, 411–425. https://doi.org/10.1007/978-3-540-74796-3.
40. Porcel, C., & Herrera-Viedma, E. (2010). Dealing with incomplete information in a fuzzy linguistic recommender system to disseminate information in university digital libraries. *Knowledge-Based Systems, 23*(1), 32–39. https://doi.org/10.1016/j.knosys.2009.07.007.
41. Puerta, A. R. (1997). A model-based interface development environment. *IEEE Software, 14*(4), 40–47. https://doi.org/10.1109/52.595902.
42. Puerta, A., & Eisenstein, J. (2002a). XIML: A common representation for interaction data. In *Proceedings of the 7th International Conference on Intelligent User Interfaces—IUI'02* (pp. 214–215). https://doi.org/10.1145/502716.502763.
43. Puerta, A., & Eisenstein, J. (2002b). XIML: A common representation for interaction data. In *Proceedings of the 7th International Conference on Intelligent User Interfaces* (pp. 214–215).
44. Reinecke, K., & Bernstein, A. (2011). Improving performance, perceived usability, and aesthetics with culturally adaptive user interfaces. *ACM Transactions on Computer-Human Interaction, 18*(2), 1–29. https://doi.org/10.1145/1970378.1970382.
45. Rogers, Y., Paay, J., Brereton, M., Vaisutis, K. L., Marsden, G., & Vetere, F. (2014). Never too old: Engaging retired people inventing the future with MaKey MaKey. In *Proceedings of the SIGCHI Conference on Human Factors in Computing Systems* (pp. 3913–3922).
46. Schulert, A. J., Rogers, G. T., & Hamilton, J. A. (1985). ADM—a dialog manager. *ACM SIGCHI Bulletin, 16*(4), 177–183.
47. Shen, J., & Pantic, M. (2013). HCIwedge2 framework: A software framework for multimodal human–computer interaction systems. *IEEE Transactions on Cybernetics, 43*(6), 1593–1606. https://doi.org/10.1109/TCYB.2013.2271563.
48. Singh, A., & Wesson, J. (2009). Evaluation criteria for assessing the usability of ERP systems. In *Proceedings of the 2009 Annual Research Conference of the South African Institute of Computer Scientists and Information Technologists* (pp. 87–95).
49. Solovey, E. T. E., Lalooses, F., Girouard, A., Jacob, R. J. K., Chauncey, K., Weaver, D. … Schermerhorn, P. (2011). Sensing cognitive multitasking for a brain-based adaptive user interface. In *Proceedings of the 2011 Annual Conference on Human Factors in Computing Systems—CHI'11, January*, (p. 383). https://doi.org/10.1145/1978942.1978997.
50. Sun, Y., Ding, X., Lindtner, S., Lu, T., & Gu, N. (2014). Being senior and ICT: A study of seniors using ICT in China. *Human Factors in Computing Systems, 1*(1), 3933–3942. https://doi.org/10.1145/2556288.2557248.
51. Szekely, P. (1994). User interface prototyping: Tools and techniques. In *Workshop on Software Engineering and Human–Computer Interaction* (pp. 76–92).
52. Szekely, P., Luo, P., & Neches, R. (1992a). Facilitating the exploration of interface design alternatives. In *Proceedings of the SIGCHI Conference on Human Factors in Computing Systems—CHI'92*, (pp. 507–515). https://doi.org/10.1145/142750.142912.
53. Szekely, P., Luo, P., & Neches, R. (1992b). Facilitating the exploration of interface design alternatives: The HUMANOID model of interface design. In *Proceedings of the SIGCHI Conference on Human Factors in Computing Systems* (pp. 507–515).
54. Szekely, P., Sukaviriya, P., Castells, P., Muthukumarasamy, J., & Salcher, E. (1996). Declarative interface models for user interface construction tools: The MASTERMIND approach. In *Proceedings of the IFIP TC2/WG2.7 Working Conference on Engineering for Human–Computer Interaction* (pp. 120–150). http://portal.acm.org/citation.cfm?id=645348.650690.
55. Topi, H., Lucas, W. T., & Babaian, T. (2005). Identifying usability issues with an ERP implementation. In *ICEIS* (pp. 128–133).

56. Vanderdonckt, J. (1995). Knowledge-based systems for automated user interface generation: The TRIDENT experience. In *Proceedings of the CHI*, March. Retrieved from http://citeseerx.ist.psu.edu/viewdoc/download?doi=10.1.1.41.8207&rep=rep1&type=pdf.

57. Vlissides, J. M., & Linton, M. A. (1990). Unidraw: A framework for building domain-specific graphical editors. *ACM Transactions on Information Systems (TOIS)*, 8(3), 237–268.

58. Wells, L. K., & Travis, J. (1996). *LabVIEW for everyone: Graphical programming made even easier*. Prentice-Hall, Inc.

59. Wiecha, C., Bennett, W., Boies, S., Gould, J., & Greene, S. (1995). ITS: A tool for rapidly developing interactive applications. *Readings in Human–Computer Interaction (Second Edition)*, 8(3), 373–389. https://doi.org/10.1016/B978-0-08-051574-8.50039-X.

60. Wilson, D. A., Rosenstein, L. S., & Shafer, D. G. (1990). *Programming with MacApp*. Addison-Wesley Longman Publishing Co., Inc.

61. Xiong, W., & Qi, H. (2010). A extended TOPSIS method for the stochastic multi-criteria decision making problem through interval estimation. In *2010 2nd International Workshop on Intelligent Systems and Applications (ISA)* (pp. 1–4).

62. Zanden, B. Vander, & Myers, B. A. (1990). Automatic, look-and-feel independent dialog creation for graphical user interfaces. In *Proceedings of the SIGCHI Conference on Human Factors in Computing Systems*, April, (pp. 27–34). https://doi.org/10.1145/97243.97248.